Advances in Sustainable Food Packaging Technology

Innovations in Agricultural and Biological Engineering

Advances in Sustainable Food Packaging Technology

Edited by

Megh R. Goyal, PhD

Preeti Birwal, PhD

Anand Kumar Pandey, PhD

Ritesh Balaso Watharkar, PhD

AAP APPLE ACADEMIC PRESS

First edition published 2024

Apple Academic Press Inc.
1265 Goldenrod Circle, NE,
Palm Bay, FL 32905 USA

760 Laurentian Drive, Unit 19,
Burlington, ON L7N 0A4, CANADA

CRC Press
2385 NW Executive Center Drive,
Suite 320, Boca Raton FL 33431

4 Park Square, Milton Park,
Abingdon, Oxon, OX14 4RN UK

Apple Academic Press exclusively co-publishes with CRC Press, an imprint of Taylor & Francis Group, LLC

Library and Archives Canada Cataloguing in Publication

CIP data on file with Canada Library and Archives

Library of Congress Cataloging-in-Publication Data

CIP data on file with US Library of Congress

ISBN: 978-1-77491-394-9 (hbk)
ISBN: 978-1-77491-395-6 (pbk)
ISBN: 978-1-00339-524-9 (ebk)

ABOUT THE BOOK SERIES: INNOVATIONS IN AGRICULTURAL AND BIOLOGICAL ENGINEERING

Under this book series, Apple Academic Press Inc. is publishing book volumes over a span of 8–10 years in the specialty areas defined by the American Society of Agricultural and Biological Engineers (<asabe.org>). Apple Academic Press Inc. aims to be a principal source of books in agricultural and biological engineering. We welcome book proposals from readers in areas of their expertise.

The mission of this series is to provide knowledge and techniques for agricultural and biological engineers (ABEs). The book series offers high-quality reference and academic content on agricultural and biological engineering (ABE) that is accessible to academicians, researchers, scientists, university faculty and university-level students, and professionals around the world.

Agricultural and biological engineers ensure that the world has the necessities of life, including safe and plentiful food, clean air and water, renewable fuel and energy, safe working conditions, and a healthy environment by employing knowledge and expertise of the sciences, both pure and applied, and engineering principles. Biological engineering applies engineering practices to problems and opportunities presented by living things and the natural environment in agriculture.

ABE embraces a variety of the following specialty areas (<asabe.org>): aquaculture engineering, biological engineering, energy, farm machinery and power engineering, food, and process engineering, forest engineering, information, and electrical technologies, soil, and water conservation engineering, natural resources engineering, nursery, and greenhouse engineering, safety, and health, and structures and environment.

For this book series, we welcome chapters on the following specialty areas (but not limited to):

- Academia to industry to end-user loop in agricultural engineering;
- Agricultural mechanization;
- Aquaculture engineering;
- Biological engineering in agriculture;

- Biotechnology applications in agricultural engineering;
- Energy source engineering;
- Farm to fork technologies in agriculture;
- Food and bioprocess engineering;
- Forest engineering;
- GPS and remote sensing potential in agricultural engineering;
- Hill land agriculture;
- Human factors in engineering;
- Impact of global warming and climatic change on agriculture economy;
- Information and electrical technologies;
- Irrigation and drainage engineering;
- Nanotechnology applications in agricultural engineering;
- Natural resources engineering;
- Nursery and greenhouse engineering;
- Potential of phytochemicals from agricultural and wild plants for human health;
- Power systems and machinery design;
- Robot engineering and drones in agriculture;
- Rural electrification;
- Sanitary engineering;
- Simulation and computer modeling;
- Smart engineering applications in agriculture;
- Soil and water engineering;
- Micro-irrigation engineering;
- Structures and environment engineering;
- Waste management and recycling;
- Any other focus areas.

For more information on this series, readers may contact:

Megh R. Goyal, PhD, PE
Book Series Senior Editor-in-Chief: Innovations in Agricultural and Biological Engineering

OTHER BOOKS ON AGRICULTURAL AND BIOLOGICAL ENGINEERING BY APPLE ACADEMIC PRESS, INC.

Management of Drip/Trickle or Micro Irrigation
Megh R. Goyal, PhD, PE, Senior Editor-in-Chief
Evapotranspiration: Principles and Applications for Water Management
Megh R. Goyal, PhD, PE and Eric W. Harmsen, PhD, Editors

Book Series: Research Advances in Sustainable Micro Irrigation
Senior Editor-in-Chief: Megh R. Goyal, PhD, PE

- **Volume 1:** Sustainable Micro Irrigation: Principles and Practices
- **Volume 2:** Sustainable Practices in Surface and Subsurface Micro Irrigation
- **Volume 3:** Sustainable Micro Irrigation Management for Trees and Vines
- **Volume 4:** Management, Performance, and Applications of Micro Irrigation Systems
- **Volume 5:** Applications of Furrow and Micro Irrigation in Arid and Semi-Arid Regions
- **Volume 6:** Best Management Practices for Drip Irrigated Crops
- **Volume 7:** Closed Circuit Micro Irrigation Design: Theory and Applications
- **Volume 8:** Wastewater Management for Irrigation: Principles and Practices
- **Volume 9:** Water and Fertigation Management in Micro Irrigation
- **Volume 10:** Innovation in Micro Irrigation Technology

Book Series: Innovations and Challenges in Micro Irrigation
Senior Editor-in-Chief: Megh R. Goyal, PhD, PE

- Engineering Interventions in Sustainable Trickle Irrigation: Water Requirements, Uniformity, Fertigation, and Crop Performance
- Management Strategies for Water Use Efficiency and Micro Irrigated Crops: Principles, Practices, and Performance

- Micro-Irrigation Engineering for Horticultural Crops: Policy Options, Scheduling, and Design
- Micro-Irrigation Management: Technological Advances and Their Applications
- Micro-Irrigation Scheduling and Practices
- Performance Evaluation of Micro-Irrigation Management: Principles and Practices
- Potential of Solar Energy and Emerging Technologies in Sustainable Micro-Irrigation
- Principles and Management of Clogging in Micro-Irrigation
- Sustainable Micro-Irrigation Design Systems for Agricultural Crops: Methods and Practices

Book Series: Innovations in Agricultural and Biological Engineering
Senior Editor-in-Chief: Megh R. Goyal, PhD, PE

- Advanced Research Methods in Food Processing Technologies
- Advances in Food Process Engineering: Novel Processing, Preservation and Decontamination of Foods
- Advances in Green and Sustainable Nanomaterials: Applications in Energy, Biomedicine, Agriculture, and Environmental Science
- Advances in Sustainable Food Packaging Technology
- Analytical Methods for Milk and Milk Products, 3-volume set:
  - Volume 1: Sampling Methods, Chemical and Compositional Analysis
  - Volume 2: Physicochemical Analysis of Concentrated, Coagulated and Fermented Products
  - Volume 3: Microbiological Analysis, Isolation, and Characterization
- Biological and Chemical Hazards in Food and Food Products: Prevention, Practices, and Management
- Bioremediation and Phytoremediation Technologies in Sustainable Soil Management, 4-volume set:
  - Volume 1: Fundamental Aspects and Contaminated Sites
  - Volume 2: Microbial Approaches and Recent Trends
  - Volume 3: Inventive Techniques, Research Methods, and Case Studies
  - Volume 4: Degradation of Pesticides and Polychlorinated Biphenyls
- Dairy Engineering: Advanced Technologies and Their Applications

- Developing Technologies in Food Science: Status, Applications, and Challenges
- Emerging Technologies in Agricultural Engineering
- Engineering Interventions in Agricultural Processing
- Engineering Interventions in Foods and Plants
- Engineering Practices for Agricultural Production and Water Conservation: An Interdisciplinary Approach
- Engineering Practices for Management of Soil Salinity: Agricultural, Physiological, and Adaptive Approaches
- Engineering Practices for Milk Products: Dairyceuticals, Novel Technologies, and Quality
- Enzyme Inactivation in Food Processing: Technologies, Materials, and Applications
- Field Practices for Wastewater Use in Agriculture: Future Trends and Use of Biological Systems
- Flood Assessment: Modeling and Parameterization
- Food Engineering: Emerging Issues, Modeling, and Applications
- Food Process Engineering: Emerging Trends in Research and Their Applications
- Food Processing and Preservation Technology: Advances, Methods, and Applications
- Food Technology: Applied Research and Production Techniques
- Functional Dairy Ingredients and Nutraceuticals: Physicochemical, Technological, and Therapeutic Aspects
- Handbook of Research on Food Processing and Preservation Technologies, 5-volume set:
    - Volume 1: Nonthermal and Innovative Food Processing Methods
    - Volume 2: Nonthermal Food Preservation and Novel Processing Strategies
    - Volume 3: Computer-Aided Food Processing and Quality Evaluation Techniques
    - Volume 4: Design and Development of Specific Foods, Packaging Systems, and Food Safety
    - Volume 5: Emerging Techniques for Food Processing, Quality, and Safety Assurance
- Modeling Methods and Practices in Soil and Water Engineering
- Nanotechnology and Nanomaterial Applications in Food, Health, and Biomedical Sciences

- Nanotechnology Applications in Agricultural and Bioprocess Engineering: Farm to Table
- Nanotechnology Applications in Dairy Science: Packaging, Processing, and Preservation
- Nanotechnology Horizons in Food Process Engineering, 3-volume set:
  - Volume 1: Food Preservation, Food Packaging and Sustainable Agriculture
  - Volume 2: Scope, Biomaterials, and Human Health
  - Volume 3: Trends, Nanomaterials, and Food Delivery
- Nonthermal Plasma Technology for Food and Food Products Editors: R. Mahendran, PhD, Preeti Birwal, and Megh R. Goyal, PhD
- Novel and Alternative Methods in Food Processing: Biotechnological, Physicochemical, and Mathematical Approaches
- Novel Dairy Processing Technologies: Techniques, Management, and Energy Conservation
- Novel Processing Methods for Plant-Based Health Foods: Extraction, Encapsulation and Health Benefits of Bioactive Compounds
- Novel Strategies to Improve Shelf-Life and Quality of Foods: Quality, Safety, and Health Aspects
- Phytochemicals and Medicinal Plants in Food Design: Strategies and Technologies for Improved Healthcare
- Processing of Fruits and Vegetables: From Farm to Fork
- Processing Technologies for Milk and Milk Products: Methods, Applications, and Energy Usage
- Quality Control in Fruit and Vegetable Processing: Methods and Strategies
- Scientific and Technical Terms in Bioengineering and Biological Engineering
- Soil and Water Engineering: Principles and Applications of Modeling
- Soil Salinity Management in Agriculture: Technological Advances and Applications
- State-of-the-Art Technologies in Food Science: Human Health, Emerging Issues and Specialty Topics
- Sustainable and Functional Foods from Plants
- Sustainable Biological Systems for Agriculture: Emerging Issues in Nanotechnology, Biofertilizers, Wastewater, and Farm Machines
- Sustainable Green Nanomaterials: Synthesis, Characterization, and Engineering Applications

- Sustainable Nanomaterials for Biomedical Engineering: Impacts, Challenges, and Future Prospects
- Sustainable Nanomaterials for Biosystems Engineering: Trends in Renewable Energy, Environment, and Agriculture
- Technological Interventions in Dairy Science: Innovative Approaches in Processing, Preservation, and Analysis of Milk Products
- Technological Interventions in Management of Irrigated Agriculture
- Technological Interventions in the Processing of Fruits and Vegetables
- Technological Processes for Marine Foods, From Water to Fork: Bioactive Compounds, Industrial Applications, and Genomics
- The Chemistry of Milk and Milk Products: Physicochemical Properties, Therapeutic Characteristics, and Processing Methods

ABOUT SENIOR EDITOR-IN-CHIEF

Megh R. Goyal, PhD, PE

Retired Professor in Agricultural and Biomedical Engineering, University of Puerto Rico, Mayaguez Campus; Senior Acquisitions Editor, Biomedical Engineering and Agricultural Science, Apple Academic Press, Inc.

Megh R. Goyal, PhD, PE, is a Retired Professor in Agricultural and Biomedical Engineering from the General Engineering Department in the College of Engineering at the University of Puerto Rico–Mayaguez Campus. During his professional career, he has worked as a Soil Conservation Inspector; Research Assistant at Haryana Agricultural University and Ohio State University; Research Agricultural Engineer/Professor at the Department of Agricultural Engineering of UPRM; and Professor of Agricultural and Biomedical Engineering in the General Engineering Department of UPRM. He spent a one-year sabbatical leave in 2002–2003 at the Biomedical Engineering Department of Florida International University, Miami, USA.

Dr. Goyal was the first agricultural engineer to receive a professional license in agricultural engineering from the College of Engineers and Surveyors of Puerto Rico. In 2005, he was proclaimed the "Father of Irrigation Engineering in Puerto Rico for the Twentieth Century" by the American Society of Agricultural and Biological Engineers, Puerto Rico Section, for his pioneering work on microirrigation, evapotranspiration, agroclimatology, and soil and water engineering.

During his professional career of 52 years, he has received many awards, including Scientist of the Year, Membership Grand Prize for the American Society of Agricultural Engineers Campaign, Felix Castro Rodriguez Academic Excellence Award, Man of Drip Irrigation by the Mayor of Municipalities of Mayaguez/Caguas/Ponce and Senate/Secretary of Agriculture of ELA, Puerto Rico, and many others. He has been recognized as one of the experts "who rendered meritorious service for the development of [the] irrigation sector in India" by the Water Technology Centre of Tamil Nadu Agricultural University in Coimbatore, India, and ASABE who bestowed on him the 2018 Netafim Microirrigation Award.

Dr. Goyal has authored more than 200 journal articles and edited more than 100 books. AAP has published many of his books, including *Management of Drip/Trickle or Micro Irrigation; Evapotranspiration: Principles and Applications for Water Management;* ten-volume set on *Research Advances in Sustainable Micro Irrigation.* Dr. Goyal has also developed several book series with AAP, including Innovations in Agricultural & Biological Engineering (with over 60 titles in the series to date); Innovations and Challenges in Micro Irrigation; and Innovations in Plant Science for Better Health: From Soil to Fork.

Dr. Goyal received his BSc degree in Engineering from Punjab Agricultural University, Ludhiana, India, and his MSc and PhD degrees from the Ohio State University, Columbus, Ohio, USA. He also earned a Master of Divinity degree from the Puerto Rico Evangelical Seminary, Hato Rey, Puerto Rico, USA.

ABOUT THE EDITORS

Preeti Birwal, PhD
Scientist (Processing and Food Engineering),
Department of Processing and Food Engineering,
College of Agricultural Engineering and Technology,
Punjab Agricultural University, Ludhiana, Punjab, India

Preeti Birwal, PhD, is working as a Scientist (processing and food engineering) in the Department of Processing and Food Engineering at the College of Agricultural Engineering and Technology at Punjab Agricultural University, Ludhiana, Punjab, India. She is currently working in the area of nonthermal food preservation, fermented beverages, food packaging, and technology of millet-based beer. She has served at Jain (Deemed to be University), Bangalore, as a member of the board of examiners and placements. She has participated at several national and international conferences and seminars and has delivered lectures as a resource person on doubling farmers' income through dairy technology training sponsored by the Directorate of Extension, Ministry of Agriculture and Farmers Welfare, Government of India. Dr. Birwal has published research papers, edited books, book chapters, popular articles, conference papers, abstracts, and editorial opinions. She is advising several MTech scholars in food technology and has successfully guided five postgraduate students for their dissertation work. She also serves as an external examiner for various Indian state agricultural universities and as an editor and reviewer of several journals. Dr. Birwal has been named outstanding reviewer of the month by the online journal *Current Research in Nutrition and Food Science.* She has successfully completed AUTOCAD 2D & 3D certification. She is a life member of IDEA. She graduated with a degree in Dairy Technology from ICAR–National Dairy Research Institute, Karnal, India; a master's degree in Food Process Engineering and Management from NIFTEM, Haryana; and PhD (Dairy Engineering) from ICAR–NDRI, Bangalore, India. She has received several fellowships from MHRD, Nestle India, GATE, and UGC-RGNF.

Anand Kumar Pandey, PhD
Assistant Professor of Biotechnology Engineering, Institute of Engineering and Technology, Bundelkhand University, Jhansi, Uttar Pradesh, India.

Anand Kumar Pandey, PhD, is working as an Assistant Professor in the Department of Biotechnology Engineering at the Institute of Engineering and Technology, Bundelkhand University, Jhansi, Uttar Pradesh, India.

He received his doctorate degree from IIT BHU Varanasi, India. He is currently working on biodegradation of cellulose and other toxic hydrocarbons with the potential to cause environmental hazards. He has great expertise in bioinformatics and has contributed immensely to the field of drug discovery by his research. His wide research area also includes discovery of food nutraceuticals with the potential to treat diseases (like neurodegenerative disorders, cancer, viral, and bacterial infections), animal cell culture, diseased animal modeling, tissue engineering, and immunology. He was ranked among the top 5% in a Cell Culture Technologies course organized by NPTEL and funded by the Ministry of HRD, Government of India.

By his revolutionary research skills, Dr. Pandey has published more than 30 research articles in international peer-reviewed journals and has authored several book chapters. He has around 600 citations (excluding self-citations) with Google Scholar (h-index of 12). He is also an active reviewer for many international SCI journals, including *Informatics in Medicine Unlocked* (Elsevier), *Food Measurement and Characterization* (Springer), *Metagene* (Elsevier), *Network Modeling Analysis in Health Informatics and Bioinformatics* (Springer), *Neuroscience* (Elsevier), *Medicinal Chemistry Research* (Springer), *Journal of Biomolecular Structure and Dynamics* (Taylor & Francis), *Medical Hypothesis* (Elsevier), *British Journal of Cancer* (Nature), *Letters of Drug Design and Discovery* (Bentham Science), *Anti-Cancer Agents in Medicinal Chemistry* (Bentham Science), etc. He is currently a member of the Indian Crystallographic Association (ICA), Indian Academy of Neurosciences (IAN), Indian society of Remote Sensing (ISRS), Society for Bioinformatics & Biological Sciences (SBBS), and Association of Food Scientists & Technologists India (AFSTI).

Ritesh Balaso Watharkar, PhD
Assistant Professor, College of Agriculture and Food Business Management, MIT-ADT University, Rajbaugh, Solapur, Maharashtra, India

Ritesh Balaso Watharkar, PhD, is currently working as an Assistant Professor in the College of Agriculture and Food Business Management at MIT-ADT University, Maharashtra, India. He received his BTech degree in Food Technology from Dr. Balasaheb Sawant Konkan Agriculture University, Dapoli, Maharashtra, India; his MTech in Food Processing and Engineering degree is from the Karunya University, India; and his PhD in Food Engineering and Technology degree from Tezpur University (A Central University), Assam, India.

He was formerly a Senior Research Fellow under a funded project by the Ministry of Food Processing Industries, New Delhi (2013–2015), and Assistant Professor at K.K. Wagh Food Tech College (2011–2013) and at Shramshakti College of Food Technology, affiliated by Mahatma Phule Agricultural University (2019–2020).

During his professional career, he received awards such as senior research fellow under the UNESCO/China Great Wall Fellowship Program and worked for one year in the School of Biosystem Engineering and Food Science at Zhejiang University, Hangzhou, China. He was also awarded best popular article from *Krishidut Magazine*, Maharashtra, India. Currently, he is a life member of the Association of Food Scientists and Technologists, India. He has published more than 10 journal articles and attended more than 10 national and international conferences. He has also published more than 10 popular articles in the *Marathi Literacy Magazine*.

CONTENTS

CONTRIBUTORS

Shwetha Acharya
Quality Control, Krishna Dairy, Shrirampur, Maharastra, India

Prajya Arya
Sant Longowal Institute of Engineering and Technology, Longowal, Punjab, India

Aastha Bharadwaj
Department of Food Technology, Jamia Hamdard University, New Delhi, India

Preeti Birwal
Department of Processing and Food Engineering, Punjab Agricultural University, Ludhiana, Punjab, India

Savita Budhwar
Department of Nutrition Biology, School of Interdisciplinary and Applied Sciences, Central University of Haryana, Mahendergarh, Haryana, India

Dharm Dutt
Department of Paper Technology, Indian Institute of Technology Roorkee, Saharanpur, Uttar Pradesh, India

Nandini Dutta
Department of Food Technology, Faculty of Engineering and Technology, Jain University (Deemed-to-be University), Karnataka, India

Sana Fatma
Food Processing and Management, Banaras Hindu University, DDU Kaushal Kendra, RGSC, Varanasi, Uttar Pradesh, India

P. G. Gauthami
Department of Food Technology, FET, Jain (Deemed-to-be) University, Bengaluru, Karnataka, India

Pavan M. Gundu
Department of Food Technology, FET, Jain (Deemed-to-be) University, Bengaluru, Karnataka, India

Mahendra Gunjal
Department of Food Technology and Nutrition, School of Agriculture, Lovely Professional University, Phagwara, Punjab, India

Sukirti Joshi
Centre for Rural Development and Technology, IIT Delhi, New Delhi, India

Kamaljit Kaur
Department of Food Science and Technology, Punjab Agricultural University, Ludhiana, Punjab, India

Ramandeep Kaur
Department of Food Science and Technology, Punjab Agricultural University, Ludhiana, Punjab, India

Hina Khan
Department of Paper Technology, Indian Institute of Technology, Roorkee, Uttar Pradesh, India

Anita Kumari
Department of Nutrition Biology, School of Interdisciplinary and Applied Sciences, Central University of Haryana, Mahendergarh, Haryana, India

Ashwani Kumar
Department of Nutrition Biology, School of Interdisciplinary and Applied Sciences, Central University of Haryana, Mahendergarh, Haryana, India

Pradyuman Kumar
Sant Longowal Institute of Engineering and Technology, Longowal, Punjab, India

Yogesh Kumar
National institute of Food Technology Entrepreneurship and Management (NIFTEM), Sonepat, Haryana, India

Chaitradeepa G. Mestri
Department of Food Technology, M. S. Ramaiah University of Applied Sciences, Bengaluru, Karnataka, India

Anand Kumar Pandey
Department of Biotechnology Engineering, Institute of Engineering and Technology, Bundelkhand University, Jhansi, India

Deepika Pawar
Department of Nutrition Biology, School of Interdisciplinary and Applied Sciences, Central University of Haryana, Mahendergarh, Haryana, India

Preetha Radhakrishnan
Department of Food Process Engineering, School of Bioengineering, College of Engineering and Technology, SRM Institute of Science and Technology, SRM Nagar, Kattankulathur, Chennai, Tamil Nadu, India

Sharad Raghuvanshi
Department of Paper Technology, Indian Institute of Technology, Roorkee, Uttar Pradesh, India

Shruthy Ramesh
Department of Food Process Engineering, School of Bioengineering, College of Engineering and Technology, SRM Institute of Science and Technology, SRM Nagar, Kattankulathur, Chennai, Tamil Nadu, India

Soumya Rathore
Department of Food Technology, Harcourt Butler Technical University, Kanpur, Uttar Pradesh, India

J. K. Sahu
Centre for Rural Development and Technology, IIT Delhi, New Delhi, India

Sonia Saini
Department of Nutrition Biology, School of Interdisciplinary and Applied Sciences, Central University of Haryana, Mahendergarh, Haryana, India

Vaishali Saroha
Department of Paper Technology, Indian Institute of Technology, Roorkee, Uttar Pradesh, India

Nitya Sharma
Centre for Rural Development and Technology, IIT Delhi, New Delhi, India

Sujata Pandit Sharma
Department of Life Sciences, SBSR, Sharda University, Greater Noida, Uttar Pradesh, India

S. Supreetha
Department of Food Technology, Faculty of Engineering and Technology,
Jain University (Deemed-to-be University), Karnataka, India

Kiran Verma
National institute of Food Technology Entrepreneurship and Management (NIFTEM),
Sonepat, Haryana, India

Shalja Verma
Biochemical Engineering and Biotechnology, Indian Institute of Technology, New Delhi, India

Ritesh Balaso Watharkar
College of Management, Food and Agriculture Management Department, MIT Art,
Design and Technology University, Loni Kalbhor, Pune, Maharashtra, India

Gaurav Yadav
Department of Biotechnology, Maulana Abul Kalam Azad University of Technology,
Haringhata, Nadia, West Bengal, India

ABBREVIATIONS

ABS	acrylonitrile butadiene styrene
ABS/PC	acrylonitrile butadiene styrene/polycarbonate
ADP	adenosine diphosphate
AF4	asymmetric flow field-flow fractionation
AFM	atomic force microscopy
Ag	silver
AMP	adenosine monophosphate
ATP	adenosine triphosphate
Au	gold
Aw	water activity (aw)
BBP	butyl benzyl phthalate
BBS	4,4'-bis(2-benzoxazolyl) stilbene
BC	bacterial cellulose
BC-PPy-ZnO	bacterial cellulose polypyrrole zinc oxide
BHA	butylated hydroxyanisole
BHET	bis-(2-hydroxy ethyl) terephthalate
BHT	butylated hydroxytoluene
CAGR	compound annual growth rate
CBD	cell-wall binding domain
CFB	corrugated fiber box
CFU	colony forming unit
CMC	carboxy methyl cellulose
COI	cytochrome c oxidase
CuO-NPs	copper oxide nanoparticles
DBP	di-n-butyl phthalate
DHEP	di-2-ethylhexyl phthalate
DMA	dimethylamine
DNA	deoxyribonucleic acid
EAD	enzymatically active domain
EC	edible coatings
EF	edible film
EU	European Union
FAO	Food and Agriculture Organization
FCA	Food Contact Article

FCM	Food Contact Material
FCS	Food Contact Substance
FDA	Food and Drug Administration
FIA	flow injection analysis
FTIR	Fourier transform infrared spectroscopy
GCE	glassine carbon electrode
GC-MS	gas chromatography–mass spectrometry
GMP	good manufacturing practices
GPS	Global Positioning System
GRAS	generally recognized as safe
GTIN	global trade item number
HDPE	high-density poly-ethylene
HIPS	high-impact polystyrene
HMP	high methoxyl pectin
HNTs	halloysite nanotubes
HPMC	hydroxypropyl methylcellulose
HTC	hydrothermal carbonization
Hx	hypoxanthine
ICP-MS	Inductively Coupled Plasma-Mass Spectrometry
ICTV	International committee on taxonomy of viruses
IMP	inosine monophosphate
In	inosine
IOSP	Intelligent or Smart Packaging
ITO	indium tin oxide
LDH	lactate dehydrogenase
LDPE	low-density polyethylene
LED	light-emitting diode
LMP	low-methoxyl pectin
MALS	multi-angle light scattering
MAP	modified atmosphere packaging
MC	methyl cellulose
MgO	magnesium oxide
MHET	mono-(2-hydroxyethyl) terephthalic acid
MMT	montmorillonites
MSW	municipal solid waste
MTT	3-[4,5-dimethylthiazol-2-yl]-2,5-diphenyl-tetrazolium bromide)
MVP	modern vacuum packaging
NEMS	nano-electromechanical systems

NIAS	non-intentionally added substances
NIR	near infrared
NMBP	nonmigratory bioactive polymers
NMR	nuclear magnetic resonance
NPs	nanoparticles
PA	polyamides
PBAT	polybutylene adipate terephthalate
PBS	polybutylene succinate
PC	polycarbonate
PCDA	10,12-pentacosadiynoic acid
PCL	polycaprolactone
PDA	polydiacetylene
PE	polyethylene
PES	polyester
PES	pristine polyether sulfones membrane
PET	polyethylene terephthalate
PGH	peptidoglycan hydrolases
PHA	polyhydroxyalkanoates
PHB	polyhydroxyl butyrate
PLA	polylactic acid
PNCs	polymer nanocomposites
PP	polypropylene
PPy	poly pyrrole
PS	polystyrene
PU	polyurethanes
PUR	polyurethane
PVA	polyvinyl acetate
PVC	polyvinyl chloride
PVDC	polyvinylidene chloride
PVOH	polyvinyl alcohol
QR	quick response
RFID	radio frequency identification
RH	relative humidity
RNA	ribonucleic acid
ROS	reactive oxygen species
RSS	reduced space symbology
SED	sacrificial electron donor
SEM	scanning electron microscopy
SiO_2	silicon dioxide

SML	specific migration limit
SML (T)	total specific migration limit
TCS	time/temperature control for food safety
TEM	transmission electron microscopy
TGAA	thermogravimetric analysis
TiO_2	titanium dioxide
TiO_2-G	grapheme-titanium dioxide
TiO_2-NPs	titanium dioxide nanoparticles
TMA	trimethylamine
TTI	time-temperature indicators
TVB-N	total volatile basic nitrogen
UPC	Universal Product Code
UV	ultraviolet
WHO	World Health Organization
WTE	waste to energy
WVTR	water vapor transmission rate
XOD	xanthine oxidase
XPS	X-ray photoelectron spectroscopy
XRD	X-ray diffraction
XRF	X-ray fluorescence
ZnO	zinc oxide
ZnO-NPs	zinc oxide nanoparticles

PREFACE

The existence of life on Earth without food is beyond imagination but contaminated food and food lacking in nutrition can become a reason for numerous diseases. Safety of food is of utmost importance in the current scenario of exponentially increasing food-borne diseases. The perishable nature of food demands for its proper maintenance and storage to enhance its shelf life along with nutritive value. The exploding population and the increase in level of awareness of consumers puts forward an urgent need for development of techniques and strategies that can fulfill the current food safety requirements in a sustainable manner.

Food packaging plays a vital role in providing protection to food at every stage of production while maintaining its nutritive value. It is an effective and usually applied technique to deal with food safety issues. Conventional plastic packaging technology possess immense potential to protect food quality, but its massive accumulation leadingly contributes to environmental pollution causing death of marine as well as terrestrial life forms. Moreover, such technology does not have any added advantages.

Degradation of plastic-based packaging by physical and chemical approaches requires huge capital investment and also leads to discharge of toxic chemical in the environment. The approach of biodegradation for management of plastic waste has been extensively investigated and seems promising but the efficiency of degradation is limited.

The concept of biodegradable plastic or edible packaging has transformed the scenario and has increased hope for a better green tomorrow. Recent innovations in technologies have encouraged the development of active, intelligent, and smart packaging systems that embrace packaging potential to not only provide protection to food but can also improve the nutritive value, reduce contamination by releasing antimicrobials, and can provide real time status of food quality. They can also be used in a wide range of tracking systems which can be of great benefit for the commercial food chain. Introduction of nanotechnology to food packaging has brought an unstoppable revolution in packaging industry by incorporating scope for till date unthinkable developments. With the leading advantages, the drawbacks associated with toxicity of food due to packaging are also nontrivial.

In view of all the above aspects, it has become obligatory to create the title *Advances in Sustainable Food Packaging Technology* for this book. This book contains three parts that discuss the leading features of food storage, management, and its packaging. Part I, *Emerging and Advanced Approaches in Packaging of Foods*, elaborates on the novel food packaging and related innovations and trends. Part II, *Nanotechnology-Based Improvements in Food Packaging*, illustrates the novel nanotechnology-based application to extemporize food packaging techniques. Part III, *Biodegradation and Bio-Based Approaches for Management of Plastic Waste*, deals with management, disposal, and degradation of food waste and its packaging.

It is a compilation of knowledge of renowned authors on the abovementioned aspects. All the authors of the present compilation are pioneers in their fields and have contributed with their sole potential in making this book a great success.

We hope that this book will be greatly beneficial to the upcoming and current researchers in this field in helping them understand latest researches in the arena of food packaging. It will also be of enormous benefit to postgraduate and graduate students to gain insights about the current food packing systems while providing knowledge about conventional packaging techniques.

We would like to acknowledge our sincere thanks to all the authors of this book for their incredible contributions. We also owe special thanks to the editorial and production staff of Apple Academic Press for their enormous efforts for this book as this book would have not been possible without their devotion.

We also express our gratitude toward our family members and friends for their support and understanding while we were dedicated to the preparation of this book.

Lastly, we oblige our readers to convey their valuable suggestions for further improvement in the upcoming edition. Access to healthy and nutritious food is the right of every living creature on this Earth and we human beings should make every effort toward this holy purpose.

—Editors

PART I

EMERGING AND ADVANCED APPROACHES IN PACKAGING OF FOODS

CHAPTER 1

RECENT TRENDS IN THE FOOD PACKAGING INDUSTRY

PRAJYA ARYA and PRADYUMAN KUMAR

ABSTRACT

The various emerging ways to generate novel, attractive, and intelligent packaging have created a boom in the food industry. Along with the attractiveness, there is a requirement of major modifications, such as installation of indicators, absorbers, and emitters inside the food package to monitor the overall quality of the food product. The novelty of the food package brings many innovations in the technical front that are considered a big market growth option by marinating the food quality inside the package till it reaches the end-consumer. The concepts of analyzing the factors, which directly affect the product quality like gaseous atmosphere and smart indicators, are becoming prevalent. This novelty has changed the face of packaging from classical stabilization techniques (such as thermal treatment) to high pressure, nano packaging, and cold plasma methods in the modern era. In nano packaging, nano sensors are being used for the detection of the changes inside the package that could help in the development of antimicrobial nanocomposite films, RFID technique, barcode readers, and NMBP techniques. In this chapter, one should be able to understand the concept of packaging and novelty in the packaging system and how it is beneficial to us and nature.

1.1 INTRODUCTION

Package consists of wrapping a material made up of wood, plastic, glass, and metal around a consumable or nonconsumable item that serves to contain, describe, preserve, identify, display, and protect, and it makes the product

Advances in Sustainable Food Packaging Technology. Megh R. Goyal, Preeti Birwal, Anand K. Pandey, & Ritesh B. Watharkar, (Eds.)

marketable and keeps it clean. In this new world full of technologies and innovations, there are many new technologies to aid in the packaging of the food product and its processing. These technologies have reduced the transportation load of food, have enhanced shelf-life of food as this packaging technology might contain indicators for the quality of the food and its deterioration, to retain the nutritional and sensory quality of the produce, to reduce the need of chemical additives up to some extent.[42,54,55,59]

The increasing gain in the market of food packaging has attracted enormous scientists to develop novel methods that aid in the better management of the food shelf-life till it reaches end consumers. The new methods involve microwave processing, high pressure processing (HPP), UV light, cold plasma techniques, ozone treatment, etc. Along with these techniques, the major focus in packaging is to disinfect the packaging material to minimize the risk of contamination and food degradation at the best level. The concept of microwave processing has gained a peak as it plays a role in generating smart values from the plastic film trays that seal the package and aid in generating quick vapor. These quick vapors are adapted for reducing the cooking time of the meal and also produced ready-to-eat food, which has enhanced the market of catering and food industry.

The main motive of these processes is that food material, when inside the food package, could also be treated, and it is majorly called in-package technology of food, which further prevents food from contamination or mishandling due to posttreatment contacts. In all these treatments, packaging material exposed to too many processing situations gets affected and changes in its structure, and its mass transfer (as migration and barrier) properties occur. Generally, three types of mass transfer processes can be considered, such as permeation, migration, and scalping.

- The process of permeation occurs with aroma, flavors, gases, and water vapor. It is related to the headspace and the external side of package, which is exposed to atmosphere and it is concise about the mass transfer that took place via package membrane.
- The substances present in packaging material having low molecular weight have tendency to move toward food matrix, which is further analyzed for toxicological and legislation evaluation.
- Food quality is majorly affected by the low molecular weight hydrophobic substance and their scalping affects the mass transfer properties of the packaging material.

The concept of scalping belongs to the absorption of flavor and aroma compounds by the packaging material, which causes loss of the product quality. Generally, low-density polyethylene (LDPE) material is used in this system to minimize the absorption of flavors and aroma.[1] It could also cause damage to the package, for example, the outcome of plasticizer activity causes absorption of specific flavor compounds from solution by plastic materials.[47] In the processing of packaging material, there are several interactions that occur between the packaging material and food product. During these interactions, at times, there is generation of other undesirable compounds inside the package and could reflect health hazard. These generated molecules belong to the category of "nonintentionally added substances (NIAS)" category, for example, degradation of PVC. Detection of NIAS is now easier with the use of sophisticated instruments and they are of high importance.

The development of active and intelligent packaging can allow the usage of active and intelligent elements to absorb or to release substances into, onto, or from the packaged food, or to provide the information of the produce and package. Nowadays, food package comes with different modifications, such as dispensing caps, re-closable openings, sprays, and ways, which make the delivery of food product more convenient and usable.

There are three levels of packaging, such as: (1) primary packaging (which comes in direct contact of food like metal cans, glass bottle, laminates, and plastic pouches, usually acting as a major protective barrier); (2) secondary packaging (which contains the primary package like corrugate fiber boxes); and (3) tertiary packaging (which contains secondary package like stretch-wrapped pallet of corrugated fiber boxes). For a package, it must fulfill some criteria, for example, convenience to the end consumer (easy openings, re-closable pouches, and suitable dosing machine) and from the market point of view, it should have reasonable price, automatic packaging machine, seal-ability, and package, which could also lower the environmental stress and is easy for reuse and recycling.

The present chapter focuses on different types of packaging and their processing techniques. Also, recent trends on a wide range of nanotechnology-based applications in packaging have been illustrated.

1.2 PROCESSING METHODS FOR PACKAGING

During processing of food packaging, some important factors should be kept in mind before developing a good package for the food. Package should not react with food, package material should not leach out into the food matrix, and it should indicate about keeping quality of food and its deterioration rate. The consideration of the food processing is based on the techniques, which are used for food packaging, such as high pressure and cold plasma techniques.

1.2.1 HIGH-PRESSURE PROCESSING

Food is subjected to withstand a hydrostatic pressure in the range of 150–600 MPa or above to make the food completely stable and free of microbes and enzymes at low increase in temperature and without further addition of the preservatives. This leads to the maintenance of keeping quality of food without degrading its nutritional and organoleptic properties. As this is a cost-intensive process, therefore it is only used for high-profile foods.[37–43] In this method, there are huge changes in structural, sealing, and mechanical properties of plastic material. Many studies have been conducted related to type of packaging material (PET/LDPE/HDPE/AL/) at different temperatures and its behavior under high pressure and different types of barrier materials. These studies depicted that there was a free volume reduction due to high pressure[12] that precludes polymeric matrix diffusion while treatment and its effect vanishes when the pressure is released. Therefore, package structure can undergo deformations like delamination, wrinkles, and even holes, and this leads to decrease in barrier properties of the material.

1.2.2 COLD PLASMA TECHNOLOGY

The plasma is a partially or completely ionized state of gas and comes up as plasma oscillations in ionized gas. The condition of cold plasma could be achieved by attaining low and high temperatures by checking the equilibrium and nonequilibrium glow discharge.[10] Plasma is a matrix of negative and positive charges along with neutral particles and photons, and it exists in massive range of temperatures and densities. Phase change that occurs from solid to liquid and further to gas as enhancement in energy input beyond a certain level in gas state causes ionization of molecules to yield the plasma state. Cold plasma is a nonthermal, novel food processing technology that employs energy and reactive gases to demobilize contaminating microbes in food products, such as meat and poultry.

In the food industry, cold plasma technology is based on gas discharge technology, which is an effective, environmentally safe, economic method for critical cleaning; and vacuum ultraviolet energy is very efficient in breaking most of the organic bonds on the food surface and its matrix. The cold plasma technique can be used in the disinfection of the surface and decontamination of the microbes, which occur externally. When cold plasma flows around objects, it gives shadow effects, which might not cover all parts of a product, but it can also be used as a disinfectant of surfaces before packaging.

1.3 TYPES OF PACKAGING

1.3.1 SMART PACKAGING

Smart packaging can sense the changes in the product that are related to its internal structure and physiochemical changes. This packaging method provides entire packaging solution that observes changes in the closely related environment of food matrices (intelligent) and also monitors type of changes, which are taking place in that environment that requires some active actions.[57] The usage of sensors (bio or chemical) would also depict the quality of end product and aids in monitoring the changes during storage.[16]

Along with chemical and biosensors, smart packaging utilizes various types of sensors for observing changes in food quality, such as: by noticing pathogens, leakage, freshness, oxygen, carbon dioxide, time, temperature, and pH level.[16] The selection of a particular type of sensor depends on which type of product would be packaged like food, beverage, ready-to-serve, pharmaceutical, and similar conditions would be monitored or varied according to the product.[16]

The major follow-up of the smart packaging includes tracking and tracing the entire product throughout its life cycle to analyze every minute change related to the package inside and outside the environment. This tracking and tracing can help the manufacturer, producer, retailer, and end consumer about the product condition at any particular stage.[49]

1.3.2 ACTIVE PACKAGING

The active packaging imposes an alternative way on the traditional packaging system. The major focus area in active packaging is to introduce the concept of observing continuous changes in the food package until the usage by end consumer, for example, to maintain food quality. This technique implants components in the package, which are capable of releasing and absorbing the substance from or into the preserved food or into its surrounding environment for maintaining the shelf-life and quality of the product.[10] Active packaging is a boom for the perishable goods that include reduction in the localization activity, amount of the substance, and migration of the particle from film to food.[11]

The active packaging consists of an oxygen scavenger, ethylene scavengers, odor /flavor emitter or absorber, antioxidants, antimicrobials, etc.[49] Active packaging is the most rapidly emerging techniques, which are smart enough to observe the reaction between package and food components because a consumer requires fresh produce, with quality and safe product.[10] Active packaging system keeps a close eye on the quality of end product by monitoring its sensory, nutritional, and physiological processes that include lipid oxidation, infestation, dehydration, and degradation by microbes. This entire process aids in marinating overall keeping quality of the end product and could also be modified based on the food product.[35]

Active packaging can be applied in such a way to enhance food quality as absorbers or scavengers can remove unwanted components like ethylene, oxygen, carbon dioxide, taints, excessive water, and other specific compounds. In the active packaging, absorbers act as an inert barrier between the product, package, and external environment to provide an efficient control of surface contamination against microbial load and also to protect food of high moisture content (such as, sea food).[37]

It is the most widely used technology that includes absorbers, such as oxygen absorber, carbon dioxide absorber, ethylene absorber, humidity absorber, moisture absorber, lactose absorber, UV-light absorber, and cholesterol absorber. Also, there are self-heating and self -cooling cans as discussed in this section.

1.3.2.1 SELF-HEATING CANS

The aluminum cans are designed in such a manner that it could maintain the temperature of the food material itself. This is generally done by the self-heating and self-cooling cans. In self -heating cans, heating is from outside the chamber, and in self-cooling cans, heating is from an inner chamber. In case of heating from the outside chamber (self-heat), there are dual chambers that make an insulating zone and generate heat between the inside and outside chambers. Food package would remain in the inside chamber and the outside chamber would aid in accomplishing the exothermic reactions so as to maintain the heating of the food.

The chemicals in the outer chamber are kept separated from each other with the help of a barrier and a ring. When there is a need to heat the food material, the ring has to be pulled out to cause breakage of the barrier and to allow the chemical to mix with water and heat is released ultimately with heating of food occurred. **In case heating from inner chamber**: the heat-causing chemicals are kept in the inside chamber while food surrounds it in the outer chamber. For heating the produce, the user has to push on the base of the can to break the barrier, which acts as a separating membrane between chemical and water. The inner self-heating can is more favorable as heat dissipation is lower compared to the outer self-heating can and it also reduces excessive heating of product's exterior wall, causing discomfort for the user and is generally made up of aluminum.[2]

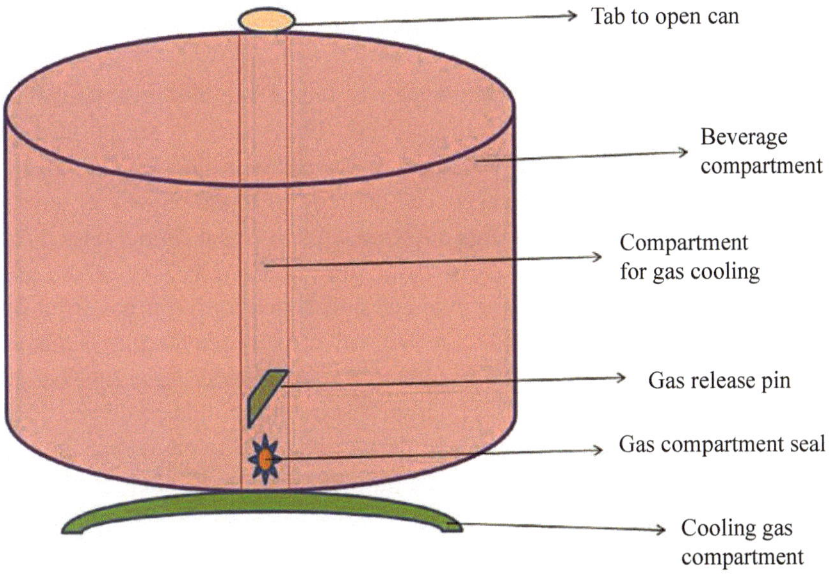

FIGURE 1.1 Representation of a self-cooling can.

1.3.2.2 SELF-COOLING CANS

The aluminum can is built in such a manner that it could minimize the heat of food products up to a maximum extent (Fig. 1.1). For minimizing the heat, there is a button that is attached to the heat exchange unit. Near the button, there is a cooling pad to contain liquid nitrogen or self-reclaimed CO_2. This would aid in quick chilling of the beverage within a minute.[2]

1.3.2.3 TYPES OF SCAVENGERS

1.3.2.3.1 Oxygen Scavengers

Oxygen scavenger aids in controlling the shelf-life of food product. The major issues related to the food are the iron content inside the food, its oxidation in the presence of oxygen in the headspace, and amount of oxygen present inside food package.[8] The amount of oxygen present would directly affect the growth and count of aerobic microbes inside the package. The oxygen scavenger has an effect on the color, nutritional and sensorial properties of a food.[61] The mechanism of working of an oxygen scavenger is based on the active component that can be a metal, organic acid, enzyme, or photosensitive dye that reacts with the entire system and can bring changes in the oxygen level.[61]

The concept of oxygen scavenger is based on the presence of iron compounds that are present in the sachets. These sachets are permeable to oxygen and utmost amount of oxygen scavenging sachets available is with 7 g of ferric oxides in the market. The demand of ferric oxide oxygen scavenger is quite often due to its high sorption capacity, as it cannot pass through the metal detector line that is its major drawback.[4] The size of the sachet would determine the oxygen scavenged and it generally ranges from 20 to 2000 cm^3 of oxygen. The major concern with oxygen scavenging is that there should be no direct contact of food material and ferric oxide that has been embedded inside the package. The iron-based compounds could possibly be replaced with the ascorbic acid or its derivatives, and to neutralize the action of ascorbic acid, there is support of alkaline compounds, such as aluminum and siliceous iron.[51]

With the increasing demand for the better shelf-life of food, different food companies are developing various types of oxygen scavengers, which work by gas flushing via modified atmosphere packaging (MAP), such as Ageless®, Cryovac OS2000, ATCO®, enzyme-based FreshPax®, OxyCatch®, and OxyGaurd®.[44,55] With the increased demand of improved version of oxygen scavengers, the iron nanoparticles usage has become prevalent, which under moisture-free conditions react with oxygen. This could also show the better view of package along with moisture control. The oxygen scavengers are majorly used for controlling the growth of microbes in the products, such as cheese, cakes, breads, pasta, coffee, beer, meat, fish, and beverages.[25]

The enzymatic oxygen scavenging systems are usually sensitive to the environmental factors like pH, water, temperature, humidity, type of substrate present, other product situations, and keeping quality. Figure 1.2 represents the interaction of the oxygen molecules with the food particles inside a food package that consists of different layers, acting as a barrier against the environment and these layers protect food as they minimize the migration of oxygen-absorbing system components into food substance.

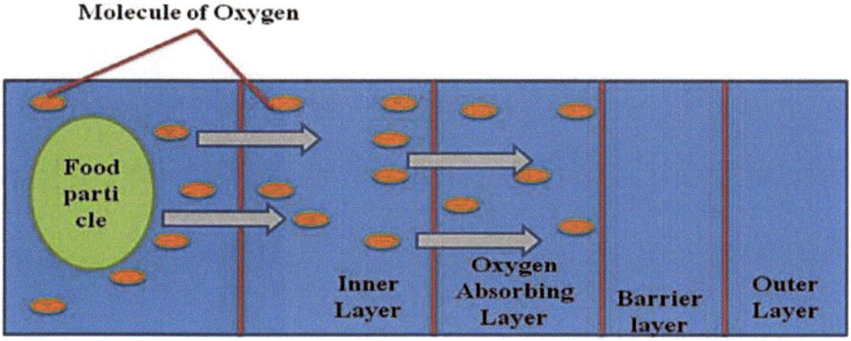

FIGURE 1.2 Representation of oxygen-absorbing system with a multi-layer active foil.

With the increasing developments in the oxygen scavenger system, a nanoscale oxygen scavenger has been reported that is based on silicon matrix with a null valent iron particle. These iron particles of nanosize exhibited an oxygen scavenging rate, which is ten times higher in testing sample polymer matrix under 100% humidity.[20]

1.3.2.3.2 Carbon Dioxide Scavengers and Emitters

The inert nature of carbon dioxide is an excellent way to preserve food products and its antimicrobial feature also add-ins to its features. The total amount of CO_2, which is available in the headspace, has a major impact on the end product quality. The usage of CO_2 emitters in the food packages may allow reduction in the package size, enhanced filling, better transport efficiency, and overall reduction in the environmental impact. The package deformation can also be prevented by the CO_2 emitter as its absorption is compensated by food product inside food package during the initial storage. Therefore, this generates a negative pressure inside the food package to aid in the enhancement of a drip loss that may create an unattractive look of the package.[63]

The presence of CO_2 also hampers the growth of microbes inside the food package. Because due to the presence of carbon dioxide, the growth of microbes gets reduced on the surface of meat products and reduction in respiration rate occurs. Carbon dioxide is more permeable than oxygen, and it may enter the food products through plastic films. Motive to enhance shelf-life of highly perishable foods could be achieved by such a system with an oxygen scavenger and a carbon dioxide emitter.

To scavenge carbon dioxide from the sachet of freshly roasted coffee beans, CO_2 absorbing sachet is used, which is made up of porous envelope containing calcium oxide along with a hydrating agent like silica gel, on which water is absorbed. This system helps, as water and calcium oxide react and produce calcium hydroxide and it further reacts with CO_2 to form calcium carbonate, and this system was developed by Multiform Desiccants Incorporated.

1.3.2.3.3 Moisture Scavenger

Moisture is one of the major causes of degradation of food material inside the package; hence, its removal becomes necessary. Due to excess moisture present inside package, a foggy layer is formed on the product. As the product respires, the accumulation of water begins inside the food package and this is also affected by temperature fluctuations and relative humidity. The major concern for the moisture comes for the meat and poultry products that contain huge amount of water and energy source that is a favorable environment for the microbes to grow. The package of meat products should be selected carefully as the accumulation of water enhances inside the package due to low permeability.[45]

There are many different emerging technologies, which aid in moisture removal, such as calcium oxide, calcium chloride, modified starch, molecular sieves, natural clay, and silica gel. Out of these methods, silica gel is commonly used as it is noncorrosive and nontoxic.

To absorb moisture, silica gel can be incorporated in the film of laminate, which has a layer for moisture absorbance that is made from a copolymer of polyester graft and a polyurethane resin, vinyl resin, and an acrylic resin. This is initially used in small sachets, such as chips, crackers, biscuits, and instant coffee. US Multisorb Technologies Inc. has used this technology for the moisture removal from the food package. There are many such technologies available in the market to provide moisture absorber for various food products.

1.3.2.3.4 Ethylene Absorber

Ethylene is a volatile molecule with a capacity to act as phytohormone, which is responsible for ripening of vegetables and fruits.[4] The control of ripening of fruits and vegetables after harvesting is always a matter of concern as they promote the degradation of end product. There is need to check the release of ethylene so as to enhance the keeping quality of end product until it reaches the end consumer via all the midway barriers. The most commonly used ethylene scavenger is potassium paramagnet, which is present on the inner matrices that are composed of silica. The change is observed as the oxidation begins and there is color change from purple to brown. This concept of color change was used to develop a nanostructured platform, which is aluminum and silica-based nanoparticles infused with potassium permanganate and indicates color change to justify the removal of ethylene.[60] Furthermore, direct usage of potassium permanganate could be harmful to the food due to severe toxicity[4,61] and it could be used in the sachet form for tomato storage for the ethylene scavenger system.[32] There are various ethylene scavengers available in the market by different companies, such as EvertFresh green Bags®, PEAKfresh®, and Ethysorb®.[4,56,61]

1.3.2.3.5 Ethanol Emitters

Ethanol emitters are used to enhance the shelf-life of food products by reducing the microbial load (especially mold growth) under the effect of ethanol. It is mainly used for bakery products, like bread and cookies. The ethanol is present in carrier material, which is further exchanged with water absorbed by the carrier material.[49] The potential of ethanol emitters generally depends on the size and type of carrier material, permeability of sachet material to ethanol and water vapor, quantity of entrapped ethanol by carrier material, packaging film permeability for ethanol, and the food material water activity. To mask the odor of alcohol present in ethanol, some sachet may also contain some flavoring agents, for example, vanilla, or other flavor like cocoa.

The ethanol emitter sachet contains a food grade ethanol encapsulated in a carrier material and its release is regulated by the water vapor permeability of the material of sachet. The ethanol existing in the material of carrier is absorbed during exchange. There are following advantages of this system:

- It eliminates the use of other methods, which utilize benzoates and sorbates, for mold growth inhibition.
- It helps in the delay of staling of bakery products
- Ethanol vapor is directly produced from ethanol-producing sachets inside the package that inhibit the direct contact of food with ethanol, thus making food safer as compared to that where ethanol spraying is done prior to food packaging.
- Few food sachets packages contain label of "Do Not Eat" that could be safely removed and discarded from the package.
- The major issue related to ethanol emitter is the controlled release of ethanol, which further creates problems in the headspace of a food package. As the ethanol-imbedded films may also require additional film to hold the ethanol and regulate its controlled release, this ultimately increases the production cost. Sometimes food may also absorb ethanol vapor, which could finally affect end consumers. It is advised to microwave or heat the ethanol containing food products to reduce its harmful effects.

1.3.2.3.6 Flavor Emitters/Absorbers

Generally, flavor enters or escapes from food matrix by reacting to its primary packaging material, which is known as flavor scalping or degradation. Flavor degradation usually occurs at high-temperature cooking or when packaging is done at high temperature. The loss of flavor could be substituted by the use of high barrier plastic that holds food flavor inside the package. At times, flavor releasing becomes necessary so that when consumer opens the food package, it should smell good and this could be achieved particularly, when the layer of heat seal of package has affinity to flavor.

Off-flavor majorly occurs due to: (a) breakdown of protein present in meat products like fish, poultry into amines, (b) oxidation of oils and fats present in food matrix, which leads to formation of aldehydes. Amines and aldehydes can be removed from the package by an appropriate usage of flavor emitter.

Few coffee manufacturers add some volatile compounds in the headspace so that when the consumer opens the package for the first time, it should smell nice and acts like a fresh product.[11] In some cases, flavor could degrade original flavor of food like in case of orange juice, as the product being pasteurized may develop some bitter flavor when limonin concentration increases up to 12 mg/kg. Further, this could be reduced by the usage of internally cellulose acetate butyrate coated plastic bottles, which will help in limonin absorption and could reduce the bitterness up to some extent.

1.3.2.4 TIME–TEMPERATURE INDICATORS

The small variation in the ambient temperature can create havoc for the food product. For close monitoring of the temperature variation in the food package, the time–temperature indicator (TTI) was developed. TTI aids in observing the food changes that took place during transportation, storage, and handling and until the end-consumption. The FDA also issued guidelines for time and temperature control of the food under time temperature control for food safety (TCS),[21] to cease and minimize the microbial load at every possible step of food processing. Even a small change duration of storage of food along with variation in temperature management could lead to food degradation at a large scale[56,58] for the entire lot of production.

The usage of TTI in food industry is possibly done due to its small size and cost-effective nature as compared to other devices. The working principle of TTI is the recognition of reaction that is irreversible in manner and mainly of enzymatic, biological, electrical, and chemical changes occurring due to exposure to temperature variation with predetermined duration of time.[17] There are three types of TTI in the market, such as[50]:

- **Entire history indicator** observes the entire product during its journey, even a minute fluctuation in its time and temperature,
- **Critical temperature indicator** reflects whether the changes in temperature occurred or not, and
- **Complete and critical TTI** observe changes of both time and temperature simultaneously as depicted.

With the enhancement of technology and science, TTI could possibly be able to represent itself as alarming devices that can provide information for the minute change inside the food package that hinders the food quality.[59]

In the active packaging system, temperature and time plays a crucial role in maintaining the end-product quality during its handling, storage, and transportation. Degradation caused by time and temperature would lead to food poisoning and food losses. The designing of these packaging systems is done in such a manner that as the deterioration of food matrix begins, change in color appearance would also be noticeable. The change indicators indicate few considerations that should be observed closely, such as: (i) change in physical and chemical characteristics of indicators; (ii) check for temper evident on indicator; and (iii) indicators should be easily activated and sensitive enough to sense minor changes.

These indicators aid in monitoring changes in food matrix by analyzing physical and chemical characteristics of indicators that should be checked. The time–temperature indicator shows temperature change inside the package, and for this many surveys are conducted.[37] For commercialization, TTI should fulfill certain requirements, such as: it should be easy to activate and use, respond quickly and accurately for the temperature fluctuations, and response must be irreversible and must correlate with actual damage to the food.

1.3.2.5 ANTIMICROBIAL FILMS

Microbial degradation is the leading cause of food degradation even inside the packed food. There are many different ways to prevent the growth of microbes through pasteurization, canning, freezing, evaporation refrigeration, and fermentation. Along with this method, there are some chemical preservatives to enhance the shelf-life further. The main antimicrobial components begin with the inactivation of microbial activities and halting their multiplication rate. The antimicrobial package must contain the chemical or natural antimicrobial agents that hinder growth of microbial load inside the package. Generally, chemical antimicrobial agents include sorbic acid, acetic acid, propionic acid, citric acid, and lactic acid, and these are categorized under acids as a preservative. Along with this, there are some other types of antimicrobial agents like bacteriocin, extract of natural plants, ethanol that is particularly effective for bacteria and out of these bacteriocin that include diolococcin, nisin, and lacticins. Among these bacteriocin and natural plant extracts came under natural antimicrobial agents.[40] There are US companies that manufacture the antimicrobial food package at the commercial level, such as: AgIONe that focus on extracts of nisin and for incorporated ingredients and films micrograde Rhone-Poulenc.[31]

Figure 1.3 shows the multi-layered film with antimicrobial properties, which consists of outer layer, then barrier layer, then layer of matrix, and finally the control layer, which lies closest to the food. In this, substance having antimicrobial nature is entrenched in the layer of matrix, which is released in a controlled manner by the control layer.

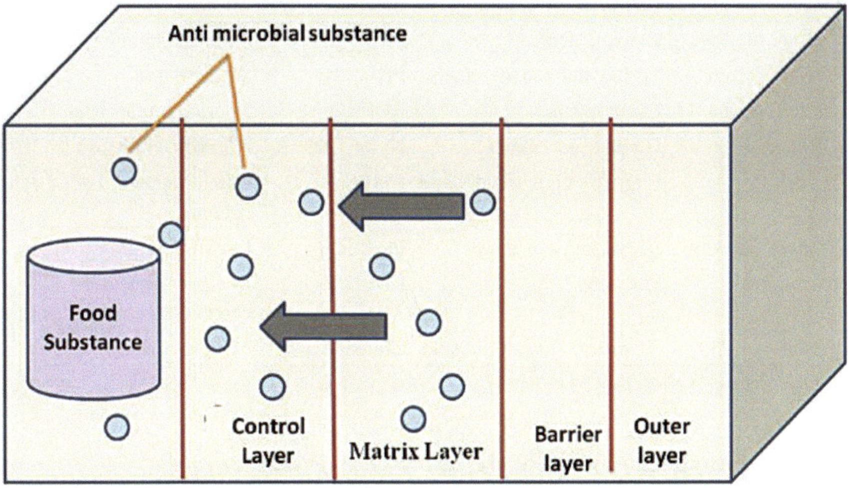

FIGURE 1.3 Representation of an antimicrobial film showing interaction between food and multi-layer active system.

The function for the release of antimicrobial substance inside the food package is always an eye-catching event for food industries, as it could also be harmful along with having its enormous benefits. The surface of antimicrobial packaging material must contain the agents that help in the migration of antimicrobial substance to the food or target and also exhibit the functional property to hinder microbial growth at a particular site for a fixed duration of time in respect of food.[24,40]

Generally, the addition of antimicrobial substances occurs in those food products, which already carry some more enzymatic and proteinaceous activity as compared to other food products like meat and meat products. The antimicrobial substance of food package has the capacity to cease the active compound in meat product and affect the microbial life (Table 1.1). There is also use of silver in antimicrobial package that aids in enhancing the shelf-life and it also contributes as an alternative to zeolite that was previously used as an antimicrobial agent.[2,34] There is example of silver zeolite, such as AgIon, Bactekiller, Zeomic, and these could be directly incorporated in the packaging polymer.[6]

TABLE 1.1 Representation of Antimicrobial Agents with Packaging Systems.

Food	Antimicrobial	Packaging material	Microorganisms
Breakfast cereal	BHT	HDPE	Molds
Cheese	Imazalil	LDPE	Molds
Culture media	Benzoic acids	PE	Total Bacteria
Culture media	Sorbic and benzoic acids	PE	*S. cerevisiae*
Culture media	Clove extract	LDPE	Microorganisms
Ground beef	Grape seed extract	Nylon, LDPE	Aerobes, Coli-forms
Ham	Eugneol, cinnamaldehyde	Chitosan	Enterobacteria, Lactic acid bacteria
Meat or chicken	Allyl isothiocyanate	PE Film	E. *coli. L. monocytogenes*
Stimulants	Parabens	LDPE	*Penicellium spp*, S. cerevisiae, *Aspergillus niger*
Water	Acetic and propionic acid	Chitosan	*Penicellium spp*

Source: Adapted from Ref. [2]

1.3.3 MODIFIED ATMOSPHERE PACKAGING (MAP)

Modified atmosphere packaging (MAP) is the most widely accepted technique for maintaining the keeping quality of the food. It is mainly done for the enhancement of shelf-life of the food product, which is attained by controlling the gaseous atmosphere inside the food package. The gaseous atmosphere inside a package consists of 78.09% nitrogen, 20.95% oxygen, 0.93% argon, and 0.03% carbon dioxide. This proportion of gasses could be modified according to the food package. This combination of gasses is highly recommended as it does not allow the permeation from the package, and selected gasses do not impose any harmful behavior on the food product.[34]

The determination of gases would also aid in determining the end-product quality by interplaying between gas and package as these control respiration rate, transpiration rate of any food product. The combined effect of these parameters will help in establishing a better shelf-life. Respiration need of the food package is controlled by gas blend, which is finely adjusted to meet specific respiration needs for each packaged food. The barrier substrates like foils, plastic films become MAP after they are formed into trays, bags, or lids and filled with fixed mixture of gasses (such as nitrogen, carbon dioxide, and oxygen). The gasses are present inside package mix and reduce loss of color, flavor, taste, odor resulting from product degradation, spoilage, and rancidity, which is caused by mold growth and other anaerobic organisms.

It is generally called for the enhancement of keeping time for sea food, prepared food, and minimally processed food. It also includes market food like pasta, baked products, and cured meat and cheese.[28]

The major usage of MAP involves, where the high-respiration food products are stored like fresh-cut fruit and vegetables. The changes in the food are noted by observing alternations in color, texture, appearance, and flavor of food.[19] Usually in the modification process, the amount of oxygen gets reduced from 20.9 to 0%, thus enhancing the amount of carbon dioxide and nitrogen. The replaced oxygen can be fulfilled with inert gas such as nitrogen and carbon dioxide, which may help in lowering pH or inhibit microbial growth. Further gas balance could be achieved by active packaging techniques like gas flushing or by planning the EMAP that is recognized as equilibrium modified atmosphere packaging and sometime also called breathable films. The modified atmosphere package has been commercialized for chilled meat packages.[28]

1.4 ROLE OF NANOTECHNOLOGY IN FOOD PACKAGING

The emerging trends of nanotechnology have been useful for food industries by incorporating novel modifications in the food. Nanotechnology deals with molecules at a nanometer scale with approximate size of 1–100 nm to create the materials, which have unique properties like color, solubility, diffusivity, strength, and optical.[23,41] The nontoxic nature, stability at high temperatures, and pressure of nano particle-based materials are a quenching trend for the food industry.[45,46]

The relationship between nanotechnology and food could be accomplished in two ways, such as food nano sensing and food nanostructured ingredients. The nano-structured food like antimicrobial agents, anticaking agents, durability of packing material, and mechanical strength could be implemented into food packaging material. While food nano sensing could be used to analyze safety evaluation and food quality,[16] the usage of nanotechnology in food enhances better precautions against plague of microbial load.[40] Along with nano sensing and nanostructured food, nanocarriers are also being utilized as delivery systems without disturbing their morphology.[16] Such a delivery system consists of properties, such as: (i) assure accessibility of target-specific rate and time, (ii) capable enough to hold lead component for a fixed duration of time in a worthy manner, and (iii) capable to extradite dynamic compound at a predefined objective site. The relationship between nanotechnology and food could be predicted in the form of emulsion, encapsulation, associated colloids, simple solutions, and biopolymer matrices with the aid of nano sensors, which can monitor the changes in microbial load.[8]

The concept of food versus microorganism is always a concern of food decay that allows more enhancements in technology to control the growth of microbes inside the food package. The nanotechnology could possibly check the growth of microbes inside food package with the aid of nano biosensors, which are capable enough of indicating the decay of food either to end consumer or to the supplier.[14,26] Nano sensors are adaptive to observe the change in the microbial growth, product degradation, fluctuation in temperature of storage area, humidity and change in environmental situation. For microbial load enhancement, these are attached to nano biosensors like nanorods, nanofibers, and thin films that are required to check and monitor the growth of microbes inside food package.[29] Along with this modification, an immunosensor could also be used for the detection of microbial contamination that has contributed toward quick and sensitive detection system. It immobilizes the protein molecule or particular antigen–antibody complex over the sensor chip or nanofilms that produce signals for the detection of a target molecule.

Furthermore, immune-sensors based on the dimetylsiloxane microfluidic system are incorporated with a particular antibody that was immobilized over alumina nano-porous membrane, which was developed for *Staphylococcus aureus and Escherichia coli* O157:H7.[53] The major use of nanoencapsulation techniques is for the release of better taste and mouthfeel, culinary balance, and retention and delivery of flavor.[38] Another case study has revealed that highly unstable and reactive plant pigments could also be nano-encapsulated for better control of their bioactive properties like by encapsulating the cyaniding-3-O-glucoside, which is present in soybean seeds H-2 subunit ferritin (rH-2) that proved to be useful in improving the thermostability and photostability.[62]

1.5 RECENT INNOVATIONS IN FOOD PACKAGING

1.5.1 RADIO FREQUENCY IDENTIFICATION TAGS (RFID)

The usage of radio frequency is focused over a chip that contains tags for storing data or information related to the food product.[30] The chip and radio waves work in a manner that it is possible to trace and retract the information of the product from both ends, that is, consumer and producer. The tracing of information occurs with the aid of antenna that transmits the signal and data is further converted into readable form and is analyzed for further understanding.[49]

The RFID system with a battery can be active or passive type. In the active type system, the battery is utilized as a power source while in the passive type system power generated from signals is received. The range of frequency is as low as (125–134 kHz) and as high as (13.56 MHz).[7] The transmission of a signal occurs via an antenna between the reader and the tag, which acts as an intermediate medium between electromagnetic waves and the server.[13] The received information is primarily analyzed by the reader from tag for further processing. The usage of an electronic chip is possibly for encoding information in the form of bits in an electronic product code.[7] The limitation with the electromagnetic waves is in the format of the distance and their effectiveness. However, in RFID, the electromagnetic waves can travel upto a range of 30 m to 4.5 m. The major use of RFID is considered in automobile industry, pharmaceutical products, bakery products, beverages, fishery products, and pets and in livestock that could possibly check the location of product by tracing microchips.

The international standards, IEC/ISO 18000 and 2916, describe more specific regulations for the use of RFID for data security and also to avoid data and monetary breaching. These microchips are further encrypted that provide a more secure way to store information about the product in this fast-growing economy. The RFID is contributing approximately 7.77 billion US$ in 2013 and its expected growth could be possibly expected to be 18.68 billion US$ by 2026 as per RFID forecast (2014–2024).[22]

1.5.2 BARCODES

Barcode is 12-digit codes, which consist of bars with some space and digits and alphabets that contain information and it is recognized as Universal Product Code (UPC). Barcode is used to track package for its in and out flows for a particular product for which it is designed. The relation of barcode and food industry is used for the tracing of food products and their information about end consumption date, price, and manufacturing details. Along with the UPC, there is a modified system that is also utilized in food industry and recognized as a portable data file (PDF), which is a two-dimensional set that carries the information and it is more informational than the UPC. The PDF is used to carry serving instruction, graphics, and nutritional composition.[50]

Both these technologies fall under the system that is required to reduce the storage space and it is called Reduced Space Symbology (RSS), which aids in coping up with storing information in small space and deals with better food packaging issues. The RSS system includes 14-digit numbers and it is mainly used as global trade item number (GTIN) and GTIN is used for loose products, such as fruits and vegetables.[49]

1.6 FOOD PACKAGING AND NONMIGRATORY BIOACTIVE POLYMERS (NMBP)

In this method of food packaging, there is migration of some bioactive compounds that is required to be hindered for the betterment of food package. The NMBP consists of polymers that have acquired some active properties of some specific biological agents that have no suppressing effects on the leaching out of active materials from polymer to substrate. There are various types of active substances, such as proteins, enzymes, peptides, chitosan, and some other biological molecules.

The NMBP can be used as antimicrobial package by bioactive agents, which hinder the survival and mechanism of microorganisms. The decreased count of microbes leads to the enhancement of shelf-life. The coating of bioactive material could be done at a particular point of entry of food or at the entire inner layer of the food package. Generally, enzymes are used as bioactive compounds as they are more efficient and also considered natural immobilizer as this supports the polymer backbone in the food package. The working concept of enzyme is based on the lock and key concept, and in NMBP, there are many enzymes, which are particular for a particular food; therefore, the selection of enzymes is always a matter of concern for halting the migration of compounds. The following are the examples of case studies, which were performed on the food and NMBP and for checking its effects, for example:

- **Milk package lactose immobilization**: the lactose in milk creates health issues so that the content of lactose should be low in lactose-intolerant persons. This special food package aids in providing less lactose by the activity of lactase enzyme and this could be done during the processing line to save time, energy, and cost as only the NMBP pack could lower the lactose content;
- **Naringinase immobilization for the citrus juice packaging material**: The major cause of bitterness in citrus products after storage is

naringin. The hydrolysis of naringin is caused by the enzyme naringinase and it aids in better storage of citrus fruit juices like grapefruit juice;

- **Reductase enzymes** are required to limit the cholesterol content of some specific food by cholesterol reductase enzyme. These are few cases that represent the NMBP and food relation in the betterment of health.

The relation of natural antimicrobial bioactive compound has been in usage since ages. The utilization of this concept is also tried in food packaging via NMBP to affect the microbial load. There are some of the majorly recognized bioactive compounds, such as chitosan and nylon.

Chitosan is produced from the deacetylation of chitin that is naturally present in the crustacean shell. The processing of chitin is further required in the form of demineralization, deproteinization and followed by deacetylation. The degree of deacetylation of chitin affects the mechanical properties and antimicrobial activity. The mechanism behind chitosan antimicrobial role is defined by the polycationic role that reacts with the microbial cell membrane that contains polyanionic compounds; it aids in the enhanced permeability of the cell membrane that causes leaching out of inner materials and hence it is generally used as the antimicrobial coating. This activity has been analyzed by many scientists against *S. typhirium, E. coli*, and *P. aeruginosa.*

The factors that affect chitosan activity are temperature, pH, and $MgCl_2$. The ions of Mg^{2+} negatively affect the chitosan activity and temperature and pH positively affect the chitosan activity.

The nylon is not that much effective as an antimicrobial factor but when irradiated with UV rays at 193 nm, the surface structure of nylon gets modified and it aids in its antimicrobial properties. It also has similar functioning like chitosan, and due to irradiation, the surface becomes catchier for microbial load.[42]

1.7 FUTURE TRENDS IN FOOD PACKAGING

With the innovation in various fields of science and technology, it has become an eye-catching view for industrialists for achieving a new and better way of packaging. There are different industries that are vigorously involved in making packaging technology more relevant for food technologists to enhance the keeping quality and also indicate the changes inside the package as described in this chapter. There are some difficulties in food packaging like heavy weight of glass container or bottles.

Glass container is more prone to damage by breaking or cracking, along with heavy weight of big metal containers that are used to store a huge amount of food. There is a need to make packaging more relevant in view of material substitution, light weight, smart labels sustainability, and biobased packaging material. In light of the material substitution, it could be understood that a glass bottle could be replaced by a PET bottle of the same size and capacity and that is less prone to damage and also is light in weight. Such a type of innovation is always a matter of concern for the industrialist that could properly functionalize its basic properties and also be a better option from the environmental point of view.[27]

1.8 SUMMARY

This chapter includes novel processing methods along with various packaging technologies like cold plasma, high-pressure processing, active and intelligent packaging, nonmigratory bioactive packaging material, and edible and biodegradable films. This chapter also includes the relationship between the nano-science and its application in food packaging. The concept of different emerging technologies could help us in understanding the new ways to develop a better way to package, store, and sell the food with more attractive and smart packaging that includes different sensors. With the development of new trends, food packaging is the most innovative way to check the growth of microbial load.

KEYWORDS

- **absorber**
- **emitter**
- **indicators**
- **modified active packaging**
- **nano-technology**
- **novel packaging**
- **scavengers**

REFERENCES

1. Adams, A.; Van Bloois, S.; Otte, B.; Caro, E.; Mekap, D.; Sandkuehler, P. Flavor Scalping by Polyethylene Sealants. *Food Pack. Shelf-life* **2019,** *21*, 1–11.
2. Altaf, U.; Kanojia, V.; Rouf, A. Novel Packaing Technology for Food Industry. *J. Pharmacogn. Phytochem.* **2018,** *7* (1), 1618–1625.
3. Álvarez-Hernández, M. H.; Artés-Hernández, F.; Ávalos-Belmontes, F.; Castillo-Campohermoso, M. A.; Contreras-Esquivel, J. C.; Ventura-Sobrevilla, J. M.; Martinez-Hernandez, G. B. Current Scenario of Adsorbent Materials Used in Ethylene Scavenging Systems to Extend Fruit and Vegetable Postharvest Life. *Food Bioprocess Technol.* **2018,** *11* (3), 511–525.
4. Arvanitoyannis, I. S.; Oikonomou, G. Active and Intelligent Packaging. In: *Modified Atmosphere and Active Packaging Technologies*; Arvanitoyannis, I. S., Ed.; CRC Press: Boca Raton, 2012; pp 628–654.
5. Appendini, P.; Hotchkiss, J. H. Review of Antimicrobial Food Packaging. *Innov. Food Sci. Emerg. Technol.* **2002**, *3*, 113–126.
6. Aziz, M.; Karboune, S. Natural Antimicrobial/Antioxidant Agents in Meat and Poultry Products as well as Fruits and Vegetables: A Review. *Crit. Rev. Food Sci. Nutr.* **2018,** *58* (3), 486–511.
7. Bibi, F.; Guillaume, C.; Gontard, N.; Sorli, B. A Review: RFID Technology Having Sensing Aptitudes for Food Industry and their Contribution to Tracking and Monitoring of Food Products. *Trends Food Sci. Technol.* **2017**, *62*, 91–103. https://doi.org/10.1016/j.tifs.2017.01.013
8. Bratovˇci´c, A.; Odobaši´c, A.; ´Cati´c, S.; Šestan, I. Application of Polymer Nanocomposite Materials in Food Packaging. Croat. *J. Food Sci. Technol.* **2015**, *7*, 86–94.
9. Brayfield, R. S.; Jassem, A.; Lauria, M. V.; Fairbanks, A. J.; Keener, K. M.; Garner, A. L. Characterization of High Voltage Cold Atmospheric Plasma Generation in Sealed Packages as A Function of Container Material and Fill Gas. *Plasma Chem. Plasma Process.* **2018**, *38*, 379–395.
10. Biji, K. B., Ravishankar, C. N., Mohan, C. O., & Srinivasa Gopal, T. K. Smart Packaging Systems for Food Applications: A Review. *J. Food Sci. Technol.* **2015,** *52*, 6125–6135.
11. Brody, A. L.; Bundy, J. A. Enzymes as Active Packaging Agents. In: *Active Food Packaging*; Rooney, M. L., Ed.; Blackie Academic & Professional: Glasgow, 1995; pp 174–192.
12. Caner, C.; Hernandez, R. J.; Harte, B. R. High-Pressure Processing Effects on The Mechanical, Barrier and Mass Transfer Properties of Food Packaging Flexible Structures: A Critical Review. *Pack. Technol. Sci.* **2004**, *17* (1), 23–29.
13. Chen, R. S.; Chen, C. C.; Yeh, K. C.; Chen, Y. C.; Kuo, C. W. Using RFID Technology in Food Produce Traceability. *WSEAS Trans. Info. Sci. App.* **2008,** *5* (11), 1551–1560.
14. Cheng, Q.; Li, C.; Pavlinek, V.; Saha, P.; Wang, H. Surface-Modified Antibacterial Tio2/Agc Nanoparticles: Preparation and Properties. *Appl. Surface Sci.* **2006,** *252*, 4154–4160.
15. Cooksey, K. Oxygen Scavenging Packaging Systems. *Encyclopedia of Polymer Science and Technology*; John Wiley & Sons, 2010; p 1. https://onlinelibrary.wiley.com/doi/full/10.1002/0471440264.pst570

16. Ezhilarasi, P. N.; Karthik, P.; Chhanwal, N.; Anandharamakrishnan, C. Nanoencapsulation Techniques for Food Bioactive Components: A Review. *Food Bioproc. Technol.* **2013,** *6*, 628–647.
17. Fang, Z.; Zhao, Y.; Warner, R. D.; Johnson, S. K. Active and Intelligent Packaging in Meat Industry. *Trends Food Sci. Technol.* **2017,** *61*, 60–71.
18. Farmer, N., Ed. *Trends in Packaging of Food, Beverages and Other Fastmoving Consumer Goods—FMCG*; Woodhead Publishing: PWN, Warszawa, 2016; p 322.
19. Finnegan, E.; O'Beirne, D. Characterising and Tracking Deterioration Patterns of Fresh-Cut Fruit Using Principal Component Analysis—Part I. *Postharv. Biol. Technol.* **2015,** *100*, 73–80.
20. Foltynowicz, Z.; Bardenshtein, A.; Sängerlaub, S.; Antvorskov, H.; Kozak, W. Nanoscale, Zero Valent Iron Particles for Application as Oxygen Scavenger in Food Packaging. *Food Pack. Shelf-Life* **2017**, *11*, 74–83.
21. Food and Drug Administration (FDA). Recommendations of the United States Public Health Service, National Technical Information Service Publication PB2013–110462, 2013. https://www.fda.gov/Food/GuidanceRegulation/RetailFoodProtection/FoodCode/ucm374275.htm
22. Ghoshal, G. Recent Trends in Active, Smart, and Intelligent Packaging for Food Products. *Food Pack. Preserv.* **2018,** 343–373.
23. Gupta, A.; Eral, H. B.; Hatton, T. A.; Doyle, P. S. Nanoemulsions: Formation, Properties and Applications. *SoftMatter* **2016**, *12*, 2826–2841.
24. Han, J. H. Antimicrobial Food Packaging: Part 1. In: *Novel Food Packaging Techniques*; Ahvenainen, R., Ed.; CRC Press: Washington, DC, 2003; p 50e70.
25. Haghighi-Manesh, S.; Azizi, M. H. Active Packaging Systems with Emphasis on Its Applications in Dairy Products. *J. Food Process Eng.* **2017,** *40*, 1–13.
26. Helmke, B. P.; Minerick, A. R. Designing A Nano-Interface in A Microfluidic Chip to Probe Living Cells: Challenges and Perspectives. *Proc. Natl. Acad. Sci. USA* **2006**, *103*, 6419–6424.
27. Hirvikorpi T.; Laine R.; Vähä-Nissi M.; Kilpi V.; Salob E.; Li W.-M.; Lindfors S.; Vartiainen J.; Kenttä E.; Nikkola J.; Harlin A.; Kostamo J. Barrier Properties of Plastic Films Coated with an Al2O3 Layer By Roll-To-Roll Atomic Layer Deposition. *Thin Solid Films* **2014,** *550*, 164–169.
28. Inns, R. Modified Atmosphere Packaging. Chapter 3. In: *Modern Processing, Packaging and Distribution Systems for Food*; Paine, F. A., Ed., Vol. 4; Blackie and Son: Glasgow, 1987; pp 36–51.
29. Jianrong, C.; Yuqing, M.; Nongyue, H.; Xiaohua, W.; Sijiao, L. Nanotechnology and Biosensors. *Biotechnol. Adv.* **2004,** *22*, 505–518.
30. Kalpana, S.; Priyadarshini, S. R.; Maria Leena, M.; Moses, J. A.; Anandharamakrishnan, C. Intelligent Packaging: Trends and Applications in Food Systems. *Trends Food Sci. Technol.* **2019,** *93*, 145–157.
31. Kerry, J. P.; O'grady, M. N.; Hogan, S. A. Past, Current and Potential Utilization of Active and Intelligent Packaging Systems for Meat and Muscle-Based Products: A Review. *Meat Sci.* **2006,** *74* (1), 113–130.
32. Köstekli, M.; Özdzikicierlev, O.; Cortés, C.; Zulueta, A.; Esteve, M. J.; Frígola, A. Role of Potassium Permanganate Ethylene on Physicochemical Properties, During Storage of Five Different Tomato Cultivars. *MOJ Food Process. Technol.* **2016,** *3* (2) 281–289.

33. Kozak, W.; Cierpiszewski, R. Opakowania Aktywne Active Packaging. *Przemysł Spożywczy* **2010**, *64* (3), 36–38.
34. Malhotra, B.; Keshwani, A.; Kharkwal, H. Antimicrobial Food Packaging: Potential and Pitfalls. *Front. Microbiol.* **2015,** *6*, 1–9.
35. Mauriello, G.; Ercolini, D.; La Storia, A.; Casaburi, A.; Villani, F. Development of Polythene Films for Food Packaging Activated with an Antilisterial Bacteriocin From *Lactobacillus curvatus* 32Y. *J. Appl. Microbiol.* **2004**, *97*, 314–322.
36. Minisrty of Agriculture, Fisheries and Food. *Time-Temperature Indicators: Research into Customer Attitudes and Behaviour*; National Consumer Council: London, 1991.
37. Millette, M.; Le Tien, C.; Smoragiewicz, W. Lacroix, M**.** Inhibition of *Staphylococcus aureus* On Beef by Nisin-Containing Modified Alginate Films and Beads. *Food Control* **2007,** *18*, 878–884.
38. Nakagawa, K. Nano- and Micro-Encapsulation of Flavor in Food Systems. Chapter 10; In: *Nano- and Microencapsulation for Foods*; Kwak, H. S., Ed.; John Wiley & Sons: Oxford, 2014; pp 249–272.
39. Pradhan, N.; Singh, S.; Ojha, N.; Srivastava, A.; Barla, A.; Rai, V.; Bose, S. Facets of Nanotechnology as Seen in Food Processing, Packaging, and Preservation Industry. *BioMed Res. Int.* **2015**, *2015* (365672), 1–17.
40. Quintavalla, S.; Vicini, L. Antimicrobial Food Packaging in Meat Industry. *Meat Sci.* **2002,** *62* (3) 373–380.
41. Rai, M.; Yadav, A.; Gade, A. Silver Nanoparticles as a New Generation of Antimicrobials. *Biotechnology* **2009,** *27*, 76–83.
42. Ramirez, R.; Saraiva, J. A.; Perez Lamela, C.; Torres, J. A. Reaction Kinetics Analysis of Chemical Changes in Pressure Assisted Thermal Processing, PATP. *Food Eng. Rev.* **2009,** *1* (1), 16–30.
43. Rani, U.; Kumar, P. Non Migratory Bioactive Polymers Pkg Part of Active Packaging System. *FandBnews.com*, 2019. http://www.fnbnews.com/Top-News/nonmigratory-bioactive-polymers-pkg-part-of-active-packaging-system-51667
44. Rastogi, N. K.; Raghavarao, K. S. M. S.; Balasubramaniam, V. M.; Niranjan, K.; Knorr, D. Opportunities and Challenges in High Pressure Processing of Foods. *Crit. Rev. Food Sci. Nutr.* **2007,** *47*, 69–112.
45. Realini, C. E.; Marcos, B. Active and Intelligent Packaging Systems for a Modern Society. *Meat Sci.* **2014,** *98*, 404–419.
46. Roselli, M.; Finamore, A.; Garaguso, I.; Britti, M. S.; Mengheri, E. Zinc Oxide Protects Cultured Enterocytes from the Damage Induced by Escherichia Coli. *J. Nutr.* **2003**, *133*, 4077–4082.
47. Sajilata, M. G.; Savitha, K.; Singhai, R. S.; Kanetkar, V. R. Scalping of Flavors in Packaged Foods. *Comprehen. Rev. Food Sci. Food Safet* **2007,** *6*, 17–35.
48. Sawai, J. Quantitative Evaluation of Antibacterial Activities of Metallic Oxide Powders (Zno, Mgo And Cao) by Conductimetric Assay. *J. Microbiol. Method* **2003**, *54*, 177–182.
49. Schaefer, D.; Cheung, W. M. Smart Packaging: Opportunities and Challenges. *Procedia CIRP* **2018,** *72*, 1022–1027.
50. Sen, C.; Das, M. Trends in Food Packaging Technology. Chapter 1; In: *Food Process Engineering: Emerging Trends in Research and Their Applications*; Meghwal, M.; Goyal M. R., Eds.; Apple Academic Press (AAP)/CRC Press, NJ, 2016; pp 1–22**.**
51. Sharma, R.; Ghoshal, G. Emerging Trends in Food Packaging. *Nutr. Food Sci.* **2018,** *48* (5), 764–779.

52. Singh, R. P. Scientific Principles of Shelf-Life Evaluation. Chapter 1; In: *Shelf-Life Evaluation of Food*; Man, D.; Jones, A. J., Eds., Vol. 1; Aspen Publishers: Gaithersburg, MD, 2000; pp 3–26.
53. Spricigo, P. C.; Foschini, M. M.; Ribeiro, C.; Corrêa, D. S.; Ferreira, M. D. Nanoscaled Platforms Based on SiO_2 And Al_2O_3 Impregnated with Potassium Permanganate Use Color Changes to Indicate Ethylene Removal. *Food Bioprocess Technol.* **2017**, *10* (9) 1622–1630.
54. Tan, F.; Leung, P. H. M.; Liud, Z.; Zhang, Y.; Xiao, L.; Ye, W.; Zhange, X.; Yi, L.; Yanga, M. Microfluidic Impedance Immunosensor for *E. Coli* O157:H7 and Staphylococcus Aureus Detection Via Antibody-Immobilized Nanoporous Membrane. *Sens. Actuat. B: Chemica* **2011,** *159*, 328–335.
55. Torres, J. A.; Sanz, P. D.; Otero, L.; Pérez Lamela, C.; Saldaña, M. D. A. Engineering Principles to Improve Food Quality and Safety by High Pressure Processing. Chapter 14; In: *Processing Effects on Safety and Quality Of Foods*; Ortega-Rivas, E., Ed.; CRC Taylor & Francis, Inc.: Boca Raton, FL, 2009; pp 379–414.
56. Torres, J. A.; Sanz, P. D.; Otero, L.; Pérez Lamela, C.; Saldaña, M. D. A. Temperature Distribution and Chemical Reactions in Foods Treated by Pressure-Assisted Thermal Processing. Chapter 15; In: *Processing Effects on Safety and Quality of Foods*; Ortega-Rivas, E., Ed.; CRC Taylor & Francis, Inc.: Boca Raton, FL, 2009; pp 415–440.
57. Vanderroost, M.; Ragaert, P.; Devlieghere, F.; De Meulenaer, B. Intelligent Food Packaging: The Next Generation. *Trends Food Sci. Technol.* **2014,** *39* (1), 47–62.
58. Wang, Y. C.; Lu, L.; Gunasekaran, S. Biopolymer/Gold Nanoparticles Composite Plasmonic Thermal History Indicator to Monitor Quality and Safety of Perishable Bioproducts. *Biosens. Bioelectron.* **2017**, *92*, 109–116.
59. Wang, S.; Liu, X.; Yang, M.; Zhang, Y.; Xiang, K.; Tang, R. Review of Time Temperature Indicators as Quality Monitors in Food Packaging. *Pack. Technol. Sci.* **2015**, *28*, 839–867.
60. Welti-Chanes, J.; Ochoa-Velasco, C. E.; Guerrero-Beltran, J. A. High-Pressure Homogenization of Orange Juice to Inactivate Pectinmethylesterase. *Innov. Food Sci. Emerg. Technol.* **2009,** *10* (4) 457–462.
61. Wyrwa, J.; Barska, A. Innovations in Food Packaging Market: Active Packaging. *Eur. Food Res. Technol.* **2017**, *243*, 1681–1692.
62. Yildirim, S.; Röcker, B.; Pettersen, M. K.; Nilsen-Nygaard, J.; Ayhan, Z.; Rutkaite, R.; Radusin, T.; Suminsk, P.; Marcos, B.; Coma, V. Active Packaging Applications for Food. *Compreh. Rev. Food Sci. Food Saf.* **2018,** *17*, 165–199.
63. Zhang, T.; Lv, C.; Chen, L.; Bai, G.; Zhao, G.; Xu, C. Encapsulation of Anthocyanin Molecules Within A Ferritin Nanocage Increases Their Stability and Cell Uptake Efficiency. *Food Res. Int.* **2014,** *62*, 183–192.

CHAPTER 2

NOVEL FOOD PACKAGING TECHNOLOGIES: APPLICATIONS AND PROSPECTS

SHRUTHY RAMESH and PREETHA RADHAKRISHNAN

ABSTRACT

The supply of healthy and safe food to consumers is one of the challenging areas in which packaging technology plays an important role. Nowadays, a lot of research is going on different types of packaging materials due to the consumer's choice for safe and eco-friendly packaging. Petroleum-based nonbiodegradable synthetic packaging materials are commonly used in the food industry, which are hazardous to the environment. Still, they are economical and have all the essential properties to make the packaging films.

2.1 INTRODUCTION

Packaging refers to the binding/wrapping of food for protection to get resistance from tampering and to provide protection from physicochemical or biological factors that can deteriorate food quality.[32] The packaging is considered a combination of art, science, and technology to protect the food products during storage, transportation, and sale.[42] Packages generally contain the details of the products, such as product name, manufacturing/expiry dates, ingredients, nutritional facts, and storage conditions.[43] Nowadays, the usage of packaged food is common due to our busy lifestyle. In the food industry, packaged food plays an essential role in hygiene and

Advances in Sustainable Food Packaging Technology. Megh R. Goyal, Preeti Birwal, Anand K. Pandey, & Ritesh B. Watharkar, (Eds.)

safety.[57] The excellent package design attracts consumers to prefer the food commodity.

Modern packaging materials (such as polymer-based packages, paper-based packages, wood, and metal-based packages) are being used worldwide. However, environmental pollution due to nondegradable synthetic polymer-based packages is one of the major issues. Therefore, many researchers and companies are coming up with new packaging techniques that will reduce the adverse effect on the environment. Mainly, the researchers are focusing on the safety and protection of the consumers by considering efficient packaging technologies. Thus, this led to the development of many packaging materials, which can extend the shelf-life of the food items. Nowadays, single-use plastic food packaging makes severe global threats in dumping as landfills. The major cause for using single-use plastic food packages is the increase in urbanization.[5] To overcome this concern, the researchers have focused on the implementation/development of food packages with recyclability and biodegradability.

Food packaging provides safe food products, helps to maintain the quality of food products, and plays an important role in customer acceptance.[42] Earlier innovations toward food packaging concepts included aseptic packaging, machine-processed packaging technology, aluminum foil-based packaging, and flexible plastic packages.[42] The packaging materials are selected based on the type of food item since food items have different tastes, flavors, texture, and color.[45] From the twentieth century onward, packaging undergoing technological advancement has come up with active and intelligent (smart) packagings. Nowadays, these technologies lead to a major role in strengthening the economy of the food packaging industries. Intelligent and active packagings provide quality food products to consumers with enhanced shelf-life and minimal product loss. The reasons behind the deterioration or loss of quality of food are microbial spoilage, metabolism, and oxidation. These issues can be overcome by innovative novel food packaging techniques, such as active packaging and intelligent or smart packaging (IOSP). Currently, the use of active packaging, IOSP, and bionano-composites has become a trend and new hope in the field of the food packaging industry with improved properties and less environmental issues.[51]

Active packaging is generally involved in the prevention/preservation of the food items packed. Active packaging includes carbon dioxide or oxygen scavenger, moisture absorbers, and antioxidant/antimicrobial agents incorporated packages.[45] Active packaging extends the shelf-life of the food items,[20] which can be improved by controlling/inhibiting microbial spoilage due to its enhanced antimicrobial or antioxidant property.[31,32,52] These findings were followed by the implementation of intelligent packaging. Intelligent packaging can able to predict/control the food conditions inside the packet itself. The main advantages of intelligent packaging technology are tracking, recording, monitoring, and identifying details of the food items packed during supply chain processing.[22] Commonly used intelligent packaging technologies are provided with gas indicators, time-temperature indicators, and radiofrequency identification (RFID).

Edible film and edible coatings are the new trends in packaging technology. They are commonly used in the food industry, especially for confectionery and perishable food items. It also reduces the pollution caused by synthetic plastic food packaging. In addition to this, the researchers have claimed that edible film and edible coatings could improve shelf-life, thereby enhancing the safety and quality of the food products. In the latest list, the nanotechnology-based packaging materials have been highlighted due to their attractive properties.[51]

Studies have also recommended the usage of renewable resource-based nanoparticles, such as nanocellulose and nano chitosan to develop the food packaging material with enhanced antioxidant, mechanical, optical, and thermal properties.[51,55]

This chapter mainly focuses on conventional and modern food packaging techniques, different types of modern packaging techniques (modified atmospheric packaging, active packaging, intelligent packaging, and edible films) and their applications, application of nanotechnology in food packaging, and future trends in food packaging.

2.2 CONVENTIONAL FOOD PACKAGING TECHNIQUES

Our ancestors have been practicing food packaging for thousands of years. Initially, primitive society practiced hunting for their food, and after hunting, they consume it as such. However, later for transportation/storage/bartering, they started preserving their food in different types of packaging in bulk quantities,[36] such as shells of nuts or fruits, leaves, bamboo stems, animal skin, the bark of trees, woven grass, etc., as shown in Figure 2.1. Later, pottery and metals came into play as a new technique for packaging.[36]

Afterward, the production and usage of paper for food packages came into use. In the early days, leaf-based packages were used for small-scale food business or local markets to wrap the food items. It is mainly used for convincing

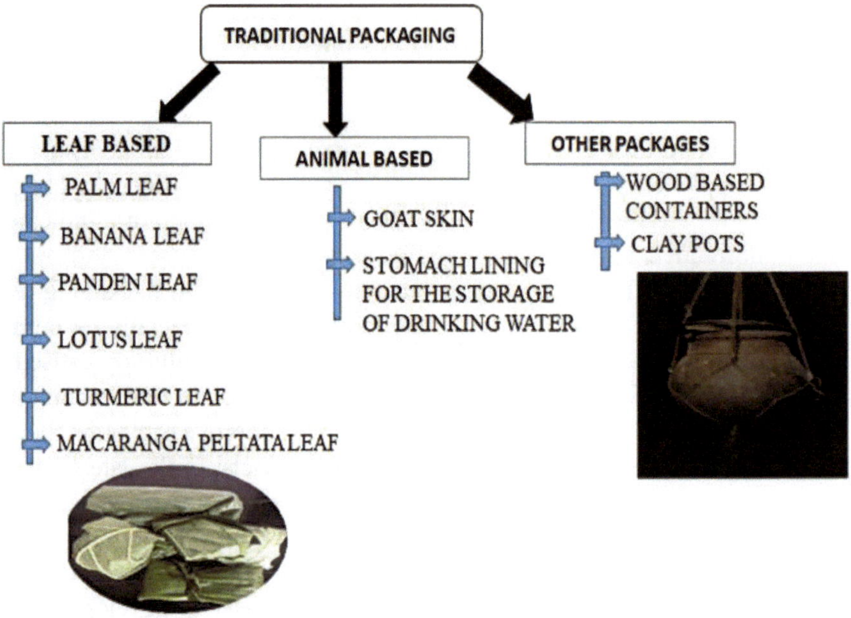

FIGURE 2.1 Classification of traditional packaging of food items.

consumers to influence the selection of food items. Paper-based packages were generally used for the packaging of medicine and other related products.[3,13] In the villages of Malaysia, since ancient times the traditional packaging has high demand because of its eco-friendly nature. The package has multilayer, including straw mats, wet clay, bamboo leaves, and packed inside bamboo baskets. The main advantage of this packaging is that it could improve shelf-life and prevent discoloration of the food.[3] Due to the usage of natural resources as packaging material, it is cost-effective, biodegradable, and will not cause any adverse effects on humans and the ecosystems.[39] Throughout history, we can see the usage of clay pots for food preservation.[36] However, traditional packaging is not effective for the long-term storage of food items due to microbial attacks, and it is the major disadvantage of traditional packaging.[3]

2.3 MODERN FOOD PACKAGING TECHNIQUES

Nowadays, vast ranges of food packaging materials are available, such as plastic, including glass, metal, ceramics, paper, or the combinations of the above materials (Table 2.1). Conventionally, packaging can be classified into three types, such as primary packaging, secondary packaging, and tertiary packaging.[25]

- In primary packaging, the food items packed are in direct contact with the packaging material.
- In secondary packaging, the primary packaged food products were collected and packed inside the packaging materials, such as wooden crates, cardboard boxes, and or in plastic-based containers. In short, after complete processing of food, it will be packed in a convenient packaging material based upon the type of food (primary packaging), and these packages will be placed inside a large or suitable secondary packaging followed by tertiary packaging.
- Tertiary packaging is used for freight transportation of a huge quantity of food items.[36] During tertiary packaging, usually plastic sleeves or shrink wraps and wooden pallets are used. Commonly, plastic-based packages are used for roadways transportation of packed food items. This plastic packaging can protect packed food items from damage and unwanted displacement.[16]

Plastic-based packages are generally made up of macromolecular polymeric materials.[52] The common types of plastic packaging combinations are thermoplastics, polystyrene, polyolefin, polyvinyl chloride (PVC), and polyethylene terephthalate (PET).[5] The generally used plastic materials for the packaging of food materials are low-density polyethylene (LDPE) and high-density polyethylene (HDPE).[20] The main disadvantage of such plastics is their nonbiodegradability and has their own limitations for preserving microbial spoilage.[30]

The consumer preference and demand for the quality/safety of foods led to the development of new packaging techniques.[11] The recent innovations in food packaging techniques, especially active packaging and intelligent packaging, led to longer shelf-life and reduced spoilage of the food items,[7,10]. The intelligent packaging includes RFID techniques, moisture absorbers, time–temperature indicators, and oxygen scavengers. At the same time, active packaging refers to antimicrobial/antioxidant packaging and modified atmospheric packaging (MAP).[22] During storage and transportation, food items (such as fruits and vegetables) get deteriorated due to microbial spoilage, oxidation, and metabolism. It will lead to the loss of quality and other nutritional factors.[58] Unhygienic practices while handling improperly packed food can lead to food-borne diseases in humans, and efficient packaging methods can overcome the above-said problems.[34]

TABLE 2.1 Types of Common Food Packaging Materials and Its Advantages.

Types of packaging material	Advantages	Food items packed	References
Plastics			
Low-density polyethylene (LDPE). High-density polyethylene (HDPE) Polyethylene terephthalate (PET)	Transparency, moldability, and recyclability based on the type of material	Milk, juice, jam, fresh fruits and vegetables, meat, fish, and so on	[45]
Polystyrene	High mechanical strength, malleable in nature, and low cost	Fresh fruits and vegetables, cheese or other processed food items, processed meat, and fish	[57]
Polyvinyl chloride (PVC)	Resistance to fat, humidity, and gases	Processed fresh fruit juice and vegetables, meat, fish, and other processed foods	[41]
Metals			
Aluminum foil	Low cost, recyclability, used at laminates during food packaging, improved barrier properties, and flexible.	Processed fish and meat, confectionary, cheese, chocolate	[52]
Thin-walled aluminum cans	Stability against steam sterilization, excellent barrier properties	Carbonated beverages, sauces, oils	[11]
Tin cans	Coatings of steel at tin can protect against corrosion. Used for the package of food items with low pH	Sauces, pickles	[25]
Others			
Glass	It can be safely used for food packaging without any leeching, impermeable to gasses and vapors, recyclable.	Milk, juice, wines, baby food, oil, liquid pharmaceuticals, and chemicals	[25]
Paper and cardboards	Cheap, recyclable, biodegradable, light weight.	Used for secondary packaging process	[57]

2.4 APPLICATIONS OF MODERN PACKAGING TECHNIQUES

Packaging the food item is an important step to protect it from damage or other environmental conditions during distribution and handling.[17] Modified atmospheric packaging (MAP) is one of the major modern packaging techniques used for food to preserve freshness and shelf-life. The selection of packaging is based on the type of food product and consumer preference.[22] The application of various modern packaging systems for the determination/ improvement of shelf-life is shown in Table 2.2.

TABLE 2.2 Application of Various Modern Packaging Systems.

Packaging system	Application	Advantages	References
Antimicrobial packaging: Nanoparticle-based antimicrobial packaging (Ag, chitosan, Cu, TiO_2, SiO_2, ZnO, Nanoclay)	Fresh foods, such as meat, fish, frozen foods, fruits, and vegetables.	Inhibit microbial spoilage	[55]
Ethanol emitters mounted packaging	Cheese, fish, high-protein food items	Reduced oxidation	[58]
Freshness indicator	Fish and meat	Freshness determination by showing change in color	[58]
Gas indicator	Beef, pickles, any food packaging system	Detect CO_2, O_2 gas levels for preventing microbial growth, prevention of dyes from leaching out	[55]
RFID	Cheese, pork, pickle, frozen food products	RFID helps to identify data such as expiry date, microbial contamination, and location	[21]
Time–temperature indicator	Juice, milk, and beverages	Monitoring food during storage and distribution, determination of the microbial quality	[24]

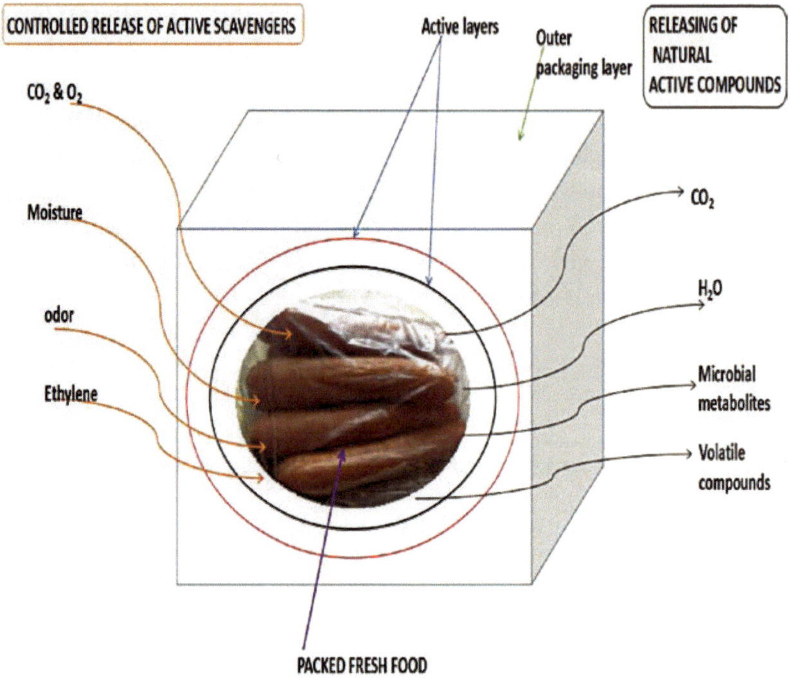

FIGURE 2.2 Pictorial representation of active packaging.

2.4.1 ACTIVE PACKAGING

The consumer prefers active packaging since it ensures prolonged shelf-life and quality/safety of food products by preventing bacterial contamination.[10,29,43] The main principle behind active packaging is the addition of the active compounds to the packaging material.[15] The materials, such as gas absorbers and emitters, will control the atmospheric conditions inside the active packaging. The controlled atmosphere and the active compounds present in the active packaging will improve the shelf-life of the food items.[43] Many studies have reported the addition of artificial antioxidants to reduce the transmission of toxic components into food items.[14] Bioactive phenol is a natural additive used in the active packaging of meat and meat products to improve the antimicrobial properties.[2] The oxygen scavengers in the packaging material are responsible for the absorption of oxygen from the atmosphere during storage and prevent the spoilage of food packed in active packaging.[48] A pictorial representation of active packaging is shown in Figure 2.2.

Cierpiszewski[6] described that the natural antimicrobials (such as bacteriocins and enzymes) can be used in active packaging due to their antimicrobial properties. During the dynamic packaging production processing, the natural antimicrobials are activated and incorporated into the packaging film by physical and chemical modifications.[32] Some examples of existing packaging applications used in the food industry are ethylene scavengers, ethanol emitters, oxygen scavengers, carbon dioxide scavengers, moisture absorbers, and temperature control packaging.[43]

2.4.1.1 BIOACTIVE PACKAGING

Bioactive packaging is a combination of biopolymer and bioactive compounds. These types of packaging have novel properties, such as shelf-life improvement, antibacterial, and antioxidant properties.[26,30]

2.4.1.1.1 Antimicrobial and Antioxidant Packaging

The usage of antimicrobial packaging is a common trend nowadays. Antimicrobial packages will reduce microbial spoilage and enhance the safety and quality of the food items. Generally, antimicrobial agents (such as metal oxides, chemicals, and natural agents) are used in packaging to prevent spoilage by microbes. The mechanism in antimicrobial-based packaging involves the slow release of antimicrobials on the surface and improves the shelf-life, quality, and safety. The functions of antimicrobial packaging include inhibition/destruction of microorganisms, improving the shelf-life, and quality of the food items. Organic substances such as essential oils, nano clays, and other organic acids are used in food packaging as an antimicrobial agent. Other inorganic antimicrobials, including silver ion (silver nanoparticles), chlorine dioxide, nisin, and metal oxides (zinc and magnesium oxide), are used for inhibiting microbial growth.[55] The study by Campos et al.[4] described the controlled release of food additives such as antimicrobials and antioxidants through the packaging material could improve the shelf-life of food items packed.

The antioxidants are used as a preservative in food packaging, which scavenge the oxygen radicals and prevent biochemical changes in food. Butylated hydroxytoluene (BHT), butylated hydroxyanisole (BHA), sulfites, benzoic acid, and nitrosamines are synthetic chemicals used as antioxidants in food packaging. However, these synthetic chemicals have carcinogenic effects and lead to safety issues among consumers at a particular level. As a replacement for synthetic chemicals, natural substitutes (such as spices, extracts of plant leaves, essential oil, chitosan, and herbs) are also used for incorporation into packaging material. These substitutes will not cause any adverse effects on consumers. Many studies describe the effective properties, including antifungal, antimicrobial, mechanical, thermal, gas barrier, and improved morphology of the packaging.[32] The distinctive property of antioxidant compounds for packaging material is to reduce/inhibit the oxidative reaction, thereby preventing the damage of tissues of fresh food items and prolonging shelf-life.

2.4.1.2 MODIFIED ATMOSPHERIC PACKAGING (MAP) AND VACUUM PACKAGING

MAP is the process of changing the internal atmosphere of the food packaging to improve its keeping quality. MAP is mainly used for fresh products, such as meat, fruits, vegetables, dairy products, and poultry. After complete sealing of the package by applying a controlled atmosphere, no further change in surrounding factors is possible.[57] During storage, the change in the modified atmosphere may occur due to the respiration of the packed item. The natural gases introduced inside the package are carbon dioxide, nitrogen, and oxygen. Carbon dioxide cannot inhibit most of the anaerobic microbes that are present inside the packaging. However, the growth of the aerobic Gram-negative bacteria and molds can be efficiently inhibited at a concentration of 5–50%. The nitrogen-infused MAP can slow down the oxidation of fat content in food products by avoiding contact with oxygen.[43] Other gasses used for the MAP of food are xenon, helium, neon, and argon.[32,51]

In the vacuum packaging, the food is placed inside the packet, then the air from the packet is removed, and the package is sealed so that air cannot enter further into the package. Vacuum packaging is mainly preferred for food items, such as cheese, paneer, and cured meat.[57] The main advantage of vacuum packaging is that it can reduce the freezer burn for meat, fatty fishes, and chicken during the storage process.[12] During vacuum packaging, the quantity of the packaging material needed to be minimum since the packaging volume is almost the same as the product volume.[6,10] In modern vacuum packaging (MVP), fresh food products (such as fruits and vegetables) are packed in stiff pouches or impermeable containers, thereby reducing the respiration rate, metabolism, and microbial spoilage during storage.[22]

2.4.1.3 EDIBLE FILMS

The edible film is used for confectionery, cheese, bakery goods, fast foods, and fruits. It provides a barrier to moisture and gas, it also acts as a vehicle for color, flavors, antioxidants, and anti-browning agents, thereby increasing the quality of the product. The most significant disadvantage of petroleum-based conventional plastic materials is their nondegradable nature. Consequently, traditional packaging materials cause ecological issues.[25] At the same time, the edible film provides better shelf-life and can improve keeping quality without any environmental hazards.[37] The edible film (EF) ought to be a sheet with a smooth texture. Interestingly, edible coatings (EC) are thin layers of eatable materials that connect to the surfaces of food items by panning, splashing, and plunging. Edible packaging overcomes the microbial spoilage of the food items, such as fresh fruits and vegetables. EF and EC can also control respiration and helps to retard dampness of the fresh food items.[40] The advantages of edible films and edible coatings[16] are as follows:

- Biodegradable and safe for edible purposes: EF and EC are safe to consume and are biodegradable. Most of the edible film and edible coatings are made from natural ingredients such as starch and proteins.
- Mechanical and physical insurance: Mechanical properties should upgrade with regard to rigidity, extension at-break, flexible modulus, pressure quality, cut quality, solidness, tearing quality, burst quality, scraped area protection grip drive, and collapsing perseverance.
- Relocation, saturation, and hindrance capacities: All boundary properties influence the film synthesis and natural conditions (relative moisture and temperature).
- Accommodation and quality protection: EF and EC can prevent dampness of food products, oxidation of food, and spoilage of the coated food products. Moreover, they provide fragrance/odor protection, enhanced oil retention property and help in the maturation of fruits/vegetables. Edible coatings or films are also providing a smooth surface; hence, it can be able to print the information regarding food products packed or coated.
- Shelf-life: The EC/EF can provide intense protection from extreme outside conditions and improves the shelf-life of the product.

2.4.2 INTELLIGENT (SMART) PACKAGING

The information regarding the safety and quality of the food item and the distribution/storage can be determined by intelligent packaging without opening the package.[23] It has recording/sensing/detecting/communicating functions, which can increase the ease of making a decision regarding the shelf-life, quality, and safety. Intelligent packaging mainly consists of RFID and indicator/sensors.[23,32] Mainly, IOSP is used to track/trace/reuse the information regarding the quality of the food items. This technique can also provide an alert regarding parameters related to shelf-life, safety, and storage condition of food products, and other food-related quires. IOSP will be in the form of small tags/labels containing small devices. These tags/labels can incorporate or be printed with the food packages. It will help to determine the storage specification and quality of the food item inside the package. The smart device inside the tag/labels commonly used in the food packages are known as indicators, and it is classified as direct and indirect indicators.[22] Direct information, such as storage condition, is proper or not, and the shelf-life of the food item packed can be detected with the help of indirect indicators.[31] This direct indicator provides the information regarding freshness (mainly), quality, and safety of the food products to consumers.[22] Many studies and reviews were addressed regarding the concept of an intelligent packaging system. Kerry et al.[23] described the advantages/efficiency of modern packaging for meat and meat products over traditional packaging. Another study on intelligent packaging reveals that it can provide an extended shelf-life of fresh food items.[27]

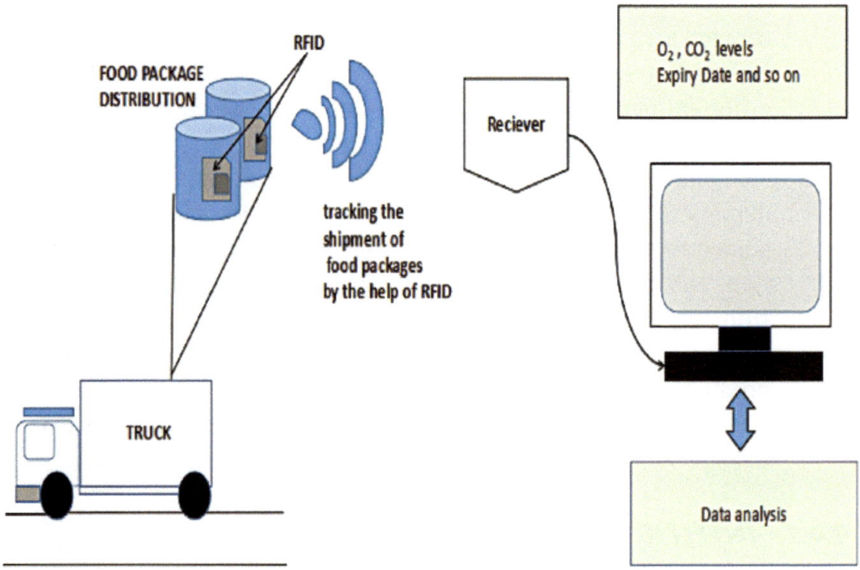

FIGURE 2.3 Working mechanism of RFID.

2.4.2.1 RFID

Bar-codes are quite commonly seen on each and every food product which we buy from the market. Bar-codes are the combination of lines arranged systematically; the lines are encoded data which can provide the information including batch number, date of manufacturing, and expiry regarding the packed products. The bar-code scanner is the device used to decode these lines.[16] RFID (Radio-frequency identifiers) is a wireless tracking system, which will provide information regarding the food item in a packaging. The working mechanism of the RFID is shown in Figure 2.3.

RFID mainly consists of three components, such as an RFID tag, RFID reader/receiver, and an antenna. The RFID tag is an important component, and it is commonly made up of plastic. The tag is embedded in between the plastic layers containing integrated circuits. The antenna can be able to stick or punch over the food product package. It transmits the data regarding food items to the reader/receiver in the form of radio waves, and a control unit inside the receiver will decode the signal. Then, a computer transforms this information into readable data, which can be stored and analyzed according to the preference. Food production organizations or distributors can monitor the traceability of food products and ensure consumers' safety. RFID can also improve the effectiveness of the supply chain management of food.[43]

2.4.2.2 TIME–TEMPERATURE INDICATORS

Intelligent packaging can control conditions, such as temperature and moisture inside the package. It also helps to sense/record/detect the food conditions inside the packaging and can improve the shelf-life and quality of the packed food item.[35] The time–temperature indicators are placed externally and internally, and they help improve the quality of the food by detecting microbial contamination, color change, and volatile compounds.[32] Kim et al.[24] used isopropyl palmitate, a diffusion-based time–temperature indicator, to determine the microbial contamination for angelica juice (unpasteurized). Later, an accurate natural indicator was developed by incorporating anthocyanin extracted from red cabbage, and the packaging film was prepared using chitosan and polyvinyl alcohol. The food item packed using this film was monitored for the detection of the deterioration of the food since anthocyanin makes the color change due to the change in pH.[46] Major commercially available time–temperature indicator, OnvuTM, is a photochromic indicator used for the analysis of change in color and pigments in food products.[24,46] Fresh-check® is a polymeric-type time–temperature indicator used to detect polymerization reaction.[38] The color change of the food while storage can be detected using Monitor markTM and Freeze WatchTM. These are microbial-type time–temperature indicators.[18] Check Point® is an enzymatic time–temperature indicator used to determine the color changes in food due to the pH variation.[18,56]

2.4.2.3 PH INDICATORS

Freshness indicators are mainly used for fresh food items, such as fruits, vegetables, milk, and frozen foods. Freshness indicators can detect the toxin formation, change of pH, microbial attack, oxidation, discoloration, unpleasant odor, and other biological reactions. For example, the original packaging attached with a freshness indicator detects leakage or uncontrolled temperature conditions. It also senses the lack of freshness, which can be identified quickly by consumers without checking the expiry date.[7] Toxin GuardTM (Toxin Alert, Ontario, Canada) is an example of a commercially available pathogen indicator in intelligent packages to detect metabolites produced by pathogens.[31,32]

2.4.2.4 MOISTURE ABSORBERS

In food packaging, moisture absorbers are used to control the water activity; thereby, these can reduce the growth of microbes and can prevent lipid oxidation. Moisture absorbers are mainly introduced to food item packages that are sensitive to moisture, including crispy products, powdered products (milk powder), candy, and sweets.[32] For dry food packages, the moisture absorbers (such as natural clays, silica gels, and calcium oxide) are incorporated into packages or introduced in the form of sachets. Internal humidity controllers can preserve poultry, meat, vegetables, and fruits. The purge absorber can remove the liquid produced/leakage from the fresh foods. Tetra packaging of six layers incorporated with moisture absorbers along with polyethylene as an outer layer can prevent the introduction of moisture from the outside and thereby preserves the food items.[1]

2.4.2.5 CO2 ABSORBER AND EMITTERS

The carbon dioxide absorbers can suppress/destroy the growth of microbes in the packages of cheese, poultry, and meat. In case of fresh fruit and vegetables packages, carbon dioxide scavengers in the form of sachets/absorbent pads are used to reduce the respiration rate. Calcium hydroxide is one of the typical carbon dioxide absorbers in the form of sachets for the preservation of food products. In modified atmospheric packaging (MAP), the usage of carbon dioxide absorbers and emitters are one of the important components for food preservation. Aday et al.[1,49] described the efficient maintenance of biochemical parameters of food products with the help of carbon dioxide absorbers. CO_2 emitters are commonly used for the preservation of dry ready-to-eat food items, confectionery food items, and perishable food items, including fish and meat. Inside the food packages, the CO_2 emitters will intently produce and liberate the CO_2 gas, which hinders the microbial spoilage, thereby conserving the safety and quality of the food item used to pack.[5] A newly reported packaging study of fish also suggests CO_2 emitters for the shelf-life extension.[32]

2.5 SCOPE OF NANOTECHNOLOGY IN FOOD PACKAGING

Nanoparticles can improve the permeability of polymer-based packaging materials.[44] In the study of Shruthy and Preetha et al.,[51] the potato cellulose nano particle-fennel oil incorporated polyvinyl chloride-based packaging film showed decreased oxygen transmission rate, thereby extending the shelf-life of raw prawns (*Penaeus monodon*) during frozen storage. Also, the addition of the nanoparticle can improve the heat resistance, mechanical properties, barrier properties, antimicrobial property and avoid UV-light transmission, and thereby enhance the keeping time of the food items inside the packaging.[5,17,33,52] The most crucial exploitation of nanotechnology is the development of intelligent and active packaging in the food industry.[44] The important applications of nanotechnology in food packaging are shown in Figure 2.4.

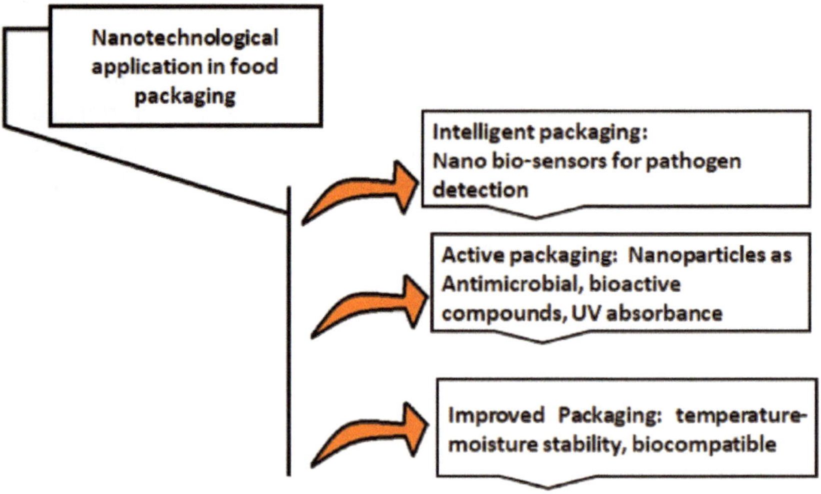

FIGURE 2.4 Applications of nanotechnology in food packaging.

The properties of intelligent packaging and active packaging have become more efficient by the intervention of nanotechnology. The incorporation of nanoparticles has enhanced thermal, mechanical, optical, biodegradability, antioxidant, and antimicrobial properties of the packaging films.[20,52,53]

Nanocomposite-based packaging for food is trending in the field of food packaging research. The combination of biodegradable biopolymers such as polyvinyl alcohol, polylactic acid, starch, and polyvinyl chloride, along with bioactive compounds, makes the nanocomposite unique. These novel nanocomposites have dynamic properties to improve the shelf-life of packed food.[37,52] For example, it has been reported that the nanocomposite prepared using nanocrystals extends the shelf-life of beer up to eighteen months.[10] Various studies have reported that the titanium dioxide nanoparticle would block the UV light and give a longer shelf-life for food.[12,54]

Nanosensors are used to locate/detect and quantify the food quality parameters by measuring its physiochemical characteristics. Mainly the sensors incorporated in packaging are used for the detection of bacterial contamination and off-flavors due to the spoilage of food items.[12,50]

2.5.1 NANOLAMINATES

Nano lamination is the process of washing/dipping/spraying of coating solution on food products to form a layer with nanometer thickness for the purpose of enhanced shelf-life and quality. The development of edible nanolaminates is made from proteins, lipids, and polysaccharides. In a study, it was reported that the protein and polysaccharides-based nanolaminates coated over food products have improved gas barrier property against carbon dioxide and oxygen. The lipid-based nanolaminates are used to coat over food products for preventing moisture absorbance. The disadvantages of lipid-based nanolaminates are: they are poor in the gas barrier and mechanical property.[9] These nanomaterials can be incorporated with active functional agents, such as antioxidants, antimicrobials, anti-browning agents, and enzymes.[8]

2.5.2 NANOCELLULOSIC MATERIALS

Food processing industry wastes are commonly used as a source of lignocellulosic materials for nano cellulose synthesis.[20,51,52] Nano-sized cellulosic materials are lightweight reinforcement materials in the polymeric matrix and also can be used as packaging material for food items.[28]

2.6 FUTURE TRENDS IN FOOD PACKAGING

The usage of active packaging enhances the shelf-life of food; however, the intelligent packaging will provide information about the safety aspects of food packed for the consumer. The development of the combination of intelligent and active food packaging will be the trend in the near future.[53]

Edible coating and edible films have satisfied the consumer's preference. Various proteins such as egg white, whey proteins, zein, keratine, gelatin, soy protein, collagen, wheat gluten, and myofibrillar proteins are commonly used by researchers for the development of edible coatings for food items. It was reported that edible films could improve the shelf-life of food products.[19,37] In addition, the usage of edible films reduces the usage of single-usage food packaging to some extent.[37] Because of these reasons, there is great potential in edible packaging material research.

Nowadays, the implementation of nanotechnology is trending in food packaging research. The recent research and development can provide an efficient, eco-friendly alternative to traditional plastic.[19] Nanocellulose/fiber-based packaging material provides improved mechanical, thermal, optical, and gas barrier properties.[20,53] The usage of agricultural waste/renewable resources for the preparation of packaging materials such as nanocellulose cellulose and nano chitosan-based packaging film provides value addition to agriculture/food processing industry waste.[20,22] Biopolymer-based packaging will also play an important role in the future due to its eco-friendly nature.

2.7 SUMMARY

More and more researchers have come up with new packaging techniques and their applications in the food industry. Innovations in packaging can improve the safety and nutritional value of food products. However, the application's success will depend on consumer acceptance of the packaging material. Advances in active and intelligent packaging can play an important role in enhancing the shelf-life of food products. Similarly, nanotechnological innovations can offer dynamic properties to the packaging material, and at the same time, the safety evaluation of nanomaterial for packaging is mandatory for its wide acceptance. In the current scenario, the demand for processed food is increasing. Hence, the development of innovative packaging material can uplift global marketing and also can ensure safety and quality food for the consumers.

ACKNOWLEDGMENTS

The authors acknowledge School of Bioengineering, SRM Institute of Science and Technology (SRMIST); Prof. C. Muthamizchelvan, V.C. (Engineering and Technology), and Dr. M. Vairamani, Chairperson, School of Bioengineering for financial and technical support.

KEYWORDS

- **active packaging**
- **antimicrobial packaging**
- **eco-friendly**
- **food packaging**
- **intelligent packaging**
- **modified atmospheric packaging**
- **packaging technologies**

REFERENCES

1. Aday, M. S.; Caner, C. The Applications of Active Packaging and Chlorine Dioxide for Extended Shelf-life of Fresh Strawberries. *Pack. Technol. Sci.* **2011**, *24* (3), 123–136.
2. Agriopoulou, A. Active Packaging for Food Applications. *EC Nutr.* **2016,** *6* (1), 86–87.
3. Ahmed, I.; Lin, H.; Zou, L.; Brody, A. L.; Li, Z.; Qazi, I. M.; Liangtao, L. A Comprehensive Review on the Application of Active Packaging Technologies to Muscle Foods. *Food Control* **2017**, *82* (1), 163–178.
4. Campos, C. A.; Gerschenson, L. N.; Flores, S. K. Development of Edible Films and Coatings with Antimicrobial Activity. *Food Bioprocess Technol.* **2011,** *4* (2), 849–875.
5. Chaudhry, Q.; Scotter, M.; Blackburn, J.; Ross, B.; Boxall, A.; Castle, L.; Aitken, R.; Watkins, R. Applications and Implications of Nanotechnologies for The Food Sector. *Food Addit. Contamin. Part A* **2008**, *25* (1), 241–258.
6. Cierpiszewski, R. Opakowaniaaktywneiinteligentne. *Active Intel. Pack. Wydawnictwo Uniwersytetu Ekonomicznego w Poznaniu, Poznań* **2016**, *70* (1), 98–99.
7. Dari, L.; Nenguwo, N.; Afari-Sefa, V. Evaluation of Packaging Liners in Wooden and Plastic Crates for Handling Tomatoes. *J. Postharv. Technol.* **2018**, *6* (1), 36–40.
8. Dasgupta, N.; Ranjan, S.; Mundekkad, D.; Ramalingam, C.; Shanker, R.; Kumar, A. Nanotechnology in Agro-Food: From Field to Plate. *Food Res. Int.* **2015**, *69* (2), 381–400.
9. Decher, G.; Schlenoff, J. B., Eds. *Multilayer Thin films: Sequential Assembly of Nanocomposite Materials*, 2nd edn.; Wiley-VCH: Weinheim-Germany, 2003; p 543.
10. Dobrucka, R.; Cierpiszewski, R. Active and Intelligent Packaging Food-Research and Development—A Review. *Polish J. Food Nutr. Sci.* **2014**, *64* (1), 7–15.
11. Dobrucka, R.; Cierpiszewski, R.; Korzeniowski, A. Intelligent Food Packaging—Research and Development. *LogForum* **2015**, *11* (1), 7–14.
12. Fang, Z.; Zhao, Y.; Warner, R. D.; Johnson, S. K. Active and Intelligent Packaging in Meat Industry. *Trends Food Sci. Technol.* **2017**, *61*, 60–71.
13. Faridah, S.; Rahah, H.; Anna, D.; Shahren, A.; Zaidi, A.; Shahri Bdul, R. Traditional Wood Wrapping and Packaging: Culture and Identity of the Sariban *Malay Sarawak. Glo. Bus. Entrepren.* **2018,** *2* (1), 1–7.
14. Farmer, N. *Trends in Packaging of Food, Beverages and Other Fast Moving Consumer Goods*; PWN: Warszawa, 2016;pp 4–7.
15. Gomez-Estaca, J.; Lopez-de-Dicastillo, C.; Hernandez-Munoz, P.; Catala, R.; Gavara, R. Advances in Antioxidant Active Food Packaging. *Trends Food Sci. Technol.* **2014**, *35*, 42–51.
16. Han, J. H. Intelligent Packaging for Food Products. Chapter 8; In: *Innovations in Food Packaging*; Han, J. H., Ed.; Ed. Elsevier Academic Press: Amsterdam, 2014; pp 171–209.
17. He, X.; Deng, H.; Hwang, H. The Current Application of Nanotechnology in Food and Agriculture. *J. Food Drug Anal.* **2019**, *27* (2), 1–2.
18. Heising, J. K.; Boekel, M. A. J. S. V.; Dekker, M. Simulations on the Prediction of Cod (Gadusmorhua) Freshness from an Intelligent Packaging Sensor Concept. *Food Pack. Shelf-Life* **2015**, *3* (1), 47–55.
19. Ishrat, M.; Nayik, G. A.; Dar., S. M.; Nanda, V. Novel Food Packaging Technologies: Innovations and Future Prospective. *J. Saudi Soc. Agric. Sci.* **2018,** *17* (5), 454–462.

20. Jancy, S.; Shruthy, R.; Preetha, R. Fabrication of Packaging Film Reinforced with Cellulose Nanoparticles Synthesized from Jack Fruit Non-Edible Part Using Response Surface Methodology. *Int. J. Biol. Macromol.* **2019**, *142*, 63–72.
21. Jedermann, R.; Lang, W. Semi-Passive RFID and Beyond: Steps Towards Automated Quality Tracing in the Food Chain. *Int. J. Radio Freq. Identification Technol. App.* **2007**, *2*, 247–259.
22. Jia-Wei, H.; Luis, R. G.; Jian-Ping, Q.; Xin-Ting, Y. Food Packaging: A Comprehensive Review and Future Trends. *Comprehen. Rev. Food Sci. Food Saf.* **2018,** *17* (4), 860–877.
23. Kerry, J. P.; O'Grady, M. N.; Hogan, S. A. Past, Current and Potential Utilization of Active and Intelligent Packaging Systems for Meat and Muscle Based Products: A Review. *Meat Sci.* **2006**, *74*, 113–130.
24. Kim, J. U.; Ghafoor, K.; Ahn, J.; Shin, S.; Lee, S. H.; Shahbaz, H. M.; Shin, H. H.; Kim, S.; Park, J. Kinetic Modeling and Characterization of A Diffusion-Based Time-Temperature Indicator (TTI) For Monitoring Microbial Quality Of Non-Pasteurized Angelica Juice. *LWT—Food Sci. Technol.* **2016**, *67*, 143–150.
25. Kirwan, M. J.; Strawbridge, J. W. Plastics in Food Packaging. In: *Food Packaging Technology*. Coles, R., McDowell, D., Kirwan, M. J., Eds.; Blackwell Publishing, CRC Press: Finland, 2003; pp 174–240.
26. Lagaron, J. M. Biodegradable and Sustainable Plastics as Essential Elements in Novel Bioactive Packaging Technologies. In: *First Conference on Biodegradable Polymers for Packaging Applications*; PIRA International: Leatherhead, 2005.
27. Lee, S. Y.; Lee, S. J.; Choi, D. S.; Hur, S. J. Current Topics in Active and Intelligent Food Packaging for Preservation of Fresh Foods. *J. Sci. Food Agric.* **2015**, *95*, 2799–2810.
28. Leistritz, F. L.; Hodur, N. M.; Senechal Stowers, M. D.; McCalla, D.; Saffron, C. M. *Biorefineries Using Agricultural Residue Feedstock in the Great Plains.* AAE Report 07001 Working Paper, Agricultural Experiment Station, North Dakota State University, Fargo North Dakota, 2007.
29. Lesiów, T.; Kosiorowska, M. Opakowaniaaktywneiinteligentne w przetwórstwiemiêsa. *Czêœæ I. GospodarkaMiêsna* **2006,** *3* (12), 1–18.
30. Lopez-Rubio, A.; Almenar, E.; Hernandez-Munoz, P.; Lagaron, J. M.; Catala, R. Overview of Active Polymer - Based Packaging Technologies for Food Applications. *Food Rev. Int.* **2004**, *20* (5), 357–387.
31. Ma, Q.; Du, L.; Wang, L. Tara Gum/Polyvinyl Alcohol-Based Colorimetric NH3 Indicator Films Incorporating Curcumin for Intelligent Packaging. *Sens. Actuators B Chem.* **2017**, *244*, 759–766.
32. Mahindra, P.; Mridula, D.; Angeson, H. Recent Developments in Food Packaging Technologies. *Beverage Food World* **2019**, *46* (1), 21–25.
33. Mihindukulasuriya, S. D. F.; Lim, L. T. Nanotechnology Development in Food Packaging: A Review. *Trends Food Sci. Technol.* **2014**, *40* (1), 149 167.
34. Morris, J. How Safe Is Our Food? *Emerg. Infect. Dis.* **2011,** *17*, 126–128.
35. Muhammad, S.; Da-Wen, S.; Zhiwei, Z. Recent Developments in Intelligent Packaging for Enhancing Food Quality and Safety. *Crit. Rev. Food Sci. Nutr.* **2018**, *58* (15), 2650–2662.
36. Muhizam, M.; Sumetha, N.; Jason, T.; Hardy, S. A. S.; Jasni, D. Looking Back to The Past: Revival of Traditional Food Packaging. In *2nd Regional Conference on Local Knowledge (KEARIFAN TEMPATAN)*, 15–16 October, Jerejak Island Rainforest Resort, Penang, 2012; pp 1–18.

37. Namratha, S.; Sreejit, V.; Preetha. R. Fabrication and Evaluation of Physiochemical Properties of Probiotic Edible Film Based on Pectin-Alginate-Casein Composite. *Int. J. Food Sci. Technol.* **2020,** *55* (4), 1497–1505.
38. Nopwinyuwong, A.; Trevanich, S.; Suppakul, P. Development of a Novel Colorimetric Indicator Label for Monitoring Freshness of Intermediate-Moisture Dessert Spoilage. *Talanta* **2010**, *81*, 1126–1132.
39. Nur Hanani, Z. A.; Roos, Y. H.; Kerry, J. P. Use and Application of Gelatin as Potential Biodegradable Packaging Materials for Food Products. *Int. J. Biol. Macromol.* **2014**, *27* (2), 1–17.
40. O'Sullivan, A.; Shaw, N. B.; Murphy, S. C.; van de Vis, J. W.; van Pelt-Heerschap, H.; Kerry, J. P. Extraction of Collagen from Fish Skins and Its Use in the Manufacture of Biopolymer Films. *J. Aquat. Food Prod. Technol.* **2006**, *15* (3), 21–32.
41. Omanovic-Miklicanin, M.; Maksimovic, I.; Vinkovic-Vrcek, D.; Mulaomerovic, A. D. Application of Nanotechnology in Food Packaging. In *Conference: 5th Workshop: Specific Methods for Food Safety and Quality*, Belgrade, Sebia, 2016; pp 119–125.
42. Pal, M. Importance of Hygienic Packaging in Food Industry. In *MSc Lecture Note*. Addis Ababa University, College of Veterinary Science, Debre Zeit, Ethiopia, 2014; pp 1–11.
43. Pal, M. Nanotechnology: A New Approach in Food Packaging. *J. Food Microbiol. Saf. Hygiene* **2017**, *2* (1), 121–134.
44. Pehanich, M. Small Gains in Processing, Packaging. *Food Process.* **2006,** *3* (4), 46–48.
45. Peltzer, M.; Wagner; Jimenez, A. Migration Study of Carvacrol as A Natural Antioxidant in High- Density Polyethylene For Active Packaging. *Food Addit. Contamin. A Chem. Ann. Control Exposure Risk Assess* **2009**, *26* (1), 938–946.
46. Pereira Jr, V. A.; de Arruda, I. N. Q.; Stefani, R. Active Chitosan/PVA Films with Anthocyanins from Brassica Oleraceae (Red Cabbage) As Time-Temperature Indicators for Application in Intelligent Food Packaging. *Food Hydrocoll.* **2015,** *43*, 180–188.
47. Ramesh, S.; Radhakrishnan, P. Areca Nut Fiber Nano Crystals, Clay Nano Particles and PVA Blended Bionanocomposite Material for Active Packaging of Food. *Appl. Nanosci.* **2020**, 1–13.
48. Restuccia, D.; Spizzirri, U. G.; Parisi, O. I.; Cirillo G.; Curcio M.; Francesca, I.; Francesco, P.; Giuliana, V.; Nevio P. New EU Regulation Aspects and Global Market of Active and Intelligent Packaging for Food Industry Applications. *Food Control* **2010,** *21* (42), 1425–1435.
49. Samaneh, M.; Baghaei-Nejad, M.; Shahrokh Abadi, M. H.; Zou. Z.; Zheng, L. R. Food Quality and Safety Monitoring Using Gas Sensor Array in Intelligent Packaging. *Sens. Rev.* **2016,** *36* (2), 169–183.
50. Sekhon, B. S. Food Nanotechnology—An Overview. *Nanotechnol. Sci. App.* **2010**, *3*, 1–15.
51. Shruthy, R.; Jancy, S.; Preetha, R. Cellulose Nano Particles Synthesized from Potato Peel for the Development of Active Packaging Film for Enhancement of Shelf-Life of Raw Prawns (Penaeus Monodon) During Frozen Storage. *Int. J. Food Sci. Technol.* **2020**, 1–9.
52. Shruthy, R.; Preetha, R. Cellulose Nanoparticles from Agro-Industrial Waste for the Development of Active Packaging. *Appl. Surface Sci.* **2019**, *484*, 1274–1281.
53. Siti, H. O.; Salam, N. R. A.; Zainal, N.; Basha, R. K.; Talib, R. A. Antimicrobial Activity of TiO_2 Nanoparticle Coated Film for Potential Food Packaging Applications. *Int. J. Photoenergy* **2014,** *5* (14), 1–7.

54. Tanja I. R.; Ivan S. R.; Branka M. P.; Aleksandra R. N. Antimicrobial Nanomaterials for Food Packaging Applications. *Food Feed Res.* **2016**, *43* (2), 119–126.
55. Vu, C. H. T.; Won, K. Novel Water-Resistant UV-Activated Oxygen Indicator for Intelligent Food Packaging. *Food Chemistry*, **2013**, *140*, 52–56.
56. Wyrwa, J.; Barska, A. Innovations in the Food Packaging Market: Active Packaging—A Review Article. *Eur. Food Res. Technol.* **2017**, *243* (2), 1681–1692.
57. Zhai, X.; Shi, J.; Zou, X.; Wang, S.; Jiang, C.; Zhang, J.; Huang, X.; Zhang, W.; Holmes, M. Novel Colorimetric Films Based on Starch/Polyvinyl Alcohol Incorporated with Roselle Anthocyanins for Fish Freshness Monitoring. *Food Hydrocoll.* **2017**, *69*, 308–317.
58. Zhao, C. J.; Han, J. W.; Yang, X. T.; Qian, J. P.; Fan, B. L. Review of Computational Fluid Dynamics for Forced-Air Cooling Process. *Appl. Energy* **2016,** *168*, 314–331.

CHAPTER 3

PRESENT AND FUTURE SCOPE OF EDIBLE PACKAGING

SOUMYA RATHORE and ANAND KUMAR PANDEY

ABSTRACT

Conventional packaging methods for food materials majorly contribute to exponentially growing packaging waste and pollution at the global level. Thus, this issue needs to be dealt with great concern and demands urgent development of an alternate approach to reduce its negative impact on the environment, keeping the long shelf-life of the product, and consumer's health in mind. Different sectors like biotech, food, and pharma acknowledge edible packaging as a better option to trim down the concern. This innovation can improvise the fancy lineaments of the product such as variety, quality, safety, and consumers' convenience. The architecture of such packaging uses coatings and films of biodegradable materials such as polysaccharides, proteins, lipids, and composite films. Rice, wheat, and other grains are used prominently as sources of these materials which are vegan as well as biodegradable in nature. According to an autopsy report of "Global Edible Packaging Market, US in 2019," the demand for edible packaging trend has increased tremendously in recent decades in developed countries, especially in Europe, North America, Asia-Pacific region, and Africa. This chapter opens up a descriptive matter about the present scenario of edible packaging trends. It also gathers information about its production and processing, the compounds or polymers having the capability to serve as carriers of pertinent novel applications, and a brief discussion of the future extent and prospects of edible packaging.

Advances in Sustainable Food Packaging Technology. Megh R. Goyal, Preeti Birwal, Anand K. Pandey, & Ritesh B. Watharkar, (Eds.)

3.1 INTRODUCTION

The rising demand for ready-to-eat food products with high-grade quality, safety, and increased shelf-life accelerates the demand for newer techniques that maintain the natural properties and wholesomeness of food. Apart from innovative processing and preservation techniques, food packaging also gained immense importance, and plays an important role in providing worth to the manufacturing sectors and supply chain.[1,3]

The prime functions of food packaging are, to protect food from extraneous matter, including physical, chemical, or biological agents, to provide prolonged shelf-life, reduction in food deterioration, maintenance of food safety and quality, and to retain the beneficial effects of processing. Packaging also enhances the market value of food products by making them more attractive to consumers. The need for perpetual changes in the food packaging sector is growing due to increased consumer demands, international food market developments, technological, and legal requirements, etc.[13] The basic and important parameter in food packaging is the selection of packaging material that is suitable and appropriate for the particular food product. The efficiency of packaging material depends on its strength to maintain quality and safety, shelf-life extension ability, and economic factors. A package has to be economical and effective as per the requirement of the product, for that reason modern packaging technologies should be improved and incorporated effectively into the food supply chain.[1,3,13]

Due to continuous development in food processing technologies, distribution chain functions lead to the need for growth and development of the packaging industry. The growth in lifestyle has also increased the demand for safe packaging. Unfortunately, food packaging industries most commonly use plastic as their raw material, which is a major threat to the environment.[1,2] Plastics do not decompose easily and hence remain accumulated in the environment as waste for a long time. Non-biodegradable plastic material has contributed to a large amount of solid waste per year and it is increasing year by year exponentially. However, some plastics are biodegradable, but their rate of degradation is very slow. It is expected that by 2025, municipal solid waste will increase by 2.2 billion tonnes in which plastic will act as the prime source causing devastating conditions of environmental pollution.[69] To resolve these issues, scientists, and environmental activists have been searching for alternative packaging strategies that can be eco-friendly and biodegradable in nature and made from natural polymers.[70]

The edible packaging field is an amalgamation of materials and food sciences. The properties of materials used in edible packaging should make it evident that the material is not only safe for packaging food but is also good to be eaten up as food. Processing and food characteristics of the material are considered simultaneously to guarantee the safety and quality of food products. The use of edible, biodegradable, and renewable materials to replace (partially or totally) petroleum-based packaging materials has increased the interest in the global market for edible packaging solutions.[54,70] Edible packaging uses biopolymers such as proteins, lipids, polysaccharides, and their composites for the production of edible coatings and films. Although such packaging is not as resistant to environmental conditions which surround packed food as conventional packaging materials but can provide protection from ultraviolet (UV) light, act as a barrier against damage, enhance shelf-life, can contain agents which are bioactive, have antimicrobial activity, biodegradable, and can also provide nutrition when eaten up.[17,19,70] Numerous researches are being conducted to further improve the efficiency and nutritional quality of edible packaging which could increase their usage around the globe and thus reduce the exponential growth of plastic packaging waste being generated by food packaging.[15–17,19,70]

This chapter focuses on the detailed description of natural edible polymers used in the manufacturing of edible films and coatings. The applications of edible packaging in the various sectors of the food industry (such as minimally processed foods, dairy-based foods, meat, fish, and poultry-based products, etc.) along with regulatory outlooks and marketability of edible packaging have also been discussed.

3.2 EDIBLE FILMS AND COATINGS

Biodegradable material leads to soil conditioning upon degeneration and produces no toxic residues under suitable conditions such as temperature, moisture, and oxygen. These materials produce compounds that can break down naturally through microbes, resulting in the production of carbon dioxide, water, methane, and biomass. The main sources of biodegradable polymers are biomass feedstock and agricultural products. These materials can provide a possible solution to wastage and environmental pollution due to their tendency to get degraded via biological means.[51]

Biodegradable polymers can be divided into either edible polymers or non-edible polymers. Edible polymers can be considered as packaging material for foods as they possess no health risk to humans. They are obtained from natural ingredients and can be directly used for consumption. Not even for consumption, they are also safe for disposal because they can easily degrade in the environment, hence harmless to the environment as compared to other polymeric materials. Apart from their consumable quality and biodegradable nature, edible polymers should also contain essential properties such as good barrier properties, high strength, stability against temperature, moisture, and microbes, and low economic cost.[65] Due to the consumer's demands for safe and high-quality food and the impact of packaging waste on the environment, edible packaging is being considered a striking alternative to the food technologist and industry. Furthermore, the increasing trend of minimally processed fresh foods with edible barriers has also contributed to increasing interest in edible packaging.[17]

Edible polymers can be utilized in two forms: edible films and edible coatings. Both forms are quite similar in their composition, but the basic difference lies in their thickness. Both edible films and coatings are made from food-derived ingredients, but their manufacturing processes are quite different. Edible films are produced in the form of sheets by wet or dry process and can be used as wraps, pouches, bags, capsules, and casings.[54] Edible films have an advantage over synthetic polymers in a way that there is no requirement for disposal of the package resulting in a reduction of environmental pollution. They also enhance the esthetic quality of packaged food along with flavor and color. These films can also be used as an additional packaging material with non-edible materials.[67]

The edible coatings are produced in the form of thin layers by dipping, spraying or panning, and can be used directly on the surface of the food. They are designed in such a way that they are attached to food as its part and cannot be separated from the food.[54] Edible coatings can be used for a wide range of products (e.g., pizza, bakery fillings, and toppings) as an external coating or continuous layers between compartments/ingredients. They can even be applied on each piece of the whole product, which has not been packed individually due to practical arguments, such as fresh-cut melons, kiwis, strawberries, nuts, beans, and pears.[52] Edible films and coatings are considered Generally Recognized as Safe (GRAS) by US Food and Drug Administration (FDA) within some specific limitations. However, further research is required to enhance the properties and manufacturing methods of these films and coating to increase their applications in food packaging.

3.3 FUNCTIONS AND PROPERTIES OF EDIBLE PACKAGING

Packaging plays an important role in enhancing the market value of a particular food. There are many drawbacks related to food packaging that affect the marketing of foods like barrier properties, migration properties, structural integrity, and maintenance of flavor and color. Edible films and coatings are considered as improved packaging versions that help to combat these obstacles. They also prevent the food from physical damage, hence maintaining and improving its esthetical properties.[50] There are many examples of edible packaging that helps to improve and maintain the quality of food such as paraffin wax for coating fruits, collagen casings for sausages, etc.[49]

The characteristics of edible films and coatings vary with the specific product, its constituents, properties, and changes associated with it during processing and storage. Although edible packaging is the best alternative to reduce environmental effluence, still it needs the support of non-edible packaging material due to hygienic reasons. The unique advantages and versatility of edible packaging materials to create eco-friendly and innovative packaging to reduce waste and/or for improving the stability, safety, quality, and variety of products, and ease of consumers, are as follows[16]:

3.3.1 BARRIER TO MOISTURE

The prime consideration taken in the selection of material for edible films and coatings is its mass transfer rate and permeability between the food and the atmosphere. Moisture plays an important role in microbial growth during the storage of food which results in spoilage and low shelf-life, especially in perishable and semi-perishable foods. So the material used for the edible packaging should be a good moisture barrier because variations in packed food water activity (aw) may cause severe microbial growth, undesired textural changes, and deteriorating enzymatic and chemical reactions. Apart from the moisture permeability, water vapor penetration also affects the barrier properties. Therefore, materials having low polarity and dense-structured molecular matrixes are considered best as the moisture barrier for food packaging.[71]

3.3.2 BARRIER TO OXYGEN

Foods having high-fat content are susceptible to oxidation in the presence of oxygen and result in degradation and spoilage like autolysis in fish, discoloration in fresh meat cuts, etc. Some fruits are also prone to enzymatic reactions in the presence of oxygen, resulting in browning of fruits. Edible films made from hydrocolloid-based material are best suited as the gas barrier by controlling the permeability of gas and modifying the atmosphere of the package. They can also control the respiration rate of fruits and ethylene production during the storage period hence enhancing the shelf-life of climacteric fruits.[8]

3.3.3 BARRIER TO AROMA AND FLAVOR

Loss of volatile flavors, aroma, and development of off-flavors usually occur during the storage of foods. Sometimes, there is also a migration of off-flavors from the environment to the package during transportation, distribution, and storage. Edible material having low diffusivity and low affinity for such migration is considered good for packaging food as they have good aroma barrier properties. It helps to retain the best aroma and flavor during storage and distribution. Although studies on the aroma permeability of these edible films are comparatively limited and challenging as compared with moisture and gas transport measurement, still, many methods have been introduced to enhance the aroma barrier properties.[22,35]

3.3.4 IMPROVEMENT IN PRODUCT RELIABILITY, STRENGTH, AND HANDLING ASPECTS

Some food products are fragile in nature so it is essential to maintain their structural integrity during distribution and storage. Edible coatings can improve the strength of the package thus making the handling easy and improving product durability and reliability. The coating also prevents the browning and rotting of fruits as it minimizes epidermal cell damage. Edible coatings are very effective on enrobed frozen foods and dried foods as it improves the structural reinforcement and prevents the disintegration and breakage of dried foods. These coatings have good mechanical properties such as tensile strength, elasticity, and elongation to perform their protective functions efficiently. The only limitation that weakens the mechanical strength of edible coatings and films is the temperature rise. It is an important environmental variable affecting the tensile properties of edible films. As the temperature rises above its transition temperature, the mechanical strength of the package sharply decreases.[65]

3.3.5 IMPROVEMENT IN ESTHETICAL PROPERTIES OF FOOD

Sensory characteristic plays an important role in the acceptance and market value of food. Various attributes like color, appearance, gloss, etc., can be enhanced by the application of edible packaging. Coatings based on lipids and waxes are used to increase the gloss and shine of fruits. Ethanol-based coatings are used for confectionary and candies to make them more bright in color. Ethanol coatings possess irreversible flavor changes in some instances. An effective alternative of ethanol-based coating is a water-based coating as it is less expensive and imparts fair glossy features to a food product.[15]

3.4 CLASSIFICATION OF EDIBLE POLYMERS USED IN FILMS AND COATINGS

There are two basic prerequisites that are taken into consideration for the material used in the manufacturing of edible films and coatings—first material is edible in nature, and second, it is capable to form a persistent layer/film. Generally, polymers that fulfill these requirements are used to make edible films and coatings. The natural polymers used for such coatings can be classified according to their method of production and their origin (Fig. 3.1).[55] On the basis of the above characteristics, edible polymers are grouped into three classes[3]:

- Hydrocolloids (proteins, polysaccharides, and alginate in nature) obtained from nature or from genetically modified microorganisms,
- Lipids (fatty acids, acylglycerol, waxes) from direct biomass, and
- Composites of lipids and hydrocolloids obtained from active compounds and chemical synthesis.

Hydrocolloids are hydrophilic polymers having long chain structures and containing polar or charged functional groups, which render them soluble in water. They are generally polysaccharides or proteins in nature and capable of forming viscous dispersions or gels. Though hydrocolloid-based films are not good water vapor barriers they are very prominent in the maintenance of the sensory quality of food like color and flavor as they are waxy and opaque.[33] On the contrary, lipids are hydrophobic in nature and hence have excellent water vapor barrier properties. They are obtained from vegetable oils having carbon atoms ranging from 14 to 18.[42]

3.4.1 PROTEIN-BASED EDIBLE PACKAGING MATERIAL

Protein-based polymers can be of plant or animal origin or both. They contain amino acids as their structural and functional units. The prime requirement for film formation is the denaturation of protein so that it can form an elongated structure. Protein denaturation can be done by different treatments like heat, acid, alkali, and/or solvent treatment.[51] There are some intrinsic and extrinsic factors that can affect the physiochemical properties of protein used for the formation of films and coatings. Intrinsic factors include the composition of amino acids, electrostatic charge, hydrophobicity, and changes in amino acid structure due to heat, enzymes, pressure, and damage due to some mechanical impact. On the contrary, extrinsic factors include the processing parameters such as temperature, pressure, pH, RH, ionic strength, etc.[57] To improve its mechanical strength, films can be combined with appropriate plasticizers. This can enhance the structural integrity resulting in low fragility and more physical protection. By improving their mechanical strength, these films have good applications in the packaging of dry fruits, cashew nuts, and beans.

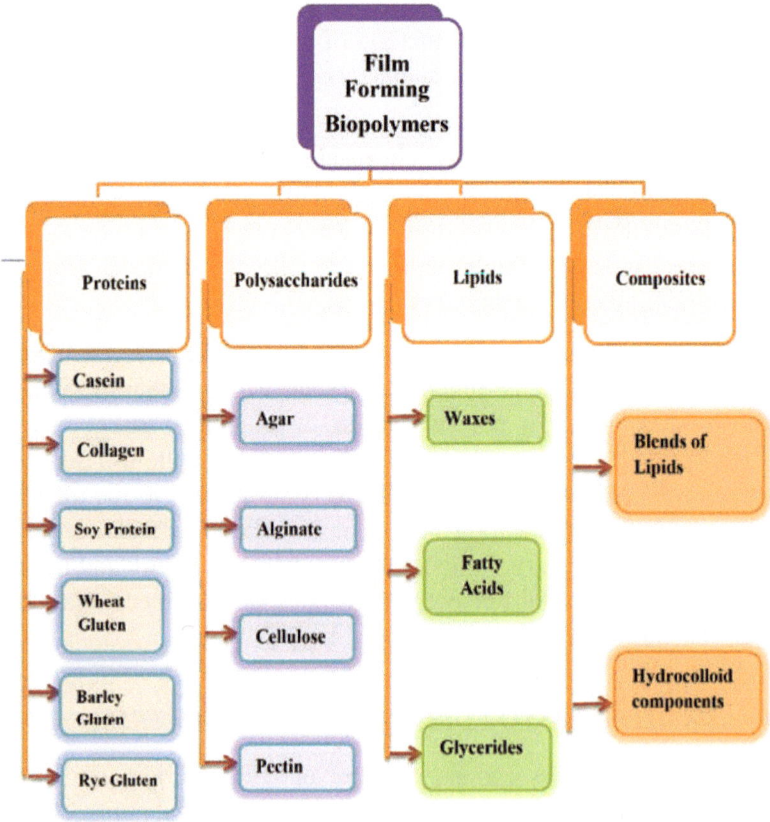

FIGURE 3.1 Natural polymeric materials used for film/coating formation.

Films and coatings made from protein-based polymers have good gas barrier properties and they are excellent barriers of oil and oxygen but they are slender in water vapor resistance. Protein-based films and coatings (Table 3.1) can also be used with the combination of non-edible material as the application of multi-layered packaging helps in reducing the environmental pressure of degradation of synthetic material.[55]

TABLE 3.1 Protein-Based Edible Films and Coatings.

Protein	Source	Functions	Applications	Reference
Calcium caseinate	Animal protein	Permeability to O_2 avoiding oxidation	Potato and apple	[71]
Corn zein	Corn kernel	Good barrier to oxygen	Candy, nuts	[73]
Gelatin	Animal protein	Provides mechanical strength, prevents oxidation, antioxidant carrier	Sausage	[11]
Soy	Vegetable protein	Excellent water and gas permeability	Apple, potato, carrots	[62]
Wheat gluten	Cereal based protein	Low oxygen perme-ability, prevents oxidation	Cheese and tomato	[68]

3.4.2 POLYSACCHARIDE-BASED EDIBLE PACKAGING MATERIAL

Polysaccharides are the only polymers that are found almost in every organism including mammals and plants, and are used for various purposes in food and non-food industries. They are comprised of mono- or disaccharides connected by glycosidic bonds and form long chain structure polymer. Polysaccharide polymers are soluble in water and thus form strenuous bonds between hydrogen and water during hydration. These hydrogens are important contributors to the formation of films and coatings. Due to such hydrogen bond formation, polysaccharides are capable of forming gels or thick solutions. These solutions or gels are further treated with treatments like drying or evaporation for the formation of a film matrix that has several positive characteristics such as tensile strength, solubility, and clarity.[37]

Polysaccharide-based films and coatings are eco-friendly and safe for food packaging due to their nontoxic nature (Table 3.2). Strong hydrogen bonds help in improving mechanical strength and provide good structural characteristics. These films are also good barriers to oil and oxygen. They can be utilized in the packaging of fruits as they can control the oxygen permeability, resulting in controlled and delayed ripening of fruits.[19] Polysaccharide polymers can be of plant origin (cellulose and its derivatives, starch, pectin, and gum Arabic), animal origin (chitin and chitosan), marine origin (alginate and carrageenan), and microbial origin (gellan and xanthan gum).

TABLE 3.2 Polysaccharide-Based Edible Films and Coatings.

Polysaccharide	Functions	Application in food	Reference
Agar	Delay growth of microorganisms	Fish, candy, and desserts	[44]
Alginate film	Used to restructure foods	Crabsticks, onion rings, cocktail berries, and meat products	[67]
Carboxymethyl Cellulose	Good resistance to oxygen, carbon dioxide	Fruits and vegetables	[7]
Carrageenan	Good mechanical properties	Sausage casings, dry solids foods, and oily foods	[72]
Chitosan	Antifungal behavior, good resistance to oxygen, fat, and oil	Apple and cherries	[33]
Hydroxypropyl cellulose	High resistant to fats and oils	Fruits and vegetables	[10]
Starch	Excellent barrier properties to O_2 and CO_2	Strawberry and tangerine	[29]

Cellulose is an organic compound of plant cells found in ample amounts in nature and linked by β-1, 4 glycosidic bonds in between D-glucose units. Although cellulose has consistent hydrogen bonding which results in low water solubility, the treatment of cellulose with alkali followed by some chemical reactions improves its water-soluble properties that help in making a film that can use for packaging.[19] Generally, three water soluble cellulose-based polymers are used for the film formation—methyl cellulose (E461, MC), carboxymethylcellulose (E466, CMC), and hydroxypropyl methylcellulose (E464, HPMC). Among all these three polymers, the best cellulose derivative is carboxy methyl cellulose having excellent barrier properties against oxygen, lipids, and high biodegradable nature with low economic value for application.

CMC is considered the best cellulose derivative for film formation as it has excellent barrier properties to oxygen, lipids, and high biodegradability with low application cost. On the contrary, hydroxypropyl methylcellulose is thermoplastic in nature and thus is very useful in moulding and extrusion. Films based on cellulose derivatives are solely transparent, odorless, flexible, and tough, but are highly resilient to oils and fats.[7]

Starch is composed of two macromolecules-amylose (20–30%) and amylopectin (70–80%) and is widely used for the production of edible films after cellulose derivatives. Starch is found in granular form in plant tubers. The various sources of starch are wheat, rice, potato, and tapioca but the most frequently employed source of starch for filmmaking is corn (maize). Although starch granules are hydrophilic in nature, they have the tendency to absorb the water during hydration. Starch granules swell during hydration and continue until a critical concentration level is achieved. This results in the formation of gel which is further treated to make a packaging film.[4] Starch polymers have features that correspond to plastic polymers such as tasteless, odorless, non-toxicity, good barriers against oxygen, lipids, and low cost. They also lower the water activity of the package resulting in retardation of microbial growth and extended shelf-life of a food product. Edible starch coatings and films are usually used in confectionery, bakery, meat products, and batters.[12]

Pectin is composed of D-galacturonic acid mainly found in the plant cell wall and used as a cementing agent, stabilizing, and thickening agent in the food industry especially in jam, jelly, and marmalade.[36] Pectin can be classified based on its degree of methylation: Low methoxyl pectin (LMP) and high methoxyl pectin (HMP). This methylation degree has a great effect on the gel formation mechanism of pectin and further impacts film strength of the film made by pectin. Pectin films alone are not so consistent for packaging but if it is used in combination with plasticizers or if they are cross-linked with some cations, they show excellent properties for food packaging. These combined films have high mechanical strength, and excellent barriers to lipid, moisture, and oxygen. The current application of these films is on minimally processed fruits and vegetables.[2]

Chitosan is not directly found in nature but it is obtained from the chitin. Chitin is the second most available polymer after cellulose present in biological materials such as the cell wall of fungi and the exoskeleton of crustaceans. Chitosan is extracted by the removal of an acetyl group from chitin in the presence of an alkali. Chitosan is highly considered for the formation of films due to some specific characteristics. The most important characteristic is that the film can be formed without any addition of additives and plasticizers and still shows excellent resistance to oxygen, fat, and microbial activity. Chitosan is not soluble in water but can get solubilized in solvents acidic in nature, such as diluted HCl, acetic acid, and formic acid. Due to their good antimicrobial and antifungal properties, these films have a wide application in fresh fruits and vegetables such as strawberries, cucumber, and apples.[31]

Alginate is a marine-based polysaccharide originating from brown algae and seaweed. It comprises of (1-4)-linkage between α-L-guluronic acid and β-D-mannuronic acid. Alginate films are formed by the gelation process by which gel or gel solution can be formed. Gel formation can be done by the addition of calcium ions which on cooling results in slow calcium release, thus forming a strong cross-linked network. The addition of calcium ions increases the water insolubility of alginate films, hence improving their barrier properties against water vapor.[32] Alginate films are also used as a preservative of food because they are good carriers of antimicrobials and antioxidants. It is also used as an encapsulating agent for food products. Another application of these films is in meat products for the prevention of dehydration in the form of sacrificing agent.[42]

Another marine-originated polysaccharide is carrageenan, which is classified on the basis of the number and position of sulfate groups, namely, κ (20%), ι (33%), and λ (41%). The film formation process includes gel formation followed by drying and solvent evaporation. The major use of carrageenan-based films and coatings are as a sacrificing agent and encapsulating agent for food flavors.[22,72] They prevent the surface dehydration of meat, poultry, and processed products. They are also useful in the retardation of lipid oxidation in meat if used in combination with other packaging materials.[42,77]

3.4.3 LIPID-BASED EDIBLE PACKAGING MATERIALS

Lipids are triglycerides of fatty acids found naturally in both plant and animal cells. Lipids can be used alone or in combination with proteins and polysaccharides for food packaging application. The competency of lipid-based films and coatings relies on some specific features such as structure and type of lipid, its chemical arrangement, hydrophobicity, and physical state.[47] Generally, wax and shellac resins are the most commonly used lipids for the production of edible films. Other materials include petroleum-based wax, essential oils and fats, acetoglyceroids, and fatty acids. Lipid-based films and coatings have excellent water vapor barrier properties due to relatively low polarity. These films also have some disadvantages such as waxy texture, being prone to rancidity, fragility, and opaqueness. However, usually, the polysaccharide is used as a polymer structure matrix to maintain these films which contribute mechanical strength.[80]

Wax is a non-polar substance that consists of functional groups such as alcohol, aldehyde, and ketone with long hydrocarbon units and has high molecular weight. It can either be natural (beeswax) or petroleum-based (paraffin wax). Paraffin is a blend of hydrocarbons attained from the distillation of crude petroleum. It is used as a coating material on fruits and cheese. On the contrary, beewax obtained from insects, that is, honey bees, is an excellent barrier to humidity. These wax coatings are very useful in improving the appearance of the food surface and also provide a barrier to moisture and gas.[12,22,42,60]

Shellac resin is secreted by a plant parasite named *Laccifer lacca*, which especially resides on big trees. It is soluble in alcohols and alkalines, translucent, yellowish brown in color, and is solid or semi-solid in nature. Shellac is used as a coating to improve the shine and gloss of the food product, however, it is not considered as GRAS by USFDA. The shellac resin coatings were used first to give a glossy appearance on fruit surfaces, it was also applied on tomatoes and green chilies, respectively. They are good moisture and water vapor barriers. These coatings are also effective in combating the pitting losses in fruits and vegetables during post-harvest handling and storage.[62]

3.5 APPLICATIONS OF EDIBLE PACKAGING (MATERIAL-BASED FILMS AND COATINGS)

The importance and value of any material whether edible or non-edible, depends on its industrial application. The continuous growth of the food industry, rising demands for processed food, and high environmental pollution due to plastic food packaging have gained the attention of technologists to develop new packaging alternatives as plastics are non-biodegradable and possess an environmental threat. Edible packaging is one of the revolutionary changes in the packaging area especially in the food industry. Here are some applications of edible packaging on food products:

3.5.1 FRUITS AND VEGETABLES: FRESH AND MINIMALLY PROCESSED

Generally, coatings that are edible in nature are used for fruits and vegetables to improve their gloss, shine, and appearance. Edible coatings also control the post-harvest metabolism by controlling the oxygen penetration that results in the slow ripening of fruits. It also provides a good barrier to moisture hence controlling the microbial activity. For fresh-cut fruits, edible coatings prevent enzymatic browning. It acts as a carrier of antioxidant and antimicrobial agents.[75] Effective use of edible films or coatings as barriers over vegetables and fruits leadingly depend on the development of film or coating that can contribute appropriate internal gas composition for different fruits/vegetables in a specific manner (Table 3.3).

TABLE 3.3 Edible Coatings for Fruits and Vegetables.

Fruit/vegetable	Type of coating	Functions	Reference
Apple	Alginate or gellan	Reduce weight loss, carrier of antioxidants, extend shelf-life	[52]
Apple slices	Starch, alginate, beewax, CMC	Reduce weight loss, increased resistance of water vapor	[12]
Avocado	Ellagic acid and candelilla wax	Improved appearance, reduce weight loss	[60]
Blueberry	Chitosan	Extend shelf-life	[77]
Citrus fruits	CMC and chitosan	Improvement of all quantified parameters	[4]
Fresh cut: baby corn asparagus, and Chinese cabbage	Glycerol and chitosan	Antimicrobial, protection against moisture	[55]
Fresh-cut carrot	Caragennan and beewax	Protection against moisture	[23]
Fresh-cut mango	cassava, starch, and glycerol	Inhibited the growth of food-borne human pathogens, increased concentration of flavor volatiles	[18]
Grapes	Shellac, beewax, and cornstarch	Extended shelf-life under refrigerated storage	[25]
Guava	Glycerol	Good water retention properties	[57]
Mushroom	Alginate	Prevention from moisture, improved appearance	[37]
Pear	Carnauba wax	Retard softening	[78]
Pumpkin	κ-Carrageenan and tapioca starch	Reduced decay and improved appearance	[30]
Tomato fruits	Chitosan	Maintenance freshness. Reduction of weight loss	[43]

3.5.2 MEAT, FISH, POULTRY, AND THEIR PRODUCTS

Meat is highly perishable food and very prone to microbial contamination and spoilage which results in low shelf-life and degraded quality. Intelligent and antimicrobial packaging has arisen as a technology that can act as a hurdle to providing food safety in recent years. Protein-based edible films and coatings are generally used for the packaging of meat and poultry products (Table 3.4). These films are very effective against moisture, fat, and microbial activity hence providing a longer shelf-life of meat with better quality. Coating made from collagen protein also helps in reducing the drip loss in meat and fish during the thawing process, especially in salmon.[59] Several polysaccharides-based films and coatings have been employed, made up of agar, chitosan, starch,and isolates of whey protein to improve the safety and shelf-life of fish.

TABLE 3.4 Edible Coatings for Meat, Fish, Poultry, and Their Products.

Product	Type of coating	Functions	Reference
Bacon	Red algae	Minimize lipid oxidation and prevent discoloration	[61]
Carp	Chitosan	Reduce fat uptake during deep frying	[80]
Cooked pork sausages	Chitosan	Barrier to lipids and extended shelf-life	[42]
Dried fish	WPI		[45]
Frozen salmon	WPC	Prevent drip loss during thawing	[40]
Minced pork	Gelatin, glycerol	Reduce weight loss and prevent discoloration	[39]
Roasted turkey	Starch, chitosan, alginate, pectin	Barrier to moisture and reduce microbial activity	[38]
Sardine fillet	Chitosan	Enhance the holding capacity of water, drip loss, and properties of texture.	[47]

3.5.3 DAIRY PRODUCTS

Milk and dairy products are an important part of the diet for both adults and children. Milk contains nutrients that are essential for the growth and development of children.[9] One of the most diversified dairy products, that is, cheese, is highly prone to microbial action. The major causes of the quality degradation of cheese are microbial contamination, loss of moisture, off-flavor development, and other esthetical changes.[14] Edible coatings rich in antimicrobial properties are required for cheese packaging (Table 3.5). It is important to notice that an increment in cross-linking by adding extra calcium is not necessary in cheese coating as it itself contains calcium.

TABLE 3.5 Edible Coatings for Dairy Products.

Product	Type of coating	Functions	Reference
Cheese	Galactomannan, whey protein isolate, and guar gum	Prevent discoloration, lowers moisture loss, and good oxygen barrier	[58]
Mozzarella cheese	Sodium alginate, chitosan, and isolate of soy protein	Possessed overall best properties while storage	[81]
Semi-hard cheese	Chitosan	Extended shelf-life at refrigerated temperature	[24]

3.6 REGULATORY APPROACHES AND MARKETABILITY

The packaging material has direct contact with the food, so it is essential to pay more attention toward the analysis and quality control measures for the safety and health of consumers. To maintain these objectives, European Union (EU) has formed regulatory approaches for the analysis of food packaging material in the 1970s.[64] These regulations are governed by US Food and Drug Administration (FDA). FDA classified food packaging material into three classes and provided clear definitions for each of them: Food Contact Material (FCM), Food Contact Substance (FCS), and Food Contact Article (FCA). European Union also collaborated with the Food and Agriculture Organization (FAO) and World Health Organization (WHO) to form the guidelines for the global acceptance of these regulations. So, FAO and WHO generated a document called Codex Alimentarius for the guidance of these regulations and legislation to be used uniformly in all the countries.[26,28]

The legislation and rules for biodegradable and edible packaging material generally depend on the country in which these products are to be sold. Each and every country falls accountable for the guarantee of the consumers' safety. Although the rules may vary with the country, it should be made mandatory that the package must be labeled with all the relevant information regarding the material such as specific characteristics of the material, processing methods, ingredients, etc. Edible films and coatings are considered intentional food additives as they are eaten with the food product directly. US FDA issued a list containing coatings and emulsions that are permitted as an ingredient for the preparation of edible packages.

These ingredients come under the category of GRAS (Generally Recognized as Safe) substances.[27]

In USA regulation, an excellent aspect of good quality food commercialization has been mentioned for fresh vegetables and fruits that have a coating, which is that labeling should be done to provide information to the consumers and they should be informed regarding their food. On the contrary, in some other countries, like Mexico, such regulation is lacking for these coating materials and there is yet a deficiency of information regarding edible bioactive, functional, and intelligent packaging. For the same reason, several consumers disapprove of the use of edible waxes or films over fruits, but if appropriate and correct information will be provided about edible films, their benefits, and their composition will definitely modify the current opinion. The formation of specific rules for these edible material types is also required to attain this goal.

Most polysaccharide, protein, and lipid-based polymers are considered GRAS by US FDA for edible packaging material. However, some of the gums such as cassia gum are still in doubt for their direct consumption.[6] The acceptability of edible polymer-based films and coatings should follow an appropriate process that determines their suitability for food formulation. The process is summarized as follows:

- If an edible polymer is used according to Current Good Manufacturing Practices and/or according to the limits designed by the USFDA, it can be considered as safe (GRAS) for food packaging.
- Any ingredient used for the edible films should be declared on the label; even if it is used in small amounts because some ingredients have the possibility of containing allergens that can be harmful to individuals.
- If the manufacturer finds an edible polymer suitable for the packaging and the polymer is not existing in the list of GRAS substances, so the manufacturer may either file a petition to FDA for the affirmation of the polymer as a GRAS or he/she may process the polymer on self-determination.
- The manufacturer may not need to establish the utilization of any edible polymer in edible films in GRAS if that polymer material already has pre-1958 clearance from FDA and hence has been priorly sanctioned.
- If an edible polymer or an ingredient cannot be declared GRAS or pass the clearance of the FDA, the manufacturer must submit a petition before its utilization as a film or coating.

The development of any product or in this case edible package, can only ~~be~~ succeed if the particulate is commercialized. For successful commercialization, three factors are taken into consideration: Economic feasibility, Public acceptance, and Regulations.[76]

The economic feasibility of any product depends upon the cost of its raw material and its manufacturing process. In the case of edible packaging, natural polymers are used that are derived from natural sources (pectin, beeswax, etc.) or waste of agriculture and industry (whey from the dairy industry). These raw materials make the edible films and coatings more economical for industrial use as well as reduce the load of solid mass from the environment by utilizing industrial waste.[5]

Public acceptance relies on the fact that "what does the consumer actually want?" Consumer demands for safe and processed food, thus, are directing attention of the nutritionist and technologist for the development of new alternatives. Packaging plays an important role in the food industry and at the same time, it also poses an environmental burden as plastics are the most widely used packaging materials. To combat these problems, edible packages have been developed to have the same properties as plastics or other regular packaging materials.[20] Linked to this part, the "regulations" are intimately related to the edible film legislation which creates the responsibility to inform the consumers regarding the function and composition of the film and establishes the requirement that the constituents must be biodegradable and edible.

3.7 FUTURE TRENDS AND PROSPECTS OF EDIBLE PACKAGING

The annual growth rate of the food packaging market is 4.4% globally till 2019 (Global Food Packaging Market 2015–2019, Technavio). According to the Economic Times of India, this growth rate is expected to rise from US$32 billion in 2015 to US$73 billion in 2022 at a compound annual growth rate (CAGR) of 18%. This data shows that the increasing demand for packaged food is the reason for such a rise in the packaging industry. The major packaging material used in the food industry is plastic as it is cheap and has good characteristic features for food packaging but it poses a big threat to the environment due to its non-biodegradable nature. So packaging materials that are easily biodegradable and/or edible in nature are gaining immense importance even at the market level. Although edible films and coatings are sometimes not so effective alone but in combination with some other additives and plasticizers they are as good as regular packaging materials.

A lot of research are going on either to improve the properties of edible packaging by utilizing some pre-treatments or with the goal of allowing for the incorporation and/or controlled release of active compounds. Recent researches have initiated using nanotechnologies to enhance the properties of edible polymers. Nanocomposites, nanoencapsulation, and nanolaminates are gaining attention to enhance the functionality, strength, and stability of edible packaging materials, though more research is still needed in this direction.

Future research should include:

- Characterization of physical and chemical properties of the solutions used for making coatings and films (composition, concentration, solubility, viscosity, density, surface tension, etc.) and the analysis of different parameters (e.g., pH, temperature, time, etc.) that affects their properties.
- Film characterization envisioned for use as coatings, in conditions similar to that of the commodity being treated. This includes vapor and gas permeability analysis, analysis of sensorial and mechanical properties, solubility, thickness, digestibility, etc.
- Metabolic reaction study which occurs in commodities and the degree towhich modification in these coatings can be done.
- Internal gas composition study of commodities that are coated and relationships between pace and internal atmosphere of physiological processes of ripening, such as metabolic reactions, respiration, and secondary compounds production while storage.
- Study of the effect of coatings on shelf-life and quality of commodities, considering all possible conditions that they could create during storage and handling.
- Optimal methods determination, for application of coatings, most favorable to gain high quality of product at least cost possible.
- Analysis of consumer acceptability for coatings.
- Investigation of effects of edible coatings on final commodity cost.

3.8 SUMMARY

The variations in the global environment are disturbing the lifestyle directly or indirectly depending upon their degree and reasons. The major cause behind this is increased environmental pollution. To combat this issue a lot of focus has been paid in the last few years to maintain and preserving environmental sustainability by using the environmentally friendly materials and biomass by industries. The packaging industry is one of them that is paying much more attention toward the use of biomass and edible polymers to reduce the environmental burden. Bioplastics will revolutionize the future generation where the unique factor of biodegradability will make them more advantageous than conventional plastics. Novel techniques of food packaging are effectively being researched to promote food quality and safety. There is a drift toward edible and environmentally friendly packaging forms. Edible packaging typically utilizes biodegradable and sustainable material that could be used for consumable coating or wrapping over the food, thus generating no waste. Several recent studies have investigated and evaluated the efficiency of edible materials as an additional value to packaged foods. Edible films and coatings are gaining more interest because of their potential to increase shelf-life, exhibit various functions and reduce the threat to the environment. Many researchers have acknowledged the perspective of edible packaging in the food and pharmaceutical industries as a substitute/collaborative addition to conventional packaging to improve the process of protection of food protection and/or recyclability of packages. Innovative diverse utilizations of edible coatings and films have been projected as new applications and also as existing technology alternatives. Research in this aspect has increased exponentially in the past few years, some issues are yet to be resolved in order to utilize them in scaling up for food packaging. As per future research is concerned, the major issue is related to the development of innovative technologies to improve the delivery properties of edible polymer as most studies have been done at a laboratory point yet. Though, further research should focus on commercial scale with the goal of providing highly accurate and precise information that can be utilized to commercialize fresh products covered with edible polymers.

KEYWORDS

- **conventional packaging**
- **edible packaging**
- **environmental pollution**
- **packaging polymers**
- **packaging waste**

REFERENCES

1. Ahvenainen, R. Active and Intelligent Packaging: An Introduction. Chapter 2; In *Novel Food Packaging Techniques*; Ahvenainen, R., Ed.; CRC Press: Boca Raton, FL, 2003; pp 5–21.
2. Arantzazu, V.; Nuria, B.; Alfonso, J.; María, C. G. Natural Pectin Polysaccharides as Edible Coatings. *Coatings* **2015,** *5*, 865–886.
3. Arismendi, C.; Chillo, S.; Conte, A.; AlessandroDel Nobile, M.; Flores, S.; Gerschenson, L. N. Optimization of Physical Properties of Xanthan Gum/Tapioca Starch Edible Matrices Containing Potassium Sorbate and Evaluation of its Antimicrobial Effectiveness. *LWT Food Sci. Technol.* **2013,** *53* (1), 290–296.
4. Arnon, H.; Granit, R.; Porat, R.; Poverenov, E. Development of Polysaccharides-Based Edible Coatings for Citrus Fruits: A Layer-By-Layer Approach. *Food Chem.* **2015,** *166*, 465–472.
5. Arvanitoyannis, I. S. Eds.; In *Irradiation of Food Commodities: Techniques, Applications, Detection, Legislation, Safety and Consumer Opinion*; 1st Ed.; Apple Academic Press, 2010; p 736.
6. Atarés, L.; Chiralt, A. Essential Oils As Additives in Biodegradable Films and Coatings For Active Food Packaging. *Trends Food Sci. Technol.* **2016,** *48*, 51–62.
7. Ghanbarzadeh, B.; Almasi, H. Physical Properties of Edible Emulsified Films Based on Carboxymethyl Cellulose and Oleic Acid. *Int. J. Biol. Macromol.* **2011,** *48*, 44–49.
8. Baldwin, E. A. Surface Treatments and Edible Coatings in Food Preservation. Chapter 21; In *Handbook of Food Preservation*; Rahman, M. S., Ed.; CRC Press: Boca Raton, FL, 2007; pp 478–508.
9. Cardador, M. J.; Gallego, M. Origin of Haloacetic Acids in Milk And Dairy Products. *Food Chem.* **2016,** *196*, 750–756.
10. Campos, C. A.; Gerschenson, L. N.; Flores, S. K. Development of Edible Films and Coatings with Antimicrobial Activity. *Food Bioproc. Technol.* **2011,** *4*, 849–875.
11. Castro, R. A.; Helena, G.; Gonzales. Recubrimiento Comestible en la Conservación de Uchuva (*Physalis peruviana* L. var Colombia). *Revista de la Asociación Colombiana de ciencia y tecnología de alimentos* **2010,** *91* (19), 16–34.
12. Chiumarelli, M.; Hubinger, M. Stability, Solubility, Mechanical and Barrier Properties of Cassava Starch – Carnauba Wax Edible Coatings to Preserve Fresh-Cut Apples. *Food Hydrocoll.* **2012,** *28* (1), 59–67.
13. Coles, R. Introduction. Chapter 1; In *Food Packaging Technology*; Coles, R., McDowell, D., Kirwan, M. J., Eds.; Blackwell Publishing: London, UK, **2003,** vol 5, pp 1–31.
14. Costa, M. J.; Maciel, L. C.; Teixeira, J. A.; Vicente, A. A.; Cerqueira, M. A. Use of Edible Films and Coatings in Cheese Preservation:Opportunities and Challenges. *Food Res. Int.* **2018,** *107*, 84–92.
15. Dangaran, K. L.; Renner-Nantz, J.; Krochta, J. M. Whey Protein-Sucrose Coating Gloss and Integrity Stabilization by Crystallization Inhibitors. *J. Food Sci.* **2006,** *71* (3), 152–157.
16. Dehghani, S.; Hosseini, S. V.; Regenstein, J. M. Edible Films and Coatings in Seafood Preservation: A Review. *Food Chem.* **2018,** *240*, 505–513.
17. Diab, T.; Biliaderis, C. G.; Gerasopoulos, D; Sfakiotakis, E. Physicochemical Properties and Application of Pullulan Edible Films and Coatings in Fruit Preservation. *J. Sci. Food Agric.* **2001,** *81*, 988–1000.

18. Dussán, S.; Torres, C.; Hleap, J. Effect of Edible Coating and Different Packaging During Cold Storage of Fresh-Cut Tommy Atkins Mango. *Inf. Technol.* **2014,** *25* (4), 123–130.
19. Erginkaya, Z.; Kalkan, S.; Üna, E. Use of Antimicrobial Edible Films and Coatings as Packaging Materials for Food Safety. In *Food Processing: Strategies for Quality Assessment*, 2014; vol 2, pp 261–295.
20. Espitia, P. J. P.; Du, W. X.; Avena-Bustillos, R. D. J.; Soares, N. D. F. F. Edible Films From Pectin: Physical-Mechanical and Antimicrobial Properties—A Review. *Food Hydrocoll.* **2014,** *35*, 287–296.
21. Shih, F. F.; Daigle, K. W.; Champagne E. T. Effect of Rice Wax on Water Vapour Permeability and Sorption Properties of Edible Pullulan Flms. *Food Chem.* **2011,** *127* (1), 118–121.
22. Fabra, M. J.; Hambleton, A.; Talens, P.; Debeaufort, F.; Chiralt, A; Voilley, A. Influence of Interactions on Water and Aroma Permeabilities of ι-Carrageenan-Oleic Acid-Beeswax Films Used For Flavor Encapsulation. *Carbohydr. Polym.* **2009,** *76* (2), 325–332.
23. Fai, A. E. C.; Alves de Souza, M. R.; Toscano de Barros, S.; Bruno, N. V.; Ferreira, M. S. L.; Branco de Andrade Gonçalves, E. C. Development and Evaluation of Biodegradable Films and Coatings Obtained From Fruit and Vegetable Residues Applied to Fresh-Cut Carrot (Daucus carota L*.). Postharvest Biol. Technol.* **2016,** *112*, 194–204.
24. Fajardo, P.; Martins, J. T.; Fuciños, C.; Pastrana, L.; Teixeira, J. A.; Vicente, A. A. Evaluation of a Chitosan-Based Edible Film as Carrier of Natamycin to Improve the Storability of Saloio Cheese. *J. Food Eng.* **2010,** *101* (4), 349–356.
25. Fakhouri, F. M.; Martelli, S. M.; Caon, T.; Velasco, J. I.; Helena, L.; Mei, I. Edible Films and Coatings Based on Starch/Gelatin: Film Properties and Effect of Coatings on Quality of Refrigerated Red Crimson Grapes. *Postharvest Biol. Technol.* **2015,** *109*, 57–64.
26. Properties, Manufacture and Application of Seaweed Polysaccharides Agar, Carrageenan and Algin. FAO, 2020; p 32. http://www.fao.org/docrep/feld/003/AB730E/AB730E03.htm (accessed Jan 10, 2020).
27. FDA (Food and Drug Administration). 2015. http://www.accessdata.fda.gov/scripts/cdrh/cfdocs/cfcfr/CFRSearch.cfm?CFRPart=175andshowFR=1andsubpartNode=21:3.0.1.1.6.3 (accessed Jan 12, 2020).
28. Food and Agriculture Organization (FAO) of the United Nations, 2016. http://www.fao.org/faowho-codexalimentarius/codex-home/es/ (accessed Feb 2, 2020).
29. Garcia, L. C.; Pereira, L. M.; de Luca Sarantopoulos, C. I. G.; Hubinger, M. D. Effect of Antimicrobial Starch Edible Coating on Shelf Life of Fresh Strawberries. *Packag. Technol. Sci.* **2012,** *25* (7), 413–425.
30. Genevois, C.; de Escalada, M.; Flores, S. Application of Edible Coatings to Improve Global Quality of Fortified Pumpkin. *Innov. Food Sci. Emerg. Technol.* **2015,** *33*, 506–514.
31. Guadalupe, I.; Olivasa Gustavo, V.; Barbosa-Canovas. Alginate–calcium films: Water Vapor Permeability and Mechanical Properties as Affected by Plasticizer and Relative Humidity. *LWT* **2008,** *41*, 359–366.
32. Guilbert, S.; Gontard, N.; Cuq, B. Technology and Applications of Edible Protective Films. *Packag. Technol. Sci.* **1995,** *8*, 339–346.
33. Nayik, G. A.; Majid, I.; Kumar, V. Developments in Edible films and Coatings for the Extension of Shelf-life of Fresh Fruits. *Am. J. Nutr. Food Sci.* **2015,** *2* (1), 16–20.

34. Hambleton, A.; Debeaufort, F.; Bonnotte, A.; Voilley, A. Influence of Alginate Emulsion-Based Films Structure on its Barrier Properties and on the Protection of Microencapsulated Aroma Compound. *Food Hydrocoll.* **2009,** *23* (8), 2116–2124.
35. Han, J. Ed.; In *Innovations in Food Packaging*, 2nd Ed.; Elsevier Academic Press: TX, USA, 2014; p 624.
36. Janjarasskul, T.; Rauch, D. J.; McCarthy, K. L.; Krochta, J. M. Barrier and Tensile Properties of Whey Protein– Candelilla Wax Film/Sheet. *LWT Food Sci. Technol.* **2014,** *56*, 377–382.
37. Jiang, Y.; Li, J.; Jiang, W. Effects of Chitosan Coating on Shelf-life of Cold-Stored Litchi Fruit at Ambient Temperature. *LWT Food Sci. Technol.* **2005,** *38*, 757–761.
38. Jiang, Z.; Neetoo, H.; Chen, H. Efficacy of Freezing, Frozen Storage and Edible Antimicrobial Coatings Used in Combination for Control of *Listeria monocytogenes* on Roasted Turkey Stored at Chiller Temperatures. *Food Microbiol.* **2011,** *28* (7), 1394–1401.
39. Kaewprachu, P.; Osako, K.; Benjakul, S.; Rawdkuen, S. Quality Attributes of Minced Pork Wrapped with Catechin–Lysozyme Incorporated Gelatin Film. *Food Packag. Shelf Life* **2015,** *3*, 88–96.
40. Kim I. H.; Yang, H. J.; Noh, B. S.; Chung, S. J.; Min, S. C. Development of a Defatted Mustard Meal-Based Composite Film and Its Application to Smoked Salmon to Retard Lipid Oxidation. *Food Chem.* **2012,** *133* (4), 1501–1509.
41. Ko, S.; Janes, M. E.; Hettiarachchy, N. S.; Johnson, M. G. Physical and Chemical Properties of Edible Films Containing Nisin and Their Action Against Listeria Monocytogenes. *J. Food Sci.* **2001,** *66* (7), 1006–1011.
42. Lekjing, S. A Chitosan-Based Coating With or Without Clove Oil Extends the Shelf-life of Cooked Pork Sausages in Refrigerated Storage. *Meat Sci.* **2016,** *111*, 192–197.
43. Limchoowong, N.; Sricharoen, P.; Techawongstien, S.; Chantha, S. An Iodine Supplementation of Tomato Fruits Coated With an Edible Film of the Iodide-Doped Chitosan. *Food Chem.* **2016,** *200*, 223–229.
44. López de Lacey, A. M.; López-Caballero, M. E.; Montero, P. Agar Films Containing Green Tea Extract and Probiotic Bacteria for Extending Fish Shelf-Life. *LWT Food Sci. Technol.* **2014,** *55* (2), 559–564.
45. Matan, N. Antimicrobial Activity of Edible Film Incorporated With Essential Oils to Preserve Dried Fish (Decapterus maruadsi). *Int. Food Res. J.* **2012,** *19* (4), 1733–1738.
46. Mehyar, G. F.; Al-Ismail, K.; Han, J. H.; Chee, G. W. Characterization of Edible Coatings Consisting of Pea Starch, Whey Protein Isolate, and Carnauba Wax and their Effects on Oil Rancidity and Sensory Properties of Walnuts and Pine Nuts. *J. Food Sci.* **2012,** *77* (2), 52–59.
47. Mohan, C. O.; Ravishankar, C. N.; Lalitha, K. V.; Srinivasa Gopal, T.K. Effect of Chitosan Edible Coating On the Quality of Double Flleted Indian Oil Sardine (*Sardinella longiceps*) During Chilled Storage. *Food Hydrocoll.* **2012,** *26* (1), 167–174.
48. Murmu, S. B.; Mishra, H. N. The Effect of Edible Coating Based on Arabic Gum, Sodium Caseinate and Essential Oil of Cinnamon and Lemon Grass on Guava. *Food Chem.* **2018,** *245*, 820–828.
49. Ncama, K.; Magwaza, L.; Mditshwa, A.; Zeray Tesfay, S. Plant-Based Edible Coatings for Managing Postharvest Quality of Fresh Horticultural Produce: A Review. *Food Packag. Shelf Life* **2018,** *16*, 157–167.

50. Nur Hanani, Z. A.; Roos, Y. H.; Kerry, J. P. Use and Application of Gelatin As Potential Biodegradable Packaging Materials for Food Products. *Int. J. Biol. Macromol.* **2014,** *71*, 94–102.
51. O'Sullivan, A.; Shaw, N. B.; Murphy, S. C.; van de Vis, J. W.; van Pelt-Heerschap, H.; Kerry, J. P. Extraction of Collagen From Fish Skins and Its Use in the Manufacture of Biopolymer Films. *J. Aquat. Food Prod. Technol.* **2006,** *15* (1), 21–32.
52. Ochoa, E.; Saucedo-Pompa, S.; Rojas-Molina, R.; Garza, H.; Charles-Rodríguez, A. V.; Aguila, C. N. Evaluation of a Candelilla Wax-Based Edible Coating to Prolong the Shelf-Life Quality and Safety of Apples. *Am. J. Agric. Biol. Sci.* **2011,** *6* (1), 92–98.
53. Oscar L. Ramos; Joao C. Fernandes; Sara I. Silva; Manuela E. Pintado; F. Xavier Malcata. Edible Films and Coatings from Whey Proteins: A Review on Formulation, and on Mechanical and Bioactive Properties. *Crit. Rev. Food Sci. Nutr.* **2012,** *52* (6), 533–552.
54. Parente Ribeiro Cerqueira, M. Â. Edible Packaging. In *Reference Module in Food Science*, 2018; pp 173–176.
55. Pitak, N.; Rakshit, S. Physical and Antimicrobial Properties of Banana Flour/Chitosan Biodegradable and Self Sealing Films Used for Preserving Fresh-Cut Vegetables. *LWT Food Sci. Technol.* **2011,** *44* (10), 2310–2315.
56. Popović, S.; Pericin, D.; Vastag, Z.; Lazić, V.; Popović, L. J. Pumpkin Oil Cake Protein Isolate Films as Potential Gas Barrier Coating. *J. Food Eng.* **2012,** *110* (3), 374–379.
57. Quezada-Gallo, J.; Gramin, A.; Pattyn, C.; Diaz Amaro, M.; Debeaufort, F.; Voilley, A. Biopolymers Used as Edible Coating to Limit Water Transfer, Color Degradation and Aroma Compound 2-pentanone Lost in Mexican Fruits. *Acta Hortic.* **2005,** *682* (3), 1709–1716.
58. Ramos, O.L.; Pereira, J.O.; Silva, S.I.; Fernandes, J.C.; Franco, M.I.; Lopes-da-Silva, J.A.; Pintado, M.E.; Malcata, F.X. Evaluation of Antimicrobial Edible Coatings from a Whey Protein Isolate Base to Improve the Shelf-life of Cheese. *J. Dairy Sci.* **2012,** *95* (11), 6282–6292.
59. Rodriguez-Turienzo, L.; Cobos, A.; Moreno, V.; Caride, A.; Vieites, J. M.; Diaz, O. Whey Protein-Based Coatings on Frozen Atlantic Salmon (Salmo salar): Influence of the Plasticiser and the Moment of Coating on Quality Preservation. *Food Chem.* **2011,** *128* (1), 187–194.
60. Saucedo-Pompa, S.; Rojas-Molina, R.; Aguilera-Carbó, A. F.; Sáenz- Galindo, A.; Garza, H. L.; Jasso-Cantú, D. Edible Film Based on Candelilla Wax to Improve the Shelf-life and Quality of Avocado. *Food Res. Int.* **2009,** *42* (4), 511–515.
61. Shin, Y. J.; Song, H. Y.; Seo, Y. B.; Song, K. B. Preparation of Red Algae Film Containing Grapefruit Seed Extract and Application for the Packaging of Cheese and Bacon. *Food Sci. Biotechnol.* **2012,** *21* (1), 225–231.
62. Shon, J. H.; Choi, Y. H. Effect of Edible Coatings Containing Soy Protein Isolate (SPI) on the Browning and Moisture Content of Cut Fruits and Vegetables. *J. Appl. Biol. Chem.* **2011,** *54* (3), 190–196.
63. Silano, V.; Rossi, L. Safety Evaluation in the European Union of Flavorings, Contact Materials, Enzymes, and Processing Aids in Food and Its Evolution Over Time. *Euro. Food Feed Law Rev.* **2015,** *10* (6), 402–432.
64. Sothornvit, R.; Krochta, J. M. Plasticizer Effect on Mechanical Properties of β-lactoglobulin Films. *J. Food Eng.* **2001,** *50* (3), 149–55.

65. Subhas C.; Shah, P. M. Edible Polymers: Challenges and Opportunities. *J. Polym.* **2014,** *2014* (427259), 1–13.
66. Ghosh, S. K.; Pal, S; Ray, S. Study of Microbes Having Potentiality for Biodegradation of Plastics. *Environ. Sci. Pollut. Res.* **2013,** *20*, 4339–4355.
67. Takala, P. N.; Salmieri, S.; Boumail, A.; Khan, R. A.; Vu, K. D.; Chauve, G.; Bouchard, J.; Lacroix, M. Antimicrobial Effect and Physicochemical Properties of Bioactive Trilayer Polycaprolactone/Methylcellulose Based Films on the Growth of Foodborne Pathogens and Total Microbiota in Fresh Broccoli. *J. Food Eng.* **2013,** *116*, 648–655.
68. Tanada Palmu, P. S.; Grosso. Development and Characterization of Edible Films Based on Gluten from Semihard and Soft Brazilian Wheat Flours. *Food Sci. Technol.* **2003,** *23* (2), 264–269.
69. What a Waste 2.0: A Global Snapshot of Solid Waste Management to 2050. The World Bank Group, 2020; p 32. https://openknowledge.worldbank.org/bitstream/handle/10986/30317/211329ov.pdf (accessed Sept 18, 2018).
70. Janjarasskul, T.; Krochta, J. M. Edible Packaging Materials. *Ann. Rev. Food Sci. Technol.* **2010,** *1*, 415–448.
71. Thian, H.; Wu, W.; Guo-Gaolun, B.; Jian, Q.; Xiang, A. Microstructure and Mechanical Properties of Glycerol Plasticized Soy Protein Plastics Containing Castor Oil. *J. Food Eng.* **2012,** *109*, 406–500.
72. Karbowiak, T.; Debeaufort, F.; Champion, D.; Voilley, A. Wetting Properties at the Surface of Iotacarrageenan-Based Edible Films. *J. Coll. Interface Sci.* **2006,** *294*, 400–410.
73. Shi, W.; Dumont, M. J. Review: Bio-Based Films from Zein, Keratin, Pea, and Rapeseed Protein Feedstocks. *J. Mater. Sci.* **2014,** *49*, 1915–1930.
74. Valencia-Chamorro, S. A.; Palou, L.; del Río, M. A.; Pérez-Gago, M. B. Antimicrobial Edible Films and Coatings for Fresh and Minimally Processed Fruits and Vegetables: A Review. *Crit. Rev. Food Sci. Nutr.* **2011,** *51* (9), 872–900.
75. Van Beilen, J. B.; Poirier, Y. Prospects for Biopolymer Production in Plants. Chapter 8; In *Green Gene Technology*; Fiechter, A., Sautter, C., Eds.; Springer: New York, 2007; pp 133–151.
76. Varela, P.; Fiszman, S. M. Hydrocolloids in Fried Foods: A Review. *Food Hydrocoll.* **2011,** *25* (8), 1801–1812.
77. Vieira, J. M.; Flores-López, M. L.; Rodríguez, D. J.; Sousa, M. C.; Vicente, A. A.; Martins, J. T. Effect of Chitosan–Aloe Vera Coating on Postharvest Quality of Blueberry (Vacciniumcorymbosum) Fruit. *Postharvest Biol. Technol.* **2016,** *116*, 88–97.
78. Xu, S.; Chen, X.; Sun, D. Preservation of Kiwifruit Coated With an Edible Film at Ambient Temperature. *J. Food Eng.* **2001,** *50* (4), 211–216.
79. Zahedi, Y.; Ghanbarzadeh, B.; Sedaghat, N. Physical Properties of Edible Emulsified Films Based on Pistachio Globulin Protein and Fatty Acids. *J. Food Eng.* **2010,** *100*, 102–108.
80. Zhang, L.; Luo, Y.; Hu, S.; Shen, H. Effects of Chitosan Coatings Enriched with Different Antioxidants on Preservation of Grass Carp (Ctenopharyngodon idellus) During Cold Storage. *J. Aquat. Food Prod. Technol.* **2012,** *21* (5), 508–518.
81. Zhong, Y.; Cavender, G.; Zhao, Y. Investigation of Different Coating Application Methods on the Performance of Edible Coatings on Mozzarella Cheese. *LWT Food Sci. Technol.* **2014,** *56* (1), 1–8.

CHAPTER 4

ROLE OF NANOMATERIALS IN EDIBLE PACKAGING SYSTEMS

SHWETHA ACHARYA, MAHENDRA GUNJAL, and PREETI BIRWAL

ABSTRACT

The applications of nanotechnology in eco-favorable coatings and encapsulation systems may have probabilities of enhancing food quality, safety, stability, and efficiency. The edible films/packaging materials are prepared by using organic and inorganic materials that are nontoxic, economical, and readily available. To improve the quality of food products, emerging techniques use edible coating as carriers of functional ingredients by integrating antibacterial, anti-browning, and nutraceuticals agents. This chapter summarizes the role of nanotechnology in edible packaging systems, different types of nanomaterials and their properties, and different methods to prepare edible films/packaging materials. The merits and demerits of this technology along with toxicity and health safety associated with nanoparticles in edible films/packaging have also been discussed.

4.1 INTRODUCTION

The consumer prefers fresh, minimally processed, nutritious, safe, and ready-to-eat food products with proper well-defined food packaging material and descriptive labels.[37] The nanotechnology intervention in packaging mainly explores three potentials, such as direct absorption into food products, incorporation in packaging material, and various applications in food processing methods.[31] The commercialized products are flourishing, which implement and also have high positive response to miscellaneous applications of

Advances in Sustainable Food Packaging Technology. Megh R. Goyal, Preeti Birwal, Anand K. Pandey, & Ritesh B. Watharkar, (Eds.)

nanotechnology.[37] Several innovative packaging developments have focused on extending the shelf life of the product and on delivering a superior quality product to the consumer. This development lacks significant advances in the packaging equipments and the incorporation of various types of sensors into the food packaging. The major types of food packaging in the small-scale food industries are glass, paper metal, CFB, and plastic containers, and the relative cost of different materials determines their adoption by local processing industries.[58] These packaging materials may have to be imported; therefore, minimum order may be an important constraint for the development of a food processing sector.[40]

Nanotechnology is a new food packaging method,[34] which can increase the shelf life of food, reduce the risk of food spoilage, ensure food safety and may also decrease food shortage, and may improve our health. Nanomaterials are classified into nanoparticles (TiO_2, Silica dioxide), nanofiber (cellulose, gelatin), and nanoplates (Silicate clay, Polymer/clay).[40] Nanotechnology in the food industry is used to identify the bacteria in food packaging, manufacture flavor color quality, and food security for raising the barrier properties.[40] Globally, nanomaterials worth about three trillion US$ in 2020 have been proposed to improve the mechanical and barrier properties of food packages along with active and intelligent packaging. Nanotechnology in food packaging is aimed to improve the main features of traditional packaging systems that include contentment (simplicity for transportation and handling), convenience (consumer responsive), preservation (avoid packaging and avoid microbial spoilage, finally long shelf-life), marketing, and communication.[59]

A mixture of nanomaterials (such as silver nanoparticles (Ag), tin nanoparticles, nanotitanium dioxide (TiO_2), and nanozinc oxide (ZnO)) as a functional additive to foodstuff package classification is based on the dimensions of their structure elements.[34] Nanocoatings help to improve packaging properties. Various types of food coatings/edible films are available in the market to protect the food from mass transfer barriers. These edible films of nanocoatings range from 3 to 5 nm, which can increase the shelf life by protecting fruits, vegetables, cheeses, and bakery products to provide flavor, color, enzymes, antioxidants, and anti-browning compounds to the products.[73] Nanoclays as nanomaterials are also intentionally used in food packaging because of their superior mechanical/thermal/barrier properties and low cost.[50]

Biopackaging is prepared from natural raw materials, such as cellulose, starch, and chitin), and environmental synthetic polymers, such as polylactic acid and polycaprolactone. Biodegradable polymers have many advantages which include renewable, melt strength, nontoxic, thermal stability, impact strength, and permeability.[31] Biodegradable polymers also enhance their industrial use to incorporate nano-sized particles for strengthening the polymer and to improve the properties of biopolymers. Some traditional nanocrystals have applications, such as food packaging, films and coatings to extend the shelf life of fresh fruits/vegetables.[31]

Nanomaterials are used to develop superior barrier properties of packaging, which are suitable for their antimicrobial, UV protection, and the prospect of oxidation prevention properties. Antimicrobial nanomaterials have high antimicrobial activity, which helps the packaging system against the spoilage of food products.[76] Further, the most recent techniques involve edible films incorporated with nanoparticles and encapsulated compounds used for food preservation. The packaging materials must possess physical–mechanical properties, such as mechanical resistance, barrier properties, and antimicrobial properties.[40]

This chapter focuses on the role of nanomaterials in edible packaging systems.

4.2 DIFFERENT TYPES OF NANOMATERIALS USED IN EDIBLE FILMS/PACKAGING

Nanomaterials can improve the properties of food packaging, such as antimicrobial, UV protection activity, and the possibility of oxidation prevention. Antimicrobial nanomaterials are made of Ag, TiO_2, ZnO, and MgO. Nanoparticles are very suitable agents for antimicrobial active packaging systems due to their high antimicrobial activity.[6] TiO_2 nanoparticles are safe to the human body and are permitted as food additives in food contact material in food packaging.[11]

Edible films used for cheese, fruits, confectionery, bakery goods, and fast foods provide a barrier to moisture and gas, and act as vehicles to color, flavor, and oxidants; and serve as anti-browning agents. Edible films also have increased the shelf-life of mango puree reinforced with nanocellulose.[8,28] Nanomaterials are included in polymers to act as film-forming for food packaging to get better physical–mechanical properties by acting as fillers.[50] Biological activity of edible films against food-borne pathogens and microbial spoilage may involve the use of biopolymer-based materials for food preservation. The food-grade biopolymers and edible films are formed by good packaging biodegradability and nutritional value in many roles like antimicrobial and antioxidant when used as additives.[34]

The edible film has an advantage of edible packaging materials over synthetic film to reduce environmental pollution. The barriers including edible films and coatings can decrease the complications and improve the packaging materials, which are compared with the traditional methods and eco-friendly packaging materials and replaced by such synthetic polymer films. New materials are being developed and considered by scientists, many from abundant natural sources that have conventionally been regarded as desecrated materials.[35] Different types of nanomaterials used in edible films/packaging are given as follows:

- Hydrocolloids[50,76]: Starches,[50,76,79] chitosan,[12,24,51] guar gum;[24]
- Lipids[28,31]: Paraffin wax,[31] protein film,[37,64] gelatin gel,[24,28] corn zein film,[24,28,83]
- Proteins: Soy protein films,[24,62,78] collagen protein film,[14] gelatine,[40,41,56]
- Other materials[24,35,37,45,65,72]: Synthetic polymers and natural polymers, composite films.[45,65]

4.3 GENERAL CHARACTERISTICS AND PROPERTIES OF EDIBLE FILMS

Edible films are used to protect food products from spoilage and unfavorable condition, which increases their shelf-life, and accordingly, edible films and coating vary to the requirements of foods during storage condition.[57,77,82]

4.3.1 CHARACTERISTICS OF EDIBLE FILMS (TABLE 4.1)

TABLE 4.1 Characteristic of Edible Films.

Characteristics	Description
Cost	Low cost
Desirable additives	Coloring agents, nutrients, flavor, antioxidants, antimicrobial agents, and vitamins
Manufacture	Easily manufacture and economically feasible
Nature	Nontoxic, nonallergic, fully digestible, and biodegradable
Packaging characteristics	Mechanically stable against damage during transportation, handling, and storage of food materials
Sensory characteristics	Good odor, flavor, taste, and appearance

4.3.2 PROPERTIES OF EDIBLE FILMS

The properties of edible film depend on the type of film-forming materials in use and molecular weight.[82]

4.3.2.1 ORGANOLEPTIC PROPERTIES

The organoleptic properties for edible films and coatings must be clear, transparent, odorless, tasteless, etc. Examples are coatings of sugar and chocolate layers or candies. Edible films are fortified with vitamins and minerals and nutrients to increase the shelf-life and to decrease the chance of defects by foreign matter.[82] This nanomaterial helps to preserve the advantageous concentration of coloring substances, flavor, sweetness, and saltiness. Also, it gives nutritional enhancement without destroying the reliability of the food product.[39,77,80]

4.3.2.2 BIODEGRADABILITY

Biodegradation is based on the type of polymer used, structure, molecular weight, crystalline structure, and additives. The polymer depends upon the material which is fully biodegradable nontoxic.[8] To maintain their edibility and biodegradability, all film components should be food-grade ingredients and biodegradable.[2] There are many quality factors, such as visual quality, texture, flavor, edible color printing, and other marketing-related factors which are contributed.[8]

4.3.2.3 MECHANICAL PROPERTIES

Mechanical properties are based upon the tensile strength which is comparable to polyethylene and PET. The mechanical properties depend on the fruit film composition. The fruit films are stronger than vegetable films due to their higher ratios of dietary fibers to total sugars.[8] Mechanical properties should be optimized regarding tensile strength, compression strength, puncture strength, burst strength, abrasion resistance, adhesion force.

4.3.2.4 WATER VAPOR TRANSFER

For any edible films, the WVTR must be low protective barriers to boundary moisture replacement for short-term utilization. Biodegradable films, such as lipids and wax coatings that have good WVTR barriers are evaluated with starch and protein-based materials due to low water vapor barrier properties. Edible films have low dehydration, moisture absorption and oxidation of ingredients, aroma loss, frying oil absorption, ripening/aging, and microbial deterioration of food products.[8,56,57] The purpose of edible films is to concern the lipid compounds, such as animal and vegetable fats that have tremendous moisture and gas barrier properties which are based on textural and organoleptic inconvenience due to oxidation which leads to rancidity and a waxy taste.[27]

4.3.2.5 CONTROL OF GAS EXCHANGE

Edible films depend on resources which are made up of remarkable physicochemical properties. All edible films have good potential barrier properties at different compositions and environmental conditions, still, the food product is safe and nontoxic.[8,77] Nowadays, food industries utilize edible films and coatings for food ingredients in the form of capsules, microcapsules, soluble strips, flexible pouches, and on hard particles.[57,82]

4.4 METHODS OF PACKAGING OF EDIBLE FILMS

4.4.1 NANOCOATING

Spraying, immersion, and encapsulation techniques are easy to apply for edible nanocoatings for food products.[83] This frequently consists of eco-friendly materials and does not require to be reduced from food before utilization.[82] Currently, nanocoating films present nanostructures in which natural ingredients which have antimicrobial and antioxidant action are included to amplify the valuable effects on the food product quality.[78]

In this study, the nanomaterials are based on biopolymer edible film made by treating the suspension with high-pressure homogenization and studied the effect of glycerol as a plasticizing agent by differing its concentration, such as 23%, 33%, and 44% w/w, the dry basis of powder and observed that the inverse relationship was found in between glycerol concentration and moisture content of the film. As glycerol content increases, the water content decreased at various levels of water activity and also showed that the increase in the glycerol content increases the barrier properties of the film. The food products depend on the edible films and coatings which are used during processes.[8,33]

The dispersed systems are covered by the food products, which produce a firm matrix that will perform as an edible film and coating. The new techniques in applying edible films on food products include dipping, spreading, spraying, and wrapping. The edible film formulations are added and dried directly on the exterior part of the food, whereas the other material is poured into a tray and dried which is integrated into the food product that possesses hydrophilic and hydrophobic properties , which has a resemblance to polar and nonpolar sites.[16] So, emulsions are classified into two types based on their component, for example, oil/water or water/oil due to the diffuse phase communicated to the second component, individual in lower concentration in the emulsion.[62] During the process of formulation of edible films, the oil/water systems have additional stability and can suspend lipophilic antimicrobial components.[83]

4.4.1.1 HALLOYSITE NANOTUBES

A halloysite nanotube (HNTs) is made from natural nanomaterial. HNTs are technologically advanced as functional nanocapsules in such a way that it has the capability of being loaded by many materials.[35]

Nanomaterial is arranged by a biopolymer film made up of pectin matrix is modified with HNTs holding peppermint oil, HNTs exterior surface with cucurbit aril particles in order to improve the affinity of the nanofiller toward peppermint oil.

In agricultural produce handling, wastes are used to synthesize the nanocomposites as resources.[56] Biopolymer film combined with banana starch nanoparticles was invented using blend cross-linking to brand an enhanced agent.[81]

The pectin extracted from pomegranate peels is formulated by films with a casting method that is fabricated by carboxymethyl cellulose, and the manufactured nanocomposite films contain ZnO-NPs, which is superior in the UV-light blocking of single-layer nanoclay.[52,70]

4.4.1.2 TIO_2 NANOPARTICLES

TiO_2 is utilized as an antimicrobial agent as it is cheap, nontoxic, and photostable and metal oxide nanomaterials show thermo permanency property and water gets attracted due to their antimicrobial activity.[55,64] TiO_2 is one of the best antimicrobial agents which has the capabilities of polyethylene-based film that is standard to the biocide success of TiO_2 in the flaw of bacteria and the accumulated TiO_2 compressed the penetration diffusivity, but did not disturb the gas barrier as well as reduced the visual transparency. TiO_2 developed the visual transparency and improved the flexible strength and contact angle and deteriorated the water vapor penetrability properties and reduced the elongation at break of the film.[55]

4.4.2 NANOLAMINATES

Nanolaminate films consisted of two or more layers with nanolayer whose interlinkage can be physical and chemical techniques for their synthesis. This technique permits surface lamination with various nanofilms layers based on several nanomaterials.[14,47] Later, these films use compounds, such as proteins, polysaccharides, colloidal molecules, and charged lipids that provide and improve the both physical and chemical properties. The active compounds, such as antimicrobials, anti-browning, and antioxidants agents can be incorporated into the films[23] to increase the quality and shelf life of packaged food products.[47]

ZnO nanoparticles are inorganic metal oxides that have been manufactured by chemical methods.[64] Different techniques, such as thermal decomposition, hydrothermal, and precipitation are functional to ZnO nanoparticles for assembly. The incorporation of ZnO nanoparticles has been amplified safely in food packaging because of their antimicrobial activities, UV blocking, and low-cost compared with the Ag nanoparticles.[55]

4.4.3 POLYMER/CLAY NANOCOMPOSITES

Clays and silicate resources are filler-stable at low-cost and nontoxic in nature. Clays, kaolinite, montmorillonite, and saponite materials have been used for PNC applications.[80] Polymer and clay nanocomposites are also characterized as auspicious classes in food packaging due to the improvement in the mechanical and barrier properties of edible packaging materials among the matrix polymers; and a small quantity of nanoclay as nanofiller is merged in the polymer matrix.[47]

The central process of nanocomposite materials is melt compounding and in situ polymerization. A nanocomposite system is made up of polar polymer, such as polyamide with nanoclay and nonpolar polymer (e.g., polyolefin) as a compatible stabilizer.[80] The commercial properties of the constituents are contingent on the distribution of the nanoparticles and their morphology. The dispersal of the exfoliated systems shows a better barrier, thus, clays can be used as improved polymers, gas, and water vapor barrier properties.

Food packaging materials depend on clay nanocomposites that provide extended shelf-life, are light in weight, heat resistant, and unbreakable. In a nanocomposite with clays, conveyance in diffusing the molecules is thwarted due to opaque clays. Thus, the nonlinear path raises the distasteful gas dispersal length, which donates to the maintenance of shelf-life of speedy spoiled foods.[80]

4.4.4 NOVEL ENCAPSULATION TECHNIQUES

Encapsulation is a process of encasing a bioactive compound or a nutrient in nanoparticles of the structure to create a thermodynamic and physical barrier that protects it from adverse conditions (heat, humidity, oxidation, chemical reactions, and other extreme conditions) while allowing it to be released when needed.[6,41] In the incorporation of materials to make films, different strategies are employed,[47] which are given as follows:

- When active chemicals are exposed to unfavorable conditions, they must maintain their stability and viability.
- Enhancing the compatibility of the packaging polymer with the bioactive ingredient (increased miscibility of lipophilic compounds with hydrophilic biopolymeric materials).
- Improving the compound's availability.
- Specific stimuli are used to modulate the release of these molecules.

Different organic systems, such as emulsions were used in the development of active packaging solutions based on the natural antibacterial and antioxidant compounds, as well as encapsulation strategies to stabilize the compounds in the food packaging material. For example, curcumin is a lipophilic chemical compound and has been encapsulated in composite films and it showed antibacterial and antioxidant effects.[81]

HNTs have been used to encapsulate carvacrol, which is found in essential oils and is compounded into LDPE films. Later, it increased the volatile carvacrol molecules' thermostability, and it served as carriers thus allowing the volatile component to be integrated into the polymers through a high-temperature melt compounding technique.[70] When compared with LDPE films with carvacrol without carriers, this preserved its antibacterial efficacy following high-temperature treatment, and the encapsulation delayed the rate of diffusion of carvacrol from the films.[81] These film characteristics had substantial impact on the duration of storage, extending it by 6 weeks. Although carvacrol encapsulation had no effect on direct antibacterial activity, it did have a positive effect on the film's structure resulting in a higher crystalline order level. It also improved the film's mechanical characteristics and delayed the release of carvacrol.[1,6,76]

Bioactive compounds are employed in films in accordance with the desired properties of the packaging system and the food to which these films will be applied. As a result, the following are the most prevalent methods of integrating phytochemicals into film-forming solutions produced by agro-based polymers:

- Using inherently bioactive biopolymeric ingredients produced from agricultural products and by-products to make the film.[47]
- During the filmmaking process directly mixing bioactive chemicals with biopolymers.[41]
- Keeping the compounds bound inside of the wall material, developing nanoparticles, encapsulates them. Later, in the filmmaking stage, the nanoparticles are mixed with the biopolymer in a physical and direct manner.[44]
- Bioactive compounds are encapsulated and are kept trapped inside a wall material, generating micro or nanoparticles.
- During the filmmaking process, micro or nanoparticles are sprayed onto the biopolymer to incorporate them.[44]

Chemical interactions among bioactive compounds and biopolymers are affected by their chemical and structural properties, such as flexibility, concentration, and pH; and are influenced by the bioactive compound's characteristics and its concentration added to the film-forming solution.[71]

Different methods of thymol nanoemulsions, an antimicrobial essential oil, have been utilized to encapsulate substances, such as mixing gelatine with soy lecithin.[44]

Microencapsulation protects sensitive food components, prevent nutritional losses, and maintains flavors by packing food items in the form of micro- and nanoparticles. Flavor encapsulation uses a variety of techniques including spray drying, spray chilling, extrusion, and freeze-drying. The spray drying encapsulation technique is a common method for foods. During the processing, it depends on strong retention of the core materials, notabe volatiles, as well as a small amount of oil for both volatiles and nonvolatiles.[1]

Antimicrobial drugs are encapsulated in core-shell nanofibers, cyclodextrins, and liposomes for stable incorporation in the packaging materials. The encapsulating substance and the microencapsulation process employed for their creation, the size and form of the nanoparticles produced are diverse. The cost, quality of the wall material, material to be encapsulated, application, and mechanisms all affect the selection of microencapsulation technology.[6]

4.5 APPLICATIONS OF NANOMATERIALS AS PACKAGING MATERIALS IN VARIOUS FOODS ITEMS

The safety of food products is very important; so, food packaging plays an important role in the manufacturing of these products. The main principle of food packaging is to stop the contamination, to inhibit the enzymes activity, to reduces weight loss, to prevent spoilage, and to increase the shelf-life of various food products.[30] There are different types of nanomaterials used in food wrapping to reduce the loss of moisture, temperature, and nutritional value of food items, barrier properties, mechanical strength, and flexibility. Nanomaterials also help to improve antimicrobial and antioxidant properties to protect from ultraviolet light and to improve biodegradability of packaging materials.[42,60]

The most common method in edible film/coating is layer-by-layer deposition.[74] The applications of nanomaterials are found in meat, dairy, bakery, fruits, and vegetables industry to protect the sensory properties, oxidation, enzymatic browning of food products. Table 4.2 presents applications and effects of nanomaterials being used in edible films/coatings for different types of food products.

TABLE 4.2 Application and Effects of Nanomaterials in Edible Films/Coatings of Different Food Products.

Food item	NM used	Effects on food item	Reference
Apricot	Pectins and modified layered double hydroxide (LDH)-salicylate coatings with glycerol additives	Improved the barrier properties and shelf life of apricots	[26]
Beef	Chitosan, gelatin, and glycerol	Reduced weight loss, lipid oxidation during storage, and preserves color	[9]
Cheese	Whey protein concentrate, glycerol, and Guar gum	Shows antifungal properties and increased shelf life of cheese	[29]
Chicken breast fillets	Whey protein isolate with oregano and clove essential oils	Shelf life was increased and show antimicrobial effects	[21]
Chicken meat	Chitosan films, peanut skin, and pink pepper residue	No differences observed in color, pH, and total mesophilic counts, and maintains the quality of chicken meat	[69]
Cooked cured ham	Chitosan, essential oils.	Reduces the microbial load, prevents chemical changes, and extends the shelf life	[63]

TABLE 4.2 *(Continued)*

Food item	NM used	Effects on food item	Reference
Kabanosy dry sausage	Beeswax coating with glycerol	Reduced weight loss during storage	[75]
Lamb meat	Whey protein isolate, cellulose nanofiber, titanium dioxide, and rosemary essential oil	Maintains physicochemical properties, sensory characteristics, nutritional values, and extends the shelf life during storage	[67]
Mango fruit	Two types of materials: cellulose and carnauba wax coating	Increased shelf life, delayed ripening and softening, reduced decay, and weight loss of mango fruit	[4]
Mashed potato	Corn zein, hydroxypropyl methyl cellulose or methyl cellulose	Reduced moisture loss and fat uptake of mashed potato. Methyl cellulose shows good barrier properties compared to other materials	[49]
Otolithesruber fish fillets	Zinc oxide, essential oil.	Increased shelf life up to 16 days	[32]
Rainbow trout	Chitosan and cinnamon oil	Maintain good quality characteristics, inhibits lipids oxidation, and extend the shelf life during the refrigerated storage	[34]
Semi-hard regional cheese	Galactomannan and chitosan	Prevents the moisture loss and reduced microbial load	[10]
Smoked salmon	Calcium alginate Oyster and hen lysozyme, nisin	Reduced the microbial growth at refrigerated temperatures	[13]
Strawberry	Nano-chitosan and copper	Reduced weight loss and firmness; Delay the respiration rate; Increased anthocyanin concentration; No effect on consumer acceptability	[19]
Turkey deli meat	Silver nanoparticles, zinc oxide nanoparticles, oregano oil, and rosemary oil	Silver nanoparticles films are more effective than zinc oxide nanoparticles and rosemary oil. All prepared films show good antibacterial properties against microbes	[38]

4.6 MERITS AND DEMERITS OF USING NANOMATERIALS IN EDIBLE FILMS/PACKAGING SYSTEMS

The edible films and coatings are frequently used in meat and meat processed products, fruits, vegetables, seafoods, dairy products, bakery, confectionary products, and fried products for the preservation of food products. Each food product has different attributes, so, it must be protected, reduced, and enhanced to maintain its quality.[6] The use of nanomaterials in edible food packaging has increased.[3] The edible films/packaging provides protection and barrier with enhanced qualities and food safety. The edible films/packaging materials are eaten along with the packaged food products. There is no need to dispose of the package and films can help to reduce environmental pollution.

The films are made from renewable and edible materials and therefore, they are decomposed rapidly compared with polymeric materials. The film contains various types of materials to improve the organoleptic properties, such as flavor, color, and sweetness of packaged food products.[7] The use of antimicrobial nanomaterials in edible films/packaging materials helps to improve the shelf life and to inhibit the growth of microorganisms in food items.[21,38] The edible films show good moisture barrier properties, and thus can prevent microbial growth, textural changes, and enzymatic activity of food products.[26,36] The nanomaterials contain different types of bioactive compounds, and thus can release antimicrobial and antioxidant agents and vitamins to the specific food matrix.[74] The edible films and coatings may have an effect on allergic reactions, product safety, cost of manufacturing, workers, and consumers, and lack of information.[15]

4.7 TOXICITY AND HEALTH SAFETY ASSOCIATED WITH NANOPARTICLES IN EDIBLE FILMS/PACKAGING

The major issue with nanomaterials is whether these are safe to use or not. Basically, edible films/packaging materials are made up of natural ingredients; and therefore, there is no major risk to human beings. However, using nanomaterials in the edible films/packaging requires consideration of several factors, such as adverse health effects on humans and their impact on environment.[1,43] Nanomaterials can have a variety of toxic effects due to their nature of compounds or materials, particle shape and size, and surface condition. The oxidative stress, inflammatory response, and genotoxic effects are common side effects that have been documented in various experimental studies. The potential side effects of nanomaterials depend on how much material was used for preparation.[48] The existing research on the toxicity associated with nanomaterials has revealed that nano-based materials have toxicological effects not only on living cells but also on numerous environmental ecosystems.[18]

The nanomaterials enter into our body by ingestion, inhalation, and cutaneous exposure, once they come in contact with biomolecules, such as proteins, fats, carbohydrates, these are mixed with body fluids among cells, tissues, and blood.[46,20] The various research studies on TiO and Ag nanoparticles indicated that these materials can penetrate the bloodstream and can accumulate in organs due to their insolubility.[61] Meat products, which contain liver and spleen, are majorly conscientious for nanomaterials dispersion, facilitating their transit from the intestine to the circulatory system.[17]

The edible films/packaging are from different types of materials that must be generally recognized as safe (GRAS) for the intended use or approved by the FDA or U.S. Pharmacopoeia/National Formulary and also follow good manufacturing pPractices (GMP).[36] The different types of edible packaging materials are made from natural (eggs, fish, milk (casein and whey), wheat, peanuts, and soy proteins) and synthetic (silver nanoparticle, nanocellulose, titanium nanoparticle, nano-starch, and carbon nanotubes) materials; and these materials must provide all information to the consumers according to Food Allergen Labelling and Consumer Protection Act of 2004.[22,36]

4.8 SUMMARY

Food packaging can improve food quality, food safety, and traceability. The recent novel packaging methods are nanopackaging, improved, active, and smart packaging. Many nanomaterials, nanoparticles, and additives with good characteristics and properties can increase the physicomechanical strength, water, and oxygen barrier properties of packaging. Safety and environmental issues of nanotechnology implementation to substitute nondegradable polymer materials are to avoid the risks of nanomaterials in the food packaging sector.

KEYWORDS

- **edible films**
- **edible packaging**
- **food packaging**
- **food quality**
- **nanomaterials**

REFERENCES

1. Acharya, S.; Jakeer, S.; Shilpa, P.; Andhale, S. Review: Flavor Encapsulation by Spray Drying Technique. *Int. J. Chem. Stud.* **2021,** *9* (1), 1836–1840.
2. Aramouni, F. M.; Abu-Ghoush, M. H. Physicochemical and Sensory Characteristics of No-Bake Wheat-Soy Snack Bars. *J. Sci. Food Agric.* **2010,** *91* (1), 44–51.
3. Arora, A.; Padua, G. W. Nanocomposites in Food Packaging. *J. Food Sci.* **2010,** *75* (1), 43–49.
4. Baldwin, E. A.; Burns, J. K.; Kazokas, W.; Brecht, J. K.; Hagenmaier, R. D.; Bender, R. J.; Pesis, E. D. N. A. Effect of Two Edible Coatings with Different Permeability Characteristics on Mango (*Mangifera indica* L.) Ripening During Storage. *Postharvest Biol. Technol.* **1999,** *17* (3), 215–226.
5. Becerril, R.; Nerín, C.; Silva, F. Encapsulation Systems for Antimicrobial Food Packaging Components: An Update. *Molecules* **2020,** *25* (5), 1134–1140.
6. Bizymis, A. P.; Tzia, C. Edible Films and Coatings: Properties for the Selection of the Components, Evolution through Composites and Nanomaterials, and Safety Issues. *Crit. Rev. Food Sci. Nutr.* **2021,** *2021*, 1–16.
7. Bourtoom, T. Edible Films and Coatings: Characteristics and Properties. *Int. Food Res. J.* **2008,** *15* (3), 237–248.
8. Brokaut, N.; Hirye, P.; Acharya, S. Edible Films Made from Apple Peel. *Int. J. Food Sci. Nutr.* **2021,** *6* (2), 91–94.
9. Cardoso, G. P.; Dutra, M. P.; Fontes, P. R.; Ramos, A. D. L. S.; de Miranda Gomide, L. A.; Ramos, E. M. Selection of A Chitosan Gelatin-Based Edible Coating for Color Preservation of Beef in Retail Display. *Meat Sci.* **2016,** *114*, 85–94.
10. Cerqueira, M. A.; Sousa-Gallagher, M. J.; Macedo, I.; Rodriguez-Aguilera, R.; Souza, B. W.; Teixeira, J. A.; Vicente, A. A. Use of Galactomannan Edible Coating Application and Storage Temperature for Prolonging Shelf-life of Regional Cheese. *J. Food Eng.* **2010,** *97* (1), 87–94.
11. Chaudhary, P.; Fatima, F.; Kumar, A. Relevance of Nanomaterials in Food Packaging and Its Advanced Future Prospects. *J. Inorg. Organomet. Polym.* **2020,** *30*, 5180–5192.
12. Cuero, R. G.; Osufi, G.; Washington, A. *N*-Carboxymethyl Chitosan Inhibition of Aflatoxin Production: Role of Zinc. *Biotechnol. Lett.* **1991,** *13*, 441–444.

13. Datta, S.; Janes, M. E.; Xue, Q. G.; Losso, J.; La Peyre, J. F. Control of Listeria Monocytogenes and *Salmonella Anatum* on the Surface of Smoked Salmon Coated with Calcium Alginate Coating Containing Oyster Lysozyme and Nisin. *J. Food Sci.* **2008,** *73* (2), 67–71.
14. Decher, G.; Schlenoff, J. B. Eds.; In *Multilayer Thin Films: Sequential Assembly of Nanocomposite Materials*; John Wiley & Sons: New York, 2006; p 1112.
15. Dhall, R. K. Advances in Edible Coatings for Fresh Fruits and Vegetables: A Review. *Crit. Rev. Food Sci. Nutr.* **2013,** *53* (5), 435–450.
16. Díaz-Montes, E.; Castro-Muñoz, R. Edible Films and Coatings as Food-Quality Preservers: An Overview. *Foods* **2021,** *10* (2), 249–255.
17. Dimitrijevic, M.; Karabasil, N.; Boskovic, M.; Teodorovic, V.; Vasilev, D.; Djordjevic, V., Cobanovic, N. Safety Aspects of Nanotechnology Applications in Food Packaging. *Procedia Food Sci.* **2015,** *5*, 57–60.
18. Dos Santos, C. A.; Ingle, A. P.; Rai, M. The Emerging Role of Metallic Nanoparticles in Food. *Appl. Microbiol. Biotechnol.* **2020,** *104* (6), 2373–2383.
19. Eshghi, S.; Hashemi, M.; Mohammadi, A.; Badii, F.; Mohammadhoseini, Z.; Ahmadi, K. Effect of Nanochitosan-Based Coating with and Without Copper Loaded on Physicochemical and Bioactive Components of Fresh Strawberry Fruit (*Fragaria X Ananassaduchesne*) During Storage. *Food Bioproc. Technol.* **2014,** *7* (8), 2397–2409.
20. Farhoodi, M. Nanocomposite Materials for Food Packaging Applications: Characterization and Safety Evaluation. *Food Eng. Rev.* **2016,** *8* (1), 35–51.
21. Fernández-Pan, I.; Carrión-Granda, X.; Maté, J. I. Antimicrobial Efficiency of Edible Coatings on the Preservation of Chicken Breast Fillets. *Food Control* **2014,** *36* (1), 69–75.
22. Franssen, L. R.; Krochta, J. M. Edible Coatings Containing Natural Antimicrobials for Processed Foods. In *Natural Antimicrobials for the Minimal Processing of Foods*, 2003; pp 250–262.
23. Galus, S.; ArikKibar, E.A.; Gniewosz, M.; Kra´sniewska, K. Novel Materials in the Preparation of Edible Films and Coatings—A Review. *Coatings* **2020,** *10*, 674–680.
24. Gennadios, A.; McHugh, T. H.; Weller, C. L.; Krochta, J. M. Edible Coating and Films Based on Protein. In *Edible Coatings and Films to Improve Food Quality*, 1994; vol 1994, pp 201–217.
25. Glicksman, M. In *Gum Technology in the Food Industry*; Academic Press: New York, NY, 1969; p 590.
26. Gorrasi, G.; Bugatti, V. Edible Bio-Nano-Hybrid Coatings for Food Protection Based on Pectins and LDH-Salicylate: Preparation and Analysis of Physical Properties. *LWT Food Sci. Technol.* **2016,** *69*, 139–145.
27. Greener, I. K.; Fennema, O. Evaluation of Edible, Bilayer Films for Use as Moisture Barrier for Food. *J. Food Sci.* **1989,** *54*, 1400–1406.
28. Guilbert, S. Technology and Application of Edible Protective Films. *Packag. Technol. Sci.* **1995,** *8* (6), 339–346.
29. Guimarães, A.; Ramos, Ó.; Cerqueira, M.; Venâncio, A.; Abrunhosa, L. Active Whey Protein Edible Films and Coatings Incorporating *Lactobacillus Buchneri* for *Penicillium nordicum* Control in Cheese. *Food Bioproc. Technol.* **2020,** *13*, 1074–1086.
30. He, X.; Deng, H.; Hwang, H. M. The Current Application of Nanotechnology in Food and Agriculture. *J. Food Drug Anal.* **2019,** *27* (1), 1–21.

31. Hernandez, E. Edible Coating from Lipids and Resins. In: *Edible Coatings and Films to Improve Food Quality*; Krochta, J. M.; Balwin, E. A.; Niperos- Carriedo, M. O., Eds.; FAO (Technomic Publ. Co.): Rome, 1994; pp 279–303.
32.
33. Heydari-Majd, M.; Ghanbarzadeh, B.; Shahidi-Noghabi, M.; Najafi, M. A.; Hosseini, M. A New Active Nanocomposite Film Based on PLA/ZnO_2 Nanoparticle/Essential Oils for The Preservation of Refrigerated Otolithes-ruber Fillets. *Food Packag. Shelf Life* **2019,** *19*, 94–103.
34. Hood, L. L. Collagen in Sausage Casting. In *Advances in Meat Research: Collagen as a Food*; Pearson, A. M.; Dutson, T. R.; Bailey, A. J., Eds.; AVI Pub. Co.: New York, 1987; pp 109–129.
35. Huang, Y.; Mei, L.; Chen, X.; Wang, Q. Recent Developments in Food Packaging Based on Nanomaterials. *Nanomaterials* **2018,** *8* (10), 830–837.
36. Jang, S. H.; Jang, S. R.; Lee, G. M.; Ryu, J. H.; Park, S. I.; Park, N. H. *Halloy sitenano* Capsules Containing Thyme Essential Oil: Preparation, Characterization, and Application in Packaging Materials. *J. Food Sci.* **2017,** *82*, 2113–2120.
37. Janjarasskul, T.; Krochta, J. M. Edible Packaging Materials. *Ann. Rev. Food Sci. Technol.* **2010,** *1*, 415–448.
38. Kester, J. J.; Fennema, O. R. Edible Films and Coatings: A Review. *Food Technol.* **1986,** *40* (12), 47–59.
39. Khalaf, H. H.; Sharoba, A. M.; El-Tanahi, H. H.; Morsy, M. K. Stability of Antimicrobial Activity of Pullulan Edible Films Incorporated with Nanoparticles and Essential Oils and Their Impact on Turkey Deli Meat Quality. *J. Food Dairy Sci.* **2013,** *4* (11), 557–573.
40. Kinsella, J. E.; Damodaran, S.; German, B. Physiochemical and Functional Properties of Isolated Proteins with Emphasis on Soy Proteins. In *New Protein Foods: A Study of a Treatise*; Altschul, A. M.; Wilcke, H. L., Eds.; Academic Press: New York, 1981; pp 107–179.
41. Kour, H. Advances in Food Packaging – A Review. *Stewart Postharvest Rev.* **2013,** *9* (4), 1–7.
42. Krochta, J. M. Edible Protein Films and Coatings. In *Food Proteins and Their Applications*; Damodaran, S.; Paraf, A., Eds., CRC Press: Boca Raton, 1997; pp 529–549.
43. Kuswandi, B. Environmentally Friendly Food Nano-Packaging. *Environ. Chem. Lett.* **2017,** *15* (2), 205–221.
44. Li, H.; Shi, H.; He, Y.; Fei, X.; Peng, L. Preparation and Characterization of Carboxymethyl Cellulose-Based Composite Films Reinforced by Cellulose Nanocrystals Derived from Pea Hull Waste for Food Packaging Applications. *Int. J. Biol. Macromol.* **2020,** *164*, 4104–4112.
45. Li, X.; Yang, X.; Deng, H.; Guo, Y.; Xue, J. Gelatin Films Incorporated with Thymol Nano-emulsions: Physical Properties and Antimicrobial Activities. *Int. J. Biol. Macromol.* **2020,** *150*, 161–168.
46. Lieberman, E. R.; Guilbert, S. G. Gas Permeation of Collagen Films as Affected by Cross-Linkage, Moisture and Plasticizer Content. *J. Polym. Sci. Symp.* **1973,** *41*, 33–43.
47. Maisanaba, S.; Pichardo, S.; Puerto, M.; Gutierrez-Praena, D.; Camean, A. M.; Jos, A. Toxicological Evaluation of Clay Minerals and Derived Nanocomposites: A Review. *Environ. Res.* **2015,** *138*, 233–254.

48. Majeed, K.; Jawaid, M.; Hassan, A.; Abu Bakar, A.; Abdul Khalil, H. P. S.; Salema, A. A.; Inuwa, I. Potential Materials for Food Packaging from Nanoclay/Natural Fibers Filled Hybrid Composites. *Mater. Design* **2013,** *46*, 391–410.
49. Malakar, A.; Kanel, S. R.; Ray, C.; Snow, D. D.; Nadagouda, M. N. Nanomaterials in the Environment, Human Exposure Pathway, and Health Effects: A Review. *Sci. Total Environ.* **2020,** *2020*, 143–160.
50. Mallikarjunan, P.; Chinnan, M. S.; Balasubramaniam, V. M.; Phillips, R. D. Edible Coatings for Deep-Fat Frying of Starchy Products. *LWT Food Sci. Technol.* **1997,** *30* (7), 709–714.
51. Mark, A. M.; Roth, W. B.; Mehltretter, C. L.; Rist, C. E. Oxygen Permeability of Amylomaize Starch Films. *Food Technol.* **1966,** *20*, 75–80.
52. Mauch, F.; Hadwiger, L. A.; Bolier, T. Ethylene: Symptom, Not Signal for the Induction of Chitinase and Beta 1,3,glucanase in Pea Pods by Pathogens and Elicitors. *Plant Physiol.* **1984,** *76*, 607–611.
53. Neethirajan, S.; Ragavan, V.; Weng, X.; Chand, R. Biosensors for Sustainable Food Engineering: Challenges and Perspectives. *Biosensors* **2018,** *8*, 23–30.
54. Nogueira, G.; Oliveira, R.; Velasco, J.; Fakhouri, F. Methods of Incorporating Plant-Derived Bioactive Compounds into Films Made with Agro-Based Polymers for Application as Food Packaging: A Brief Review. *Polymers* **2020,** *12* (11), 25–18.
55. Ojagh, S. M.; Rezaei, M.; Razavi, S. H.; Hosseini, S. M. H. Effect of Chitosan Coatings Enriched with Cinnamon Oil on the Quality of Refrigerated Rainbow Trout. *Food Chem.* **2010,** *120* (1), 193–198.
56. Oliveira, T. I.; Zea-Redondo, L.; Moates, G. K.; Wellner, N.; Cross, K.; Waldron, K. W.; Azeredo, H. M. Pomegranate Peel Pectin Films as Affected by Montmorillonite. *Food Chem.* **2016,** *198*, 107–112.
57. Orsuwan, A.; Sothornvit, R. Development and Characterization of Banana Flour Film Incorporated with Montmorillonite and Banana Starch Nanoparticles. *Carbohydr. Polym.* **2017**, *174*, 235–242.
58. Osman, E.; Aykut, O. General Characteristics of Edible Films. *J. Food Biotechnol. Res.* **2018,** *2* (1), 25–30.
59. Paine, F. A.; Paine, H. Y. In *Handbook of Food Packaging*; Blackie Academic & Professional: London, 1992**;** vol 1992, pp 212–214.
60. Pal, M. Recent Developments in Food Packaging Technologies. *Beverage Food World* **2017,** *46* (1), 21–25.
61. Ranjan, S.; Dasgupta, N.; Chakraborty, A. R.; Samuel, S. M.; Ramalingam, C.; Shanker, R.; Kumar, A. Nanoscience and Nanotechnologies in Food Industries: Opportunities and Research Trends. *J. Nanop. Res.* **2014,** *16* (6), 1–23.
62. Rhim, J. W.; Park, H. M.; Ha, C. S. Bio-nanocomposites for Food Packaging Applications. *Prog. Polym. Sci.* **2013,** *38* (10-11), 1629–1652.
63. Romano, N.; Tavera-Quiroz, M.vJ.; Bertola, N.; Mobili, P.; Pinotti, A.; Gómez-Zavaglia, A. Edible Methylcellulose-Based Films Containing Fructo-Oligosaccharides as Vehicles for Lactic Acid Bacteria. *J. Food Res. Int.* **2014,** *64*, 560–566.
64. Ruiz-Navajas, Y.; Viuda-Martos, M.; Barber, X.; Sendra, E.; Perez-Alvarez, J. A.; Fernández-López, J. Effect of Chitosan Edible Films Added with *Thymus Moroderi* and *Thymus Piperella* Essential Oil on Shelf-life of Cooked Cured Ham. *J. Food Sci. Technol.* **2015,** *52* (10), 6493–6501.

65. Salame, M. Barrier Polymers. In *The Wiley Encyclopaedia of Packaging Technology*; Bakker, M., Ed.; John Wiley & Sons: New York, 1986; pp 48–54.
66. Salgado, P. R.; Ortiz, C. M.; Musso, Y. S.; Di Giorgio, L.; Mauri, A. N. Edible Films and Coatings Containing Bioactive. *Food Sci.* **2015**, *5*, 86–92.
67. Salvia-Trujillo, L.; Soliva-Fortuny, R.; Rojas-Graü, M. A.; McClements, D. J.; Martín-Belloso, O. Edible Nano-emulsions as Carriers of Active Ingredients: A Review. *Ann. Rev. Food Sci. Technol.* **2017,** *8*, 439–466.
68. Sani, M. A.; Ehsani, A.; Hashemi, M. Whey Protein Isolate/Cellulose Nanofiber/ TiO_2 Nanoparticle/Rosemary Essential Oil Nanocomposite Film: Effect on Microbial and Sensory Quality of Lamb Meat and Growth of Common Food Borne Pathogenic Bacteria during Refrigeration. *Int. J. Food Microbiol.* **2017,** *251*, 8—14.
69. Sarfraz, J.; Gulin-Sarfraz, T.; Nilsen-Nygaard, J.; Pettersen, M. Nanocomposites for Food Packaging Applications: An Overview. *Nanomaterials* **2020,** *11* (1), 10–15.
70. Serrano-León, J. S.; Bergamaschi, K. B.; Yoshida, C. M.; Saldaña, E.; Selani, M. M.; Rios-Mera, J. D., Contreras-Castillo, C. J. Chitosan Active Films Containing Agro-Industrial Residue Extracts for Shelf-life Extension of Chicken Restructured Product. *Food Res. Int.* **2018,** *108*, 93–100.
71. Shemesh, R.; Krepker, M.; Natan, M.; Danin-Poleg, Y.; Banin, E.; Kashi, Y.; Nitzan, N.; Vaxman, A.; Segal, E. Novel LDPE/Halloysite Nanotube Films with Sustained Carvacrol Release for Broad-Spectrum Antimicrobial Activity. *RSC Adv.* **2015,** *5*, 108–117.
72. Silva-Weiss, A.; Ihl, M.; Sobral, P. J. A.; Gómez-Guillén, M. C.; Bifani, V. Natural Additives in Bioactive Edible Films and Coatings: Functionality and Applications in Foods. *Food Eng. Rev.* **2013,** *5*, 200–216.
73. Subirade, M.; Kelly, I.; Guéguen, J.; Pézolet, M. Molecular Basis of Film Formation from a Soybean Protein: Comparison Between the Conformation of Glycinin in Aqueous Solution and in Films. *Int. J. Biol. Macromol.* **1998,** *23* (4), 241–249.
74. Subroto, E.; Indiarto, R.; Pangawikan, A. D.; Prakoso, F. Production and Characteristics of Composite Edible Films Based on Polysaccharides and Proteins. *Int. J. Proteins* **2021,** *9* (2), 60—67.
75. Suhag, R.; Kumar, N.; Petkoska, A. T.; Upadhyay, A. Film Formation and Deposition Methods of Edible Coating on Food Products: A Review. *Int. Food Res. J.* **2020,** *136*, 109–122.
76. Tyburcy, A.; Kozyra, D. Effects of Composite Surface Coating and Pre-Drying on the Properties of Kabanosy Dry Sausage. *Meat Sci.* **2010,** *86* (2), 405–410.
77. Whistler, R. L.; Daniel, J. R. Carbohydrate. In *Food Chemistry*; Fennema, O. R., Ed.; Taylor & Francis Group: London, 1985; pp 69–76.
78. Wittaya Prince, T. Edible Films and Coatings: Characteristics and Properties. *Int. Food Res. J.* **2008,** *15* (3), 69–73.
79. Wu, Y.; Luo, Y.; Wang, Q. Antioxidant and Antimicrobial Properties of Essential Oils Encapsulated in Zein Nanoparticles Prepared by Liquid–Liquid Dispersion Method. *Food Sci. Technol.* **2012,** *48*, 283–290.
80. Xu, X.Y.; Kim, K.M.; Hanna, M.A.; Nag, D. Chitosan-Starch Composite Film: Preparation and Characterization. *Ind. Crops Prod.* **2005,** *21*, 185–192.
81. Yano, K.; Usuki, A.; Okada, A. Synthesis and Properties of Polyimide-Clay Hybrid Films. *Polym. Sci. Part A Polym. Chem.* **1997,** *35*, 122–130.

82. Yousuf, B.; Qadri, O. S.; Srivastava, A. K. Recent Developments in Shelf-Life Extension of Fresh-Cut Fruits and Vegetables by Application of Different Edible Coatings: A Review. *J. Food Sci. Technol.* **2018,** *89*, 198–209.
83. Zahedi, Y.; Fathi-Achachlouei, B.; Yousefi, A. R. Physical and Mechanical Properties of Hybrid Montmorillonite/Zinc Oxide Reinforced Carboxymethyl Cellulose Nanocomposites. *Int. J. Biol. Macromol.* **2018,** *108*, 863–873.
84. Zambrano-Zaragoza, M. L.; González-Reza, R.; Mendoza-Muñoz, N.; Miranda-Linares, V.; Bernal-Couoh, T. F.; Mendoza-Elvira, S.; Quintanar-Guerrero, D. Nanosystems in Edible Coatings: A Novel Strategy for Food Preservation. *Int. J. Mol. Sci.* **2018,** *19*, 705–711.

CHAPTER 5

POTENTIAL ROLE OF DYES AND PIGMENTS IN INTELLIGENT PACKAGING

VAISHALI SAROHA, HINA KHAN, SHARAD RAGHUVANSHI, and DHARM DUTT

ABSTRACT

Intelligent packaging enables to monitor the quality of food from manufacturing to consumption. For developing an intelligent package, the study of food deterioration factors and reaction kinetics is necessary. When food has deteriorated, then the pH of food may change due to the release of gases such as CO_2 and ammonia. Generally, Arrhenius activation energy of food spoilage is taken into consideration when time and temperature-based indicators are developed. The dyes, such as bromothymol blue, methyl red, anthocyanin, and chlorophenol red have been used as indicators. These dyes are used in various types of prototypes for colorimetric indication of pH change, which may be due to enzymatic reaction, microbial growth, and diffusion of acid. Apart from dyes, other compounds, including inks and polymers that change color due to changes in the working conditions have also been developed and applied to various food products. Hence, this chapter focuses on various such prototypes that have been developed for use for intelligent packages.

5.1 INTRODUCTION

Active and intelligent packaging is one of the innovative food packaging technologies, which utilizes the interaction between the package and food

Advances in Sustainable Food Packaging Technology. Megh R. Goyal, Preeti Birwal, Anand K. Pandey, & Ritesh B. Watharkar, (Eds.)

or environment.[8] Active packaging improves the shelf life and improves or maintains the quality of a food item, whereas intelligent packaging monitors or senses the quality of food during manufacturing, transportation, and storage. Labels on the package provide intelligent functions by monitoring and giving information on the quality of food, and can be printed, incorporated into, or attached to the food package to function as an intelligent packaging system.[14] Different actors in the supply chain, including the consumers, should be able to transcribe the message from the output of sensors on the quality of food into an understandable message. The functional integrity of an active packaging system can be monitored by intelligent packaging devices that include sensors and indicators for time and temperature, microbial growth, and physical shock.

Meat and sea foods are highly perishable foods irrespective of the chilled storage system. Spoilage of meat and seafood is largely due to yeasts and bacteria (*Pseudomonas* spp. and *Shewanella putrefaciens*).[47] Biochemical, physicochemical, and microbiological transformations in refrigerated food reduce its freshness with time. Protein can be transformed into smaller compounds like amino acids by the action of proteolytic activity by the microorganisms or by hydrolysis. Oxidative deamination, decarboxylation, and desulfurization of amino acids can result in the release of gases, such as NH_3, CO_2, and H_2S.[5] Microbial growth also produces CO_2. Inflated levels of TVB-N (e.g., NH_3, DMA, and TMA) can also be used as indicators of the spoilage of food protein by microorganisms.[23]

Intelligent packaging devices are categorized based on the factor taken into consideration for responding to food deterioration. Time and temperature indicators (TTIs), freshness indicators, and oxygen indicators are different indicators used in intelligent packaging systems. These indicators are generally attached to the surface of a package and give a response based on the relevant physical, chemical, or microbial reaction.

For developing a sensor or an indicator, it is necessary to study the deterioration process of food. The response of the sensor or indicator should match the kinetics of food deterioration. Various types of colorimetric dyes of natural and artificial origins are known and can be used for the colorimetric determination of the pH. According to the literature, the dyes, such as bromothymol blue, methyl red, anthocyanin (extracted from *Brassica olearacea*), and chlorophenol red are commonly used as indicators.[40,65] Dyes and pigments show different colors with changes in the pH of the food matrix in which the dye is embedded. Dyes and pigments are used in intelligent packaging due to their ability to provide visual responses against changes in the quality of food.

Synthetic pH-sensitive dyes have been used for developing indicators with different ranges of colors, such as red (cresol, chlorophenol, phenol, methyl), blue (bromothymol, thymol, xylenol), purple (bromocresol, m-cresol), and yellow/orange (methyl, dimethyl).[37]

Anthocyanin extracted from various fruits, vegetables, and flowers has been investigated widely for developing indicators due to its ability to produce different colors in different ranges of pH.[62] Anthocyanin is a polyphenol-based flavonoid composed of glycosylated polyhydroxy and polymethoxy structures having two aromatic rings connected by a linear three-carbon chain. The color change of anthocyanin[71] is due to the structural modification as a function of pH in the form of (1) flavylium cation (red) at pH 1–3, (2) carbinol pseudobase (green) at pH 4–5, (3) quinonoidal base (purple) at pH 6–7, (4) anionic quinonoidal base (blue) at pH 7–8, and (5) chalcone (yellow) at pH 8–9.

This chapter explores the potential role of dyes and pigments used for developing various indicators for their use in the intelligent packaging of foods and food products.

5.2 TIME–TEMPERATURE INDICATOR (TTI)

Temperature is a crucial environmental factor, which determines the safety of food. Due to variations in temperature, the shelf life of a food product gets affected and food safety is more likely to get compromised. The TTI is a powerful tool for observing the history of time and temperature of food in the cold chain.[44] TTI is economical, simple, and convenient to use to record and monitor the temperature. It indicates the ramification of temperature on the quality of food from the manufacturing stage to the consumption stage.[13,54]

In the Arrhenius model, the chemical reaction rate is exponentially proportional to the inverse of absolute temperature as follows:

$$k = k_0 e^{-Ea/RT} \tag{5.1}$$

where k is the rate constant, k_0 is the pre-exponential factor, E_a is the activation energy, R is the gas constant, and T is the absolute temperature.

Activation energy (E_a) is the measure of the temperature sensitivity of the reaction. High activation energy value indicates that the reaction will go faster if the temperature is increased. However, the value of activation energy is very specific for each system and can vary with the water activity.

The activation energy of simple hydrolysis, lipid oxidation by free radical mechanism, nonenzymatic browning, and enzyme and microbial destruction is 10–20, 15–25, 20–40, and 50–150 kcal/mol, respectively.[24]

The requirement for developing a time–temperature integrator is that its response should coincide with the quality loss of food. Therefore, it is required that the TTI's temperature sensitivity activation energy should match the activation energy of food deterioration.[53,55]

Labuza et al.[25] indicated that the difference between indicator and food activation energy should be below 40 kJ/mol, which can increase the evaluation of food quality by 15%. The activation energy of the loss of food product ranges from 0 to 627.6 kJ/mol due to the diffusion-controlled process, enzymic reaction, hydrolysis, lipid oxidation, nutrient loss, nonenzymic browning, microbial growth, spore, and vegetative spore cell destruction.[25]

Based on the working principle, TTIs can be classified as physical, chemical, enzymatic, and biological systems. The pH is the vital sign of food spoilage along with the external and internal properties of a food package. Depending on the type of packaged food (meat, fish, milk, poultry, etc.) and storage conditions, various microbial activity, endogenous enzymes, and decomposition of protein change the pH of the food. TTI based on dyes and pigments generally consists of pH-responsive dye and substrate in which dye is immobilized. Due to the resonance effect by either acidic or alkaline compounds, the color of pH-sensitive dye changes. Delocalization of the negative charges or hydroxyl groups occurs due to the acidic compounds, whereas deprotonation occurs due to the alkaline compounds.[42]

Microbial deterioration of food is a very common phenomenon. Various microbes prevailing in the food or surroundings are responsible for food deterioration. These microbes will change the texture, color, and smell in the food and make it unsuitable for consumption. Microbial TTI response is directly related to the deterioration of food by microbes. *Lactobaccilus* spp., *Leuconostoc* spp., *Pseudomonas* spp., *B. thermospecta, Carnobacterium* spp., and pathogenic members of *Enteriobacteriace* are the predominant bacteria associated with the decaying of meat and pork.[9]

Lactic acid bacteria (LAB) are accountable for the spoilage of meat, milk, and many other food products. A microbial-based TTI was developed based on the metabolic activity and proliferation of *Lactobaccilus sakei.* Seven different types of TTIs were fabricated with growth media containing dyes (methyl red, chlorophenol red, bromocresol purple, a mixture of bromocresol green and chlorophenol red, resazurin, alizarin, and p-nitrophenol) and *Lactobaccilus sakei.* This TTI has an activation energy of 103.8 kJ/mol.[58] These microbial-based indicators are widely used in the modified atmosphere packaging (MAP) for minced meat and monitoring the microbiological quality of refrigerated food.[59]

At suitable temperature, the population of microbes grows exponentially with time in the food as shown in eq. 5.1. The growth of bacteria in food develops a sour flavor and the pH of the food is increased. For developing a microbial TTI, the pH indicator plays an important role. Different types of indicators show different orders of reactions for producing different colors.[19] Andretta et al. developed cassava starch films by thermocompression containing the residue of blueberries (anthocyanin). This starch film showed good mechanical properties and changes in color with changes in pH.[3]

The TTI prototype has been developed, its working principle is based on lactic acid vapor diffusion. The prototype was fabricated with agar and dye mixture (bromophenol blue and methyl red). The lactic acid-coated paper was used as an activator for TTI. Due to the effect of temperature, the vapor of lactic acid diffused through the agar matrix with time, and pH was changed resulting in a change in the color of a dye and the activation energy (E_a) varied from 50 to 54.4 kJ/mol.[65]

Chitosan/PVA film containing anthocyanin (extracted from *Brassica olearacea*) has been developed. Applicability of film to act as TTI was supported by a noticeable change in the coloration of the film when tested on pasteurized milk.[40] Recently, an intelligent and active film with starch/polyvinyl alcohol containing anthocyanin as a color indicator and limonene as an antibacterial agent has been developed. With time, the efficiency of antibacterial agents was reduced and food spoilage occurred due to the growth of bacteria leading to a change in the pH of milk. A color change in the film indicates a change in pH.[30]

Yan and colleagues developed a TTI using iodine in which the enzymatic reaction between amylase and starch had a wide range of activation energy that varied from 102.7 to 114.6 kJ/mol for food products destroyed by the microbial growth, enzyme browning, and nutrition loss.[69] Simultaneously, *Burkholderia cepacia* lipase and tricaprylin enzymatic reactions were involved to develop an enzymatic TTI system with an activation energy of 70.6 ± 11.1 kJ/mol.[20]

An enzyme diffusion-based TTI for monitoring the perishable food in cold chain logistics was developed. Due to change in temperature, a reaction between lipase and glycerol tributyrate led to a change in the pH of the TTI prototype, and the color of bromine thymol in TTI was changed to pink.[31] Furthermore, urease and carbamide enzymatic reaction-based TTI using phenol red as the indicator has been investigated and the activation energy was 23.1 ± 1.15 kJ/mol by changing the color from yellow to red.[68] Moreover, *Aspergillus niger* lipase as an enzyme component and glycerol tributyrate as a substrate along with thymol blue as a pH indicator have been used to develop a TTI. The four different types of TTIs have been developed by using different concentrations of enzyme and substrate with higher activation energy ranging from 36.1 to 42.1 kJ/mol. These indicators have been effectively used for monitoring some fruits, vegetables, and fish, where deterioration occurs due to lipid oxidation and enzymatic reactions.[67]

An ink composed of a different composition (on a weight basis) of sodium anthraquinone b-sulfonate, sodium hydrosulfite, sodium hydroxide, and hydroxypropyl cellulose will fade as a result of time and temperature indicating the freshness of food based on the redox property of the anthraquinone that has been formulated with a water-based polymeric binder. Oxygen diffuses through the acrylate layer and will oxidize the anthraquinone. This ink can be easily printed on paper and glass surfaces.[11] Recently, a time–temperature indicator ink (which can be printed by rotary printer on the cellulose substrate and protected by polypropylene self-adhesive films) has been developed. The formulation of water-based ink contains chlorophenol red as an indicator, a growth medium, and a food industry maturation mixture of bacteria.[72]

A polymer blend of TOPAS-5013 (a commercial copolymer of ethylene and norborene) and photoluminescent chromophore-BBS (4,4′-bis(2-benzoxazolyl)stilbene) has been developed, which changes the color with temperature. This polymer-based film indicator, with activation energy from 179 to 185 kJ/mol, can be used as an indicator in various food processing industries, such as Retort food processing and monitoring of bakeries and confectioneries.[27]

Polydiacetylene (PDA) is known as colorimetric polymeric material, which changes the color due to the effect of temperature, chemicals, and biological agents. Silica nanoparticles act as a template during the formation of PCDA (10,12-pentacosadiynoic acid) vesicles in an aqueous solution and increase the ordered arrangement of PCDA. Ultraviolet (UV) light has been used for the formation of silica/polydiactylene nanocomposite, which can be employed as a chromatic nanosensor.[51] Zinc oxide (ZnO) nanoparticles can also be used as nanosubstrate for the self-assembly of PCDA on its surface. The thermochromic behavior of PCDA/ZnO nanoparticles gets altered due to interfacial interaction and molecular ordering.[57]

Due to advancements in the field of nanotechnology, a colorimetric indicator of PDA/silica (SiO_2) nanocomposite has been prepared, which changes the color from blue to red with a temperature change.[38] Furthermore, blue suspension of PDA/SiO_2 nanocomposite in polyvinyl alcohol (PVOH)-based film has been reported. The higher the concentration of nanoparticles in plasticized films, the slower the process of color transition as compared with the low concentration of nanoparticles.[39]

Gold (Au) nanoparticles have been reported to perform the TTI function using gelatin as the reducing agent. This system can be directly incorporated into the food package system by housing the Au precursor gelatin nanoreactor in a biocompatible polymeric system. It irreversibly changes color when kept above 30°C.[29,63] Recently, a chitosan and gold nanoparticle-based nanocomposite has been developed, which can be integrated into various packaging materials for indicating the frozen and thermal history of foods.[64]

Biodegradable colorimetric indicator has been developed using anthocyanin incorporated in a matrix coated on card paper changes color from light violet to light yellow, as the temperature is varied from 40 to 70°C.[33] Another biodegradable pH indicator has been developed by entrapping the anthocyanin in polyelectrolyte complex formed by the interaction of chitosan and pectin.[32] Bacterial cellulose nanofiber containing anthocyanin extracted from red cabbage (*Brassica oleracea*) extract has also been reported to be used as pH indicator.[41] The TTIs developed by various researchers are given in Table 5.1.

5.3 FRESHNESS INDICATORS

Various metabolites produced during food storage are used to detect the freshness of food. Carbon monoxide, H_2S, sulfur dioxide, nitric oxide, and NH_3 are swiftly detected compounds and thus can be easily used for the quality control of fish. TVB-N contents of the fish and its products are used by various researchers to measure its freshness. Furthermore, the concentration of ammonium ions produced due to fish spoilage is correlated with the contents of TVB-N produced due to spoilage. Ammonium ions-selective electrode is used for the measurement of ammonium ion concentration.[16] Simultaneously, a colorimetric polyaniline film for monitoring the TVB-N has been developed by correlating with microbial spoilage growth. This film can be easily recycled by washing it with an acid solution.[22]

The increase in trimethyl amine concentration in the package headspace can be monitored with the colorimetric indicator film containing bromocresol green in the polymeric matrix. The response of the indicator is found to be correlated with *Pseudomonas fragi* and pH value.[7] An electrode for

TABLE 5.1 Various TTI Prototypes: Substrate, Mechanism, Activation Energy, Response, and Applications in Food Industry.

Substrate	Mechanism	Activation energy, kJ/mol	Applications	Indicator response	Reference
Amylase, starch, iodine	Reaction between amylase and starch	102.74–114.59	Enzymatic browning, microbe growth, and nutrition loss	Colorless-blue	[69]
Burkholderia cepacia lipase, tricaprylin, bromothymol blue, phenyl red, and neutral red	Reaction between Burkholderia cepacia lipase and tricaprylin	70.61 ± 11.10		Green-orange-red	[19]
Chlorophenol blue. Maturation mixture of bacteria	Change in pH	91.92	Cold chain management	Violet-purple-orange-amber-yellow	[72]
Gelatin, gold nanoparticles		___	Indicator for temperature abuse	Colorless-cherry red-blue	[29]
Lactic acid, bromothymol blue, methyl red	Lactic acid diffusion	50	Fruits and vegetables	Green-red	[65]
Lacto bacillus sakei, chemical indicator	Change in pH due to microbial growth	111.90	Meat	Red-yellow	[58]
Lipase, bromine thymol blue, glycerol triacetate	Enzymatic and diffusion-based	71.8–83.4	Cold chain logistics of food	Formation of color band	[31]
Phenol red, urease, carbamide	Chemical reaction between urease and carbamide	23.05 ± 1.15	Quality loss of diffusion-controlled	Yellow-red	[68]
Polydiacetylene/silica nanocomposite	Thermochromic transition	___	____	Blue-red	[38]

TABLE 5.1 *(Continued)*

Substrate	Mechanism	Activation energy, kJ/mol	Applications	Indicator response	Reference
PVA/chitosan film doped with anthocyanin	Change in pH	____	Milk	Red-light pink-light green-dark green	[40]
Reduced anthraquinone	Oxygen diffusion	_	Freshness of food	Red-beige	[11]
Thymol blue, glycerol tributyrate, Aspergillus niger lipase	Reaction between Aspergillus niger lipase and glycerol tributyrate	31.60–42.12	Some fruits, vegetables, and fish (spoilage due to lipid oxidation and enzymatic lipase)	Blue-yellow	[67]
TOPAS 5013 (copolymer of ethylene and norborene), 4,4′-bis(2-benzoxazolyl) stilbene (BBS)	Solid-state polymerization	179–185	Indicator of sterilization point of food and retort food, process monitoring of bakeries and confectioneries	______	[27]

monitoring the concentration of volatile amines in the aqueous phase and pH in the headspace of the package has been developed. pH electrode measures the pH, which changes with variation in temperature, whereas the conductivity electrode signal increases with an increase in volatile amines.[15]

Shukla et al. extracted anthocyanin from rose flowers and red cabbage and immobilized it in filter paper. Anthocyanin-immobilized paper was successfully tested in an ammonia atmosphere mimicking the conditions of a seafood packaging system.[49] Zhai developed a starch/polyvinyl alcohol-based indicator containing different concentrations of roselle-extracted anthocyanin. Lower concentrations containing anthocyanin showed more sensitivity to ammonia gas.[73]

Dudnyk et al. developed an indicator for the detection of seafood freshness based on pectin and red cabbage extract showing high sensitivity to gaseous amines.[10] Zhang et al. developed an indicator based on different combinations of chitosan, starch, and polyvinyl alcohol containing roselle anthocyanin for monitoring pork freshness. Starch/polyvinyl alcohol/roselle anthocyanin indicator film showed a change in color from red to green due to an increase in TVB-N concentration.[74]

K-value is often used as the fish freshness indicator. The ratio of the total amount of ATP degradation compounds to the total amount of ATP is the K-value. A sensor containing thiamazole blue, which changes color due to the redox reaction of MTT accompanying oxidation of hypoxanthine by XOD, has been developed.[66] The correlation between gold and silver electrode potential and various degradation parameters of fish spoilage such as a change in pH, microbial count, and nucleoside concentration has been studied. Liner correlation between the response of the electrode and the k_1 index of the metabolite of fish (such as Hx) has been used for monitoring the freshness of fish.[4]

A microdevice allows on-site measurement of the K, K_i, and H values. Four sensors measure the concentration of depository compounds (ATP, ADP, AMP, IMP, In, and Hx) to indicate the K, K_i, and H values.[43] A hypoxanthine sensor based on a nanocomposite of graphene–titanium dioxide (TiO_2-G) is used for assessing the freshness of meat. The excellent electrocatalytic activity was shown by fabricated nafion/XOD/TiO_2-G/glassine carbon electrode (GCE) sensor toward hypoxanthine within a range of 20–512 μM.[1]

The partial pressure of CO_2 in packaged food acts as a quality indicator of food. Food spoilage due to microbial growth produces CO_2. The concept of CO_2 as an indicator of food spoilage was validated by Jung and his coworkers by developing a chitosan-based indicator for indicating microbial spoilage by detecting CO_2.[18] Two dye mixtures having bromothymol blue and methyl red, and bromothymol blue, bromocresol green, and phenol red were investigated for the detection of CO_2 in the headspace of packaged skinless chicken breast spoilage. Bromothymol blue and methyl red gave results faster than the other. The color response was also correlated with the pattern of microbial growth.[46] Varying combinations of printing dyes (such as m-cresol blue, purple red, tributyl ammonium hydroxide, ethylcellulose, used with ethanol and butanol solvent), which can be printed on paper and polymer surfaces using an inkjet printer, have been investigated for developing the colorimetric CO_2 indicator.[75]

Metal oxide sensors are used for developing artificial olfactory system to assess egg freshness, which was based on the correlation between the artificial olfactory response and corresponding chemicals (lactic and succinic acids) and microbiological parameters.[52] Furthermore, an olfaction system using colorimetric sensor array was developed. In this system, dyes were printed on SiO_2 plate. This system was analyzed by using principal component analysis to discriminate the spoilage degree.[17]

Agarose-immobilized myoglobin detects the hydrogen sulfide concentration in the MAP of poultry meat products.[50] A bimetallic complex (Re(5-phenyl-1,10-phenanthroline)(CO_3)NCS^{-1})-(Pt(DMSO)Cl_2) acts as a good luminescent chemodosimeter. It is selective and sensitive to volatile biogenic sulfide. This complex showed the detection limits of 0.07, 0.065, and 0.05 mg per kg to H_2S, methanethiol (CH_3SH), and dimethylsulfide (CH_3SCH_3), respectively in real pork samples.[6] Koskela et al. developed a sensor, which has a high potential for large-scale production for monitoring the quality of raw poultry by detecting hydrogen sulfide with a printed sensor. This sensor provides a clear response to raw poultry at MAP in the retail store.[21]

An array of metalloporphyrin, which is sensitive to the headspace concentration of compounds formed during the storage of fish fillets, has been developed. Deviation in optical properties of layers of metalloporphyrin has been monitored using a computer screen photo-assisted sensor.[2]

A model based on Arrhenius equation measured the activation energy of 105.93 $kJmol^{-1}$ for the total aerobic count of freshwater fish.[70] Sixteen sensing materials were fabricated with different combinations involving 13 indicators and three organic supports, and these were used for making a chromogenic sensor array for monitoring the freshness of boiled marinated turkey kept in the MAP. The array was combined with the camera and Photoshop software to analyze and extract the color change.[48] Textural analysis of meat using light-scattering technique has been used to sense the freshness of meat.[28] To determine the firmness and soluble solid content, the indicators of apple freshness, spectroscopy technique using LED (light-emitting diode)-induced fluorescence was used successfully.[12]

5.4 OXYGEN INDICATORS

Oxygen plays a major role in the spoilage of food. The amount of oxygen in the package is controlled or modified by other inert gases such as nitrogen and CO_2. Oxygen can permeate through the package due to poor seal ability, sealing, pinholes, and package tampering during transportation. Food decay is accelerated due to oxygen availability. Hence, the presence of oxygen needs to be managed for the assurance of food preservation. Colorimetric oxygen sensors are cheap and allow consumers to check the presence of oxygen in the package.

Gutiérrez-Tauste et al. fabricated a methylene blue/titanium oxide nanocomposite thin film by a liquid deposition method, which can be activated by light and can be used as an oxygen indicator.[56] Further, alginate was used as a coating material for the development of a UV-activated oxygen indicator, which helps to reduce the leaching of dye in the water by 5.08 ± 0.06%.[61] UV-activated oxygen indicator in the form of water-based ink containing methylene blue as redox dye, TiO_2 as a colloidal semiconductor photocatalyst, and tartaric acid as an electron donor can be printed on a polyester film using an inkjet printer.[26] In the presence of oxygen, yellow color leucomethylene blue, integrated into the polyvinyl alcohol matrix, changes color to green and can also be used as a time-monitoring sensor.[34] Coatings of carrageen during the fabrication of colorimetric oxygen indicator are commercially available as nylon/polyethylene vacuum packaging films that reduce the leakage of redox dye (methylene blue, Azure-A, thionine).[60]

Mills have used nanocrystalline tin(iv) dioxide (SnO_2), in which redox dye methylene blue was activated by UVB light.[35]

Polymer medium containing dissolved sacrificial electron donor (SED) and semiconductor photocatalyst (TiO_2) slowed down the recovery rate of light-activated redox dye methyl blue.[36] Polyviologens have been investigated for their potential as oxygen indicators. Polyviologen's reduction under UV light was higher, whereas oxygen sensitivity was lower compared with methylene blue.[45]

5.5 SUMMARY

Recently, the use of indicators for real-time monitoring of food has increased in intelligent packaging. Various colorimetric indicators based on pH-sensitive dyes and color-changing polymers have been used. Different prototypes using polymers, enzyme–substrate reactions, and pH-sensitive colorimetric dyes have been developed with a vast range of activation energy. Dyes-based intelligent packaging material has been used to indicate food freshness and spoilage due to temperature abuse and oxygen. A few nanoparticles have also been reported to be used in the development of nanosensors by improving the properties of the sensor.

ACKNOWLEDGMENTS

Authors express their gratitude to Mahatma Gandhi Central Library, Indian Institute of Technology, Roorkee for providing all the research facilities.

KEYWORDS

- **dyes**
- **indicators**
- **intelligent packaging**
- **sensors**

REFERENCES

1. Albelda, Jasmine A.V. U.; Santos, A.; Saanciu, G. N. C.; Lia, A. Graphene-Titanium Dioxide Nanocomposite Based Hypoxanthine Sensor for Assessment of Meat Freshness. *Biosens. Bioelectron.* **2017,** *89*, 518–524.
2. Alimelli, A.; Pennazza; Santonico, M.; Paolesse, R.; Filippini, D.; D'Amico, A.; Lundstrom, I.; Natale, C. D. Fish Freshness Detection by a Computer Screen Photoassisted Based Gas Sensor Array. *Anal. Chim. Acta* **2007,** *582* (2), 320–328.
3. Andretta, R.; Luchese, C. L.; Tessaro, I. G.; Spada, J. C. Development and Characterization of pH-Indicator Films Based on Cassava Starch and Blueberry Residue by Thermocompression. *Food Hydrocoll.* **2019,** *93*, 317–324.

4. Barat, J.; M., Gil, L.; Breijo, E. G.; Aristoy, M. C.; Toldra, F.; Manez, R. M.; Soto, J. Freshness Monitoring of Sea Bream (*Sparus Aurata*) with a Potentiometric Sensor. *Food Chem.* **2008,** *108* (2), 681–688.
5. Baston, O.; Tofan, I.; Stroia, A.L.; Moise, D.; Barna, O. Refrigerated Chicken Meat Freshness. Correlation Between Easily Hydrolysable Nitrogen, pH Value and Biogenic Amines Contents. *Food Technol.* **2008,** *2*, 37–43.
6. Chow, C. F.; Ho, P.Y.; Sun, D.; Lu, Y. J.; Wong, W. L.; Tang, Q.; Gong, C. B. Development of Sensitive and Selective Food Sensors Using New Re(I)-Pt (II) Bimetallic Complexes to Detect Volatile Biogenic Sulfides Formed by Meat Spoilage. *Food Chem.* **2017,** *216*, 382–389.
7. Chun, H. N.; Kim, B.; Shin, H. S. Evaluation of A Freshness Indicator for Quality of Fish Products During Storage. *Food Sci. Biotechnol.* **2014,** *23* (5), 1719–1725.
8. Dainelli, D.; Gontard N.; Spyropoulos, D.; Zondervan-vanden Beuken, E.; Tobback, P. Active and Intelligent Food Packaging: Legal Aspects and Safety Concerns. *Trends Food Sci. Tech.* **2008,** *19* (Suppl. 1), S103–S112.
9. Dainty, R.H.; Mackey, B.M. The Relationship Between the Phenotypic Properties of Bacteria from Chill-Stored Meat and Spoilage Processes. *J. Appl. Microbiol.* **1992,** *73* (21), 103–114.
10. Dudnyk, I.; Janecek, E.; VAucher-Joset, J.; Stellacci, F. Edible Sensors for Meat and Seafood Freshness. *Sens. Actuators B* **2018,** *259*, 1108–1112.
11. Galagan, Y.; Su, W. F. Fadable Ink for Time–Temperature Control of Food Freshness: Novel New Time–Temperature Indicator. *Food Res. Int.* **2008,** *41* (6), 653–657.
12. Gao, F.; Dong, Y.; Xiao, W.; Yin, B.; Yan, C.; He, S. LED-Induced Fluorescence Spectroscopy Technique for Apple Freshness and Quality Detection. *Postharvest Biol. Technol.* **2016,** *119*, 27–32.
13. Giannakourou, M.; Koutsoumanis, K.; Nychas, G.; Taoukis, P. Field Evaluation of The Application of Time Temperature Integrators for Monitoring Fish Quality in The Chill Chain. *Int. J. Food Microbiol.* **2005,** *102* (3), 323–336.
14. Han, J. H.; Ho, C. H. L.; Rodrigues, E. T. Intelligent Packaging for Food Products. Chapter 8; In *Innovations in Food Packaging*; Han J. H. Ed.; Second Edition; New York: Elsevier Academic Press, 2013; pp 413–438.
15. Heising, J. K.; Battels, P.V.; Boekel, M.A.J.S., V.; Dekker, M. Non-Destructive Sensing of The Freshness of Packed Cod Fish Using Conductivity and pH Electrodes. *J. Food Eng.* **2014,** *124*, 80–85.
16. Heising, J. K.; Dekker, M.; Bartels, P. V.; Boekel, M. A. J. S. V. A Non-Destructive Ammonium Detection Method as Indicator for Freshness for Packed Fish: Application on Cod. *J. Food Eng.* **2012,** *110* (2), 254–261.
17. Huang, X.; Xin, J.; Zhao, J. A Novel Technique for Rapid Evaluation of Fish Freshness Using Colorimetric Sensor Array. *J. Food Eng.* **2011,** *105* (4), 632–637.
18. Jung, J.; Puligundla, P.; Ko, S. Proof-Of-Concept Study of Chitosan-Based Carbon dioxide Indicator for Food Packaging Applications. *Food Chem.* **2012,** *135* (4), 2170–2174.
19. Kim, M. J.; Jung, S. W.; Park, H. R.; Lee, S. J. Selection of an Optimum pH-Indicator for Developing Lactic Acid Bacteria-Based Time–Temperature Integrators (TTI). *J. Food Eng.* **2012,** *113* (3), 471–478.
20. Kim, W.; Park, E.; Hong, K. W. Development of A Time-Temperature Integrator System Using *Burkholderia cepacia* Lipase. *Food Sci. Biotech.* **2012,** *21* (2), 497–502.

21. Koskela, J.; Sarfraz, J.; Ihalainen, P.; Maattanen, A.; Pulkkinen, P.; Tenhu, H.; Nieminen, T.; Kilpela, A.; Peltonen, J. Monitoring the Quality of Raw Poultry by Detecting Hydrogen Sulfide with Printed Sensors. *Sens. Actuators B* **2015,** *218*, 89–96.
22. Kuswandi, B.; Jayus, Restyana A.; Abdulah, A.; Heng, L.Y.; Ahmad, M. A Novel Colorimetric Food Package Label for Fish Spoilage Based on Polyaniline Film. *Food Control* **2012,** *25* (1), 184–189.
23. Kuswandi, B.; Oktaviana, R.; Abdullah, A.; Heng, L.Y. Novel On-Package Sticker Sensor Based on Methyl Red for Real-Time Monitoring of Broiler Chicken Cut Freshness. *Packag. Technol. Sci.* **2014,** *27* (1), 69–81.
24. Labuza, T. P. Application of Chemical Kinetics to Deterioration of Foods. *J. Chem. Educ.* **1984,** *61* (4), 348–353.
25. Labuza, T. P.; Kamman, J. Reaction Kinetics and Accelerated Tests Simulation as a Function of Temperature. In *Applications of Computers in Food Research*; Saguy I. Ed.; New York: Marcel Dekker; **1983**; pp 110–129.
26. Lawrie, K., Mills, A., Hazafy, D. Simple Inkjet-Printed, UV-Activated Oxygen Indicator. *Sens. Actuators B* **2013,** *176*, 1154–1159.
27. Lee, B.; Sin, H. Polymer Based Time Temperature Indicator for High Temperature Processed Food Products. *Food Sci. Biotech.* **2012,** *21* (5), 1483–1487.
28. Li, H.; Sun, X.; Pan, W.; Kutsanedzie, F.; Zhai, J.; Chen, Q. Feasibility Study on Nondestructively Sensing Meat's Freshness Using Light Scattering Imaging Technique. *Meat Sci.* **2016,** *119*, 102–109.
29. Lim, S.; Gunasekaran, S.; Imm, J. Y. Gelatin-Templated Gold Nanoparticles as Novel Time Temperature Indicator. *J. Food Sci.* **2012,** *77* (9), N45–N49.
30. Liu, B.; Xu, H.; Zhao, H.; Liu, W.; Zhao, L.; Li, Y. Preparation and Characterization of Intelligent Starch/PVA Films for Simultaneous Colorimetric Indication and Antimicrobial Activity for Food Packaging Applications. *Carbohydr. Polym.* **2017,** *157*, 842–849.
31. Lu, L.; Zheng, W.; Lv, Z.; Tang, Y. Development and Application of Time-Temperature Indicator Used on Food During the Cold Chain Logistics. *Packag. Technol. Sci.* **2013,** *26* (S1), 80–90.
32. Maciel, V. B. V.; Yoshida, C. M. P.; Franco T. T. Chitosan/Pectin Polyelectrolyte Complex as A pH Indicator. *Carbohydr. Polym.* **2015,** *132*, 537–545.
33. Maciel, V. B. V.; Yoshida, C. M. P.; Franco, T. T. Development of Prototype of Colorimetric Temperature Indicator for Monitoring Food Quality. *J. Food Eng.* **2012,** *111* (1), 21–27.
34. Marek, P.; Velez, J. J. V.; Haas, T.; Doll, T.; Sadowski, G. Time-Monitoring Sensor Based on Oxygen Diffusion in An Indicator/Polymer Matrix. *Sens. Actuators B* **2013,** *178*, 254–262.
35. Mills, A.; Hazafy, D. Nanocrystalline SnO–Based, UVB-Activated, Colorimetric Oxygen Indicator. *Sens. Actuators B* **2009,** *136*, 344–349.
36. Mills, A.; Hazafy, D.; Lawrie, K. Novel Photocatalyst-Based Colorimetric Indicator for Oxygen. *Catal. Today* **2011,** *161*, 59–63.
37. Musso, Y. S.; Salgado, P. R.; Mauri, A. N. Gelatin Based Films Capable of Modifying Its Color against Environmental PH Changes. *Food Hydrocoll.* **2016,** *61*, 523–530.
38. Nopwinyuwong, A.; Boonsupthip, W.; Pechyen, C.; Suppakul, P. Formation of Polydiacetylene/Silica Nanocomposite as A Colorimetric Indicator: Effect of Time and Temperature. *Adv. Polym. Technol.* **2012,** *32* (S1), E724–E731.

39. Nopwinyuwong, A.; Kaisone, T.; Hanthanon, P.; Nandhivajrin, C.; Boonsupthip, W.; Pechyen, C.; Suppakul, P. Effects of Nanoparticle Concentration and Plasticizer Type on Colorimetric Behaviour of Polydiacetylene/Silica Nanocomposite as Time –Temperature Indicator. *Energy Procedia* **2014,** *56* (C), 423–430.
40. Pereira, V. A.; Arruda, I. N. Q.; Stefani, R. Active Chitosan/PVA Films with Anthocyanins from *Brassica Oleracea* (Red Cabbage) As Time – Temperature Indicators for Application in Intelligent Packaging. *Food Hydrocoll.* **2015,** *43*, 180–188.
41. Pourjavaher, S.; Almasi, H.; Meshkini, S.; Pirsa, S.; Parandi, E. Development of A Colorimetric pH Indicator Based on Bacterial Cellulose Nanofibers and Red Cabbage (*Brassica Oleracea)* Extract. *Carbohydr. Polym.* **2017,** *156*, 193–201.
42. Priyadarshi, R.; Ezati, P.; Rhim, J. W. Recent Advances in Intelligent Food Packaging Applications Using Natural Food Colorants. *ACS Food Sci. Technol.* **2021,** *1* (2), 124–138.
43. Qiu, X.; Itoh, D.; Satake, T.; Suzuki, H. Microdevice with Integrated Multienzyme Sensors for the Measurement of Pork Freshness. *Sens. Actuators B* **2016,** *235*, 535–540.
44. Realini, C. E.; Marcos, B. Active and Intelligent Packaging Systems for a Modern Society. *Meat Sci.* **2014,** *98* (3), 404–419.
45. Roberts, L.; Lines, R.; Reddy, S.; Hay, J. Investigation of Polyviologens as Oxygen Indicators in Food Packaging. *Sens. Actuators B* **2011,** *152* (1), 63–67.
46. Rukchon, C.; Nopwinyuwong, A.; Trevanich, S.; Jinkarn, T.; Suppakul, P. Development of A Food Spoilage Indicator for Monitoring Freshness of Skinless Chicken Breast. *Talanta* **2014,** *130*, 547–554.
47. Russell, S. M.; Fletcher, D. L.; Cox, N. A. Spoilage Bacteria of Fresh Broiler Chicken Carcasses. *Poult. Sci.* **1996,** *75*, 2041–2047.
48. Salinas, Y.; Ros-lis, J. V.; Vivancos, J.L.; Manez, R. M.; Aucejo, S.; Herranz, N.; Lorente, I.; Garcia, E. A Chromogenic Sensor Array for Boiled Marinated Turkey Freshness Monitoring. *Sens. Actuators B* **2014,** *190*, 326–333.
49. Shukla, V.; Kandeepan, G.; Vishnuraj, M. R.; Soni, A. Anthocyanins Based Indicator Sensor for Intelligent Packaging Application. *J. Agri. Res.* **2016,** *5* (2), 205–209.
50. Smolander, M.; Hurme, E.; Latva-Kala, K.; Luoma, T.; Alakomi, H. L.; Ahvenainen, R. Myoglobin-Based Indicators for The Evaluation of Freshness of Unmarinated Broiler Cuts. *Innov. Food Sci. Emerg. Technol.* **2002,** *3* (3), 279–288.
51. Su, Y. Preparation of Polydiacetylene/Silica Nanocomposite for Use as Chemosensor. *React. Funct. Polym.* **2006,** *66*, 967–973.
52. Suman, M.; Riani, G.; Dalcanale, E. MOS-Based Artificial Olfactory System for The Assessment of Egg Products Freshness. *Sens. Actuators B* **2007,** *125* (1), 40–47.
53. Taoukis, P. S. Modeling the Use of Time-Temperature Indicators in Distribution and Stock Rotation. Chapter 19; In *Food Process Modelling*; Tijskens, L. M. M.; Hertog, M. L. A. T. M.; Nicolaï B. M. Eds; Cambridge: Woodhead Publishing Ltd., 2001; pp 402–432.
54. Taoukis, P.S.; Labuza, T. P. Applicability of Time Temperature Indicators as Shelf-Life Monitors of Food Products. *J. Food Sci.* **1989,** *54* (4), 783–788.
55. Taoukis, P.S.; Labuza, T. P. Time Temperature Indicators (TTIs). In *Novel Food Packaging Techniques*; Ahvenainen, R. Ed.; Cambridge: Woodhead Publishing Ltd., 2003; pp 103–126.
56. Tauste, D. G.; Domenech, X.; Pastor, N. C.; Ayllon, J. A. Characterization of Methylene Blue/Tio2 Hybrid Thin Films Prepared by Liquid Phase Deposition (LPD) Method:

Application for Fabrication of Light Activated Colorimetric Oxygen Indicators. *J. Photochem. Photobiol. A* **2007,** *187* (1), 45–52.
57. Traiphol, N.; Rungruangviriya, N.; Potai, R.; Traiphol, R. Stable Polydiacetylene/Zno Nanocomposites with Two-Step Reversible and Irreversible Thermochromism: The Influence of Strong Surface Anchoring. *J. Colloid Interface Sci.* **2011,** *356* (2), 481–489.
58. Vaikousi, H.; Biliaderis C. G.; Koutsoumanis K. P. Development of Microbial Time/Temperature Indicator Prototype for Monitoring the Microbiological Quality of Chilled Foods. *Appl. Environ. Microbiol.* **2008,** *74* (10), 3242–3250.
59. Vaikousi, H.; Biliaderis, C. G.; Koutsoumanis K. P. Applicability of A Microbial Time Temperature Indicator (TTI) For Monitoring Spoilage of Modified Atmosphere Packed Minced Meat. *Int. J. Food Microbiol.* **2009,** *133* (3), 272–278.
60. Vu, C. H. T.; Won, K. Leaching-Resistant Carrageenan Based Colorimetric Oxygen Indicator Films for Intelligent Food Packaging. *J. Agric. Food Chem.* **2014,** *62* (29), 7263–7267.
61. Vu, C. H. T.; Won, K. Novel Water Resistant UV-Activated Oxygen Indicator for Intelligent Food Packaging. *Food Chem.* **2013,** *140* (1), 52–56.
62. Wallace, T. C.; Giusti, M. M. Anthocyanins-Nature's Bold, Beautiful, and Health-Promoting Colors. *Foods* **2019,** *8* (11), 550–555.
63. Wang, Y. C.; Lu L.; Gunasekaran S. Gold Nanoparticle Based Thermal History Indicator for Monitoring Low-Temperature Storage. *Microchim. Acta* **2015,** *182*, 1305–1311.
64. Wang, Y.C.; Mohan, C.O.; Guan, J.; Ravishankar, C.N.; Gunasekaran, S. Chitosan and Gold Nanoparticles- Based Thermal History Indicators for Perishable and Temperature Sensitive Products. *Food Control* **2018,** *85*, 186–193.
65. Wanihsuksombat, C.; Hongtrakul, V.; Suppakul, P. Development and Characterization of a Prototype of a Lactic Acid-Based Time-Temperature Indicator for Monitoring Food Product Quality. *J. Food Eng.* **2010,** *100* (3), 427–434.
66. Watanabe, E.; Tamada, Y.; Sato, N. H.; Development of Quality Evaluation Sensor for Fish Freshness Control Based on Ki Value. *Biosens. Bioelectron.* **2005,** *21* (3), 534–538.
67. Wu, D.; Hou, S.; Chen, J.; Sun, Y.; Ye, X.; Liu, D.; Meng, R.; Wang, Y. Development and Characterization of Enzymatic Time Temperature Indicator (TTI) Based on *Aspergillus Niger* Lipase. *LWT - Food Sci. Technol.* **2015,** *60* (2), 1100–1104.
68. Wu, D.; Wang, Y.; Chen, J.; Ye, X.; Wu, Q.; Liu, D.; Ding, T. Preliminary Study on Time-Temperature Indicator (TTI) System Based on Urease. *Food Control* **2013,** *34* (1), 230–234.
69. Yan, S.; Huawei, C.; Limin, Z.; Fazheng, R.; Luda, Z.; Hengtao, Z. Development and Characterization of a New Amylase Type Time-Temperature Indicator. *Food Control* **2008,** *19* (3), 315–319.
70. Yao, L.; Luo, Y.; Sun, Y.; Shen, H. Establishment of Kinetic Models Based on Electrical Conductivity and Freshness Indicators for The Forecasting of Crucian Carp (*Carssius Carassius*) Freshness. *J. Food Eng.* **2011,** *107* (2), 147–151.
71. Yong, H.; Wang, X.; Zhang, X.; Liu, Y.; Qin, Y.; Liu, J. Effects of Anthocyanin-Rich Purple and Black Eggplant Extracts on the Physical, Antioxidant and pH-Sensitive Properties of Chitosan Film. *Food Hydrocoll.* **2019,** *94*, 93−104.
72. Zabala, S.; Cast, J.; Martínez, C. Development of A Time and Temperature Indicator (TTI) Label by Rotary Printing Technologies. *Food Control* **2015,** *50*, 57–64.
73. Zhai, X.; Shi, J.; Zou, X.; Wang, S.; Jaing, C.; Zhang, J.; Huang, X.; Zhang, W.; Holmes M. Novel Colorimetric Films Based on Starch/Polyvinyl Alcohol Incorporated

with Roselle Anthocyanin for Fish Freshness Monitoring. *Food Hydrocoll.* **2017,** *69*, 308–317.

74. Zhang, J.; Zou, X.; Zhai, X.; Huang, X.; Jiang, C.; Holmes, M. Preparation of An Intelligent pH Film Based on Biodegradable Polymers and Roselle Anthocyanins for Monitoring Pork Freshness. *Food Chem.* **2019,** *272*, 306–312.
75. Zhang, Y.; Lim, L. T. Inkjet-Printed Colorimetric Indicators. *Talanta* **2016,** *161*, 105–113.

CHAPTER 6

ROLE OF BACTERIOPHAGES IN ACTIVE FOOD PACKAGING

KIRAN VERMA and YOGESH KUMAR

ABSTRACT

Active food packaging with bacteriophages is a novel emerging technology to reduce the load of bacteria; and thus reduce the potential for human infection. Use of bacteriophages represents a divergence from the conventional approach to food packaging. Hence, this chapter focuses on the current approach to use bacteriophages in food packaging, to reduce or eliminate bacteria from target food products and detect bacteria present in the food;; and on various methods of incorporation of bacteriophages in food packaging. Moreover, the mode of action as a microbial agent in active food packaging, morphology, and structure of bacteriophages has also been discussed. As the use of bacteriophages in food packaging may raise concerns regarding their propagation during the shelf-life of food products; therefore, limitations and safety concerns related to such packaging material containing the bacteriophages have also been illustrated.

6.1 INTRODUCTION

Foods of animal origin are excellent sources of adequate quality protein, vitamins, and minerals and provide high-quality nutrition not only to the consumer but also to microorganisms. According to WHO (2015), about 420,000 deaths and 600 million cases of foodborne illnesses occur globally by 31 key-known pathogens each year.[12]

Advances in Sustainable Food Packaging Technology. Megh R. Goyal, Preeti Birwal, Anand K. Pandey, & Ritesh B. Watharkar, (Eds.)

Food-borne illnesses pose a serious threat to human health and the economy. Diverse microorganisms that exist in food as normal microbiota (such as: *Listeria monocytogenes, Salmonella, Campylobacter, Escherichia coli, Clostridium,* and *Staphylococcus aureus*[18]) present an everlasting challenge to the food safety of consumers like natural food, which is safe for consumption. Microorganisms need optimum conditions to multiply (such as: nutrients, moisture, temperature, air, and salt). This makes it crucial to maintain such a state during the food processing and transportation that can inhibit the microbial growth and chemical reactions, which cause the spoilage of food, until the product reaches consumers safely. The best approach to dealing with this, is to either reduce the initial load or to prevent the growth of microorganisms in the food. For that purpose, active packaging materials have been used diversely in food packaging

Antimicrobial food packaging has gained significant attention as a new emerging technology to inhibit pathogenic microorganisms, making the quality and safety of food stable for longer period of time. Antimicrobial active food packaging can be applied using diverse methods to protect food products.

Traditionally antimicrobial agents were applied either by dipping or spraying for preservation of food. When antimicrobial agents come into contact with food materials, they can get diluted through diffusion or denaturation. Incorporation of the antimicrobial agent onto the surface of food packaging material has been considered more efficient, allowing the slow and longer release of antimicrobial agent efficiently. Most of the antimicrobial agents are broad-spectrum, not targeting the specific pathogens.

This novel technology allows target-specific approach to safeguard the food from specific pathogens. Pathogen-specific approaches can enhance the antimicrobial efficacy by reducing interactions with non-targeted microbes.[21] However, the indiscriminate use of antibiotics leads to a decrease in the effectiveness of antimicrobial agents, over the period of time due to rapid increase in multi-drug-resistant species. The need to control pathogens in foods and in different areas of the food industry has aroused interest in alternative methodologies to inhibit the emergence of resistant microorganisms. Bacteriophages are providing an attractive alternative to the traditional antimicrobial agent, when antimicrobial resistant bacteria are on rise. Bacteriophages make an excellent tool for the detection and inhibition of bacteria. Incorporation of bacteriophages in packaging film provides a promising tool to work against pathogenic bacteria. Bacteriophages have been used to control pathogens for more than 70 years in both medicinal and food industries because of their highly specific ability to lyse the bacteria.[10]

This chapter focuses on: (1) the use of bacteriophages in the food industry and food packaging; (2) morphology of bacteriophages and their mode of action against bacteria; and (3) the methods for incorporation of bacteriophages into food packaging.

6.2 ROLE OF BACTERIOPHAGES IN FOOD INDUSTRY

In animal food industries, pathogenic bacteria (such as: *Salmonella, E. coli, L. monocytogenes, Clostridium and Campylobacter*) are responsible for the contamination of food products. The persistent occurrence of these microorganisms in the processing area and over the surface of equipment is a leading source of contamination. According to the WHO, major cause of foodborne deaths are: *Salmonella Typhi* (approx. 52,000 deaths), non-typhoidal *Salmonella* (approx. 59,000 deaths), *Norovirus* (approx. 35,000 deaths), and Enteropathogenic *E. coli* (about 37,000 deaths).[12]

To overcome frequent bacterial foodborne pathogens (like *Salmonella, Campylobacter, Escherichia coli*, and *Listeria*), the regulatory agencies of US Department of Agriculture (DA) and Food and Drug Administration (FDA) have approved several phage-based preparations. Phage preparations are commercially used as additives in the production of ready-to-eat (RTE) foods and meats, or they are applied to live animals prior to slaughtering or added in the animal feed.[7]

Emerging technologies can attain high food safety by making food free from pathogenic as well as foodsborne illness causing microorganisms. *Salmonella* forms biofilm over the metallic surface. This pre-formed biofilm is difficult to remove by using regular disinfectants or sanitizers. Therefore, in such situations, bacteriophages come into play, and are used to control bacterial biofilms. Bacteriophages degrade the polysaccharide cell-wall of bacteria by the depolymerase enzyme.

The lysis effect of six different bacteriophage strains on *salmonella* isolated is shown in Table 6.1. These were further evaluated for effective reduction of the growth of different *Salmonella* strains,[4,10] which get attached to stainless steel surfaces under laboratory and greenhouse conditions.[10] Bacteriophage effectively reduced 3.4, 1.4, and 3.0 log CFU/cm^2 of *S. Typhimurium* strain 8243 in summer, fall/winter and spring seasons, respectively under greenhouse conditions. In comparison to this, underlaboratory conditions, bacteriophage treatment reduced up to 2.9 and 3.0 log CFU/cm^2

of attachment and slightly formed biofilm of selected top 10 *Salmonella* strains and a virulent *Salmonella Typhimurium* strain 8243, respectively.[10]

TABLE 6.1 Action of Bacteriophages Against Selected *Salmonella* Isolates.

***Salmonella* isolates**	***Salmonella* serotype**	**Bacteriophage**					
		JC1	**S5p2**	**29**	**52**	**1PB**	**VCA1**
V-RF1*,A	Typhimurium	3B	3	3	3	1	1
V-SFdO2*,A	Ouakam	2	3	3	3	1	2
V-SRF1*,A	Kentucky	0	0	2	2	1	2
V-SRF2	Hadar	1	1	0	0	0	0
V-FdM1*,A	Johannesburg	3	3	3	3	1	2
A-SRF1*,A	Schwarzengrund	0	0	0	0	0	2
V-SFdO1A	Senftenberg	2	2	3	3	1	1
V-RF2A	Typhimurium	3	3	3	3	0	2
A-SFMW1A	Typhimurium	2	2	3	3	2	3
A-SPOA	Kentucky	0	0	0	0	1	3
A-SRFA	Schwarzengrund	2	2	3	3	1	2
8243	Typhimurium	2	2	3	3	2	3

Source: Adapted from Ref. [2]

Also, bacteriophages have the functional ability to "auto dose"; meaning that they infect subsequent bacterial cells as the phage population grows. They multiply specifically in close vicinity of host bacteria. In the absence of host cells, which are targeted bacterial cells, bacteriophages are eliminated by the immune system.[7] Other bacteriophages, which are used to treat various illnesses of the human body, are used to avoid the deterioration of food products (meat, milk and dairy products) and are also used in adsorbent pads (food packaging). These pads allow for the diffusion of bacteriophages into the substrate like cellulose membrane, alginate beads and paper[8] and thus keep the food fresh by reducing the bacterial load. The various significant aspects of increasing the use of bacteriophages for effectively reducing the bacterial load are as follows[14]:

- Bacteria infected by the lytic phages are unable to regain their viability.
- Bacteriophage can be used as an indicator of food and water quality.
- Bacteriophage does not alter physico-chemical properties of food.
- Phages are self-replicating and over the course of treatment to increase efficacy.
- The high specificity toward the host pathogen.

- They do not affect the gut microbiota and nor do they change the organoleptic properties of the food products
- Thirty-eight phages from 22 to 40 refrigerated food samples were present naturally in the presence of specific pathogens and were mainly responsible for the lysis of pathogenic microbes.

6.3 BACTERIOPHAGES IN FOOD PACKAGING: METHODS OF INCORPORATION

Food packaging not only provides the barrier to the external environment, but can also safeguard the food product, if needed, which is possible with smart packaging. In smart and active packaging, it is possible to enhance the applicability of packaging material. Active food packaging incorporated with antimicrobial agents can release the antimicrobial agent gradually at a slower rate and for a longer duration.

Bacteriophages in active packaging have high rate of replication and because of that they can easily lyse the entire bacterial culture. If the generation time of a bacteria is 20 min, there will be 8 cells after one hour. On the other hand, after the duration of one hour, there may be 20–100 phages considering the replication duration of 20–40 min and the burst size (number of phages released after lysis of a bacterial cell).

Listex® is a commercially available preparation containing bacteriophages, which has been approved as processing aid in Canada, the United States of America and Netherlands. This is sprayed onto the surface of ready-to-eat meat to reduce the risk of foodborne illness. This is active against *L. monocytogenes*, manufactured by Micreos, Netherlands.[5,18]

Another commercial preparation is EcoShield®, manufactured by Intralytix Ltd, US. Bacteriophages in this preparation is active against *E. coli* O157:H7.[8] Dipping and spryaing techniues are used to apply bacteriophages to food surfaces.[18] Application of this preparation directly to food surface creates a barrier for consumer acceptance. In this case, bioactive packaging material incorporated with bacteriophages emerges as savior that would not only provide the safety from targeted pathogens but would also be more acceptable to consumers, as preparations are part of food packaging material. Table 6.2 indicates various methods that are used for the incorporation of bacteriophages having application against specific target organisms.

TABLE 6.2 Methods of Incorporation of Bacteriophages in Active Food Packaging Films.

Mode of incorporation	Type of film	Application	Ref.
Encapsulation	Whey protein film	Extended storage and release	[21]
Entrapment	Sodium alginate emulsion-based film	Antimicrobial activity	[2]
Homogenized in packaging film solution	Acetate film	Inhibition of Salmonella Typhimurium	[11]
Immobilization by encapsulation	Modified cellulose membrane	• To inhibit L. monocytogenes growth on fresh cut cantaloupe • To control L. monocytogenes on vacuum-sealed cooked turkey	[18]
Immobilization on membrane surface	• Pristine polyether sulfones membrane (PES) • PES treated with oxygen (PES-O2) • PES-O2 treated with crosslinker	• biofouling control. • antimicrobial activity.	[13]
Microfluidization and microencapsulation	Polypropylene glycol gel	Antimicrobial activity	[4]

6.3.1 BACTERIOPHAGE IMMOBILIZATION

Bacetiophage immobilization was introduced as an alternative to the application of bacteriophages as a suspension.[3] Bacetiophage immobilization is more efficient as compared to conventional antimicrobial agents, as that get reduced in concentration, due to rapid diffusion and is denatured upon contact with food constituents. Generally, immobilization is achieved by physical adsorption and chemical functionalization.[23]

- In physical adsorption, bacteriophages are incorporated on solid supports to develop bioactive packaging that can effectively control the growth of the target pathogen in food effectively. At the same time this would limit the amounts of bacteriophages on the food.

- In chemical functionalization, bacteriophages are used instead of antibodies, crosslinked to the surface of carbon screen printed electrodes for bacterial biosensing.

Presently, bacteriophages are used as antimicrobial agents, due to their self-proliferation activity, which keeps them effective as long as the host is present. Indium tin oxide (ITO) is used as a model substrates, where T4 phage is immobilized.[17] ITO is a reliable surface for functionalization with different chemical moieties, such as: amine (-NH_2), carboxylic (-COOH) and methyl (-CH_3). The ITO/T4 system inhibits the growth of *E. coli* up to 99.99% in 2h and also its efficacy is not reduced with time, unlike conventional antimicrobial agents. It was also found to have stable antimicrobial activity in food components, such as, starch and casein. Therefore, bacteriophage may be used as an antimicrobial agent in the development of food packaging.[17]

6.3.2 MICROENCAPSULATION OF BACTERIOPHAGE

Bacteriophages encapsulated in films (edible and packaging) are stabilized prior to application and released to the surrounding environment when required.[21] Encapsulation of bacteriophage is one of the emerging techniques to reduce pathogenic microorganisms in food products and it also ensures the level of antibiotic residues in food. Encapsulation is a technique in which the material to be encapsulated remains stable inside the encapsulating material to reduce losses (that occur during industrial process, gastrointestinal tract, etc.). Phages, which are incorporated by encapsulation in the films, show better storage stability at ambient temperature.[21] It is critical for antimicrobial activity of phage encapsulated films that the encapsulated phages in films are in direct contact with the target microbes that are present in the surroundings. Phage release from edible coatings upon contact with a food product can enable phage delivery to target microbes.

Bacteriophage microencapsulated in gel matrix for phage therapy has been reported to be effective over antibiotics and other antimicrobial agents due to the lack of associated development of bacterial resistance as is the case with other antimicrobial substances. Bacteriophages encapsulated within films of cellulose acetate showed activity for up to two weeks.[20]

Microfluidics is one of the distinctive equipments, which can encapsulate (micro and nanostructure) bacteriophages in polymeric microgels. Calcium-alginate acts as an encapsulation material for various biological substances, such as: cells, proteins, drugs, and enzymes due to its authenticity and

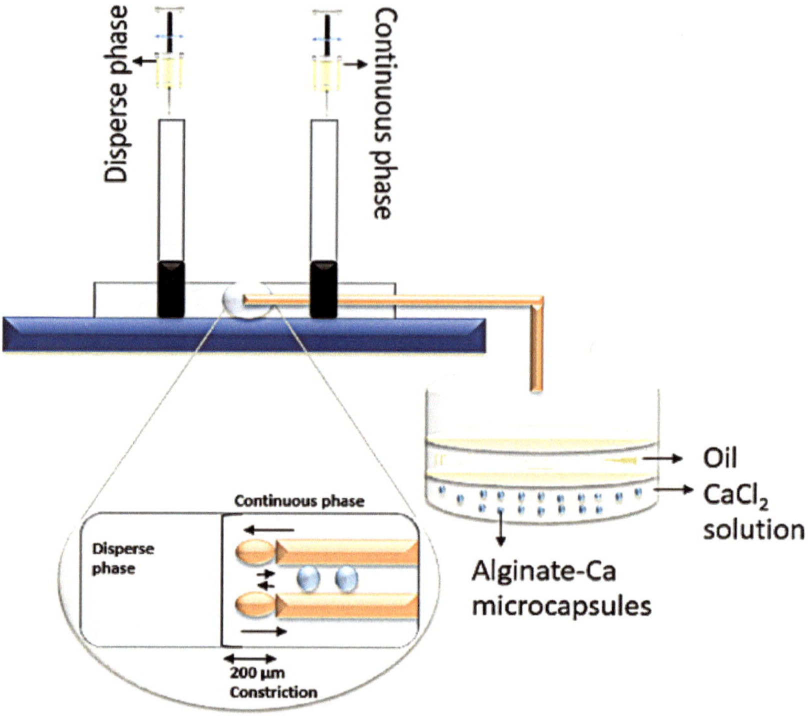

FIGURE 6.1 Microfluidic assisted alginate calcium capsules.

Source: Adapted from Ref. [4]

recognition as a GRAS (generally recognized as safe) material, which does not pose any toxicity. Microcapsules prepared by microfluid using calcium-alginate were found to be stable for 21 days in propylene glycol gel. The microcapsules added to propylene glycol gel showed antimicrobial activity by sanitizing *E. coli* O157:H7 on surface.[4] Figure 6.1 shows the formation of alginate-calcium microcapsules by microfluidics.

To have a better understanding of the action of phage on bacterial cell, it is important to perceive the basic morphology of phage.

6.4 MORPHOLOGY OF BACTERIOPHAGES

Bacteriophages are bacterial viruses that infect and release new viruses by infecting and infecting the host bacteria.[6] The word "bacteriophage" was coined by Felix d'Herelle, and it means "to eat bacteria."[1] Viral bacteriophages can cause complete lysis of a susceptible bacterial culture. They are classified as "phage" and are typically limited to a single species or even a single strain of that species. Phages are obligate intracellular parasites that multiply within bacteria by using some or all of the host's biosynthetic machinery to multiply.[1] They are found all over the place in nature and can easily be separated from faeces and sewage.

Bacteriophages occur naturally in the ecosystems with different genomes. The International Committee on Virus Taxonomy (ICTV) classifies phages based on their morphology, host specificity, and nucleic acid material (Table 6.3). ICTV has recognized 19 families of phages, which infect bacteria and archaea. Two families have RNA genomes; 8 families with circular DNA genomes and only 5 families are surrounded by an envelope; and 9 have linear genomes. Of the viral families with DNA genomes, only two have single-stranded genomes. Nine families infect bacteria only, 9 infect archaea only and *Tectiviridae* infect both bacteria and archaea.[16]

The nucleic acid may be either DNA or RNA, but not both, depending on the phage (Table 6.3). Phage nucleic acids often contain unique or modified bases that shield phage nucleic acids from nucleases that break down host nucleic acids during infection.[20,23] Simple phages can have as few as 3–5 genes, while complex phages can have over 100.[1] The nucleic acid of some explicit phages is single-stranded DNA. The majority of phages is between 24 and 200 nm in length. T4 is one of the largest phages, measuring about 200 nm in length and 80–100 nm in width. A head structure is present in all phages and can differ in size and shape (Fig. 6.2). Others are filamentous, while others are icosahedral (20 sides).

The head surrounds and protects nucleic acid by enclosing it in a protective shell. The heads of certain phages have tails attached to them. During infection, the nucleic acid moves through the tail, which is a hollow tunnel. A contractile sheath covers the T4 tail, which contracts when the bacterium is infected. A base plate and one or more tail fibres are connected to the end of the tail of phages like T4. Phages with base plate and tail fibres are involved in the phage's bacterial cell binding.[9]

6.4.1 MODE OF ACTION OF PHAGE ON BACTERIAL CELLS

Bacteriophages are viruses that infect bacteria and are responsible for inhibiting or killing susceptible bacteria. Adsorption of the phage to the bacterial cell is the first step in the infection process (Fig. 6.3). The tail fibres or an analogous structure on phages that don't have tail fibres mediate this stage.

TABLE 6.3 Classification of Bacteriophages by ICTV.

Order	Family	Morphology	Nucleic acid	Genus	Host specific
Caudovirales	*Podoviridae*	Non-enveloped, short non-contractile tail	Linear ds DNA	BPP-1, Epsilon 15, LUZ24, N4, P22, Phieco32	Bacteria (Enterobacteriaceae), archaea
	Myoviridae	Non-enveloped, long contractile tail	Linear ds DNA	T4, Mu, P1, P2, I3, SPO1, PhiH, PhiKZ	
	Siphoviridae	Non-enveloped, long non-contractile tail	Linear ds DNA	Lamda, T1, T5, L5, PsiM1, PhiC31, c2, SPbeta, N15	
Ligamenvirales	*Lipothrixviridae*	Enveloped rod-shaped	Linear ds DNA	Acidianus filamentous virus 1	Hyper thermophilic archaea
	Rudiviridae	Non-enveloped, rod shaped	Linear ds DNA	Sulfolobus islandicus rod-shaped virus 1	
Unassigned	*Ampullaviridae*	Enveloped, bottle-shaped	Linear ds DNA	Acidianus bottle-shaped virus (EF432053), species *Acidianus bottle-shaped virus*, genus *Ampullavirus*	Hyper thermophilic archaea from the genus *Acidianus*; non-lytic
	Clavaviridae	Non-enveloped, rod shaped	Circular ds DNA	Aeropyrum pernix bacilliform virus 1 (AB537968), species *Aeropyrum pernix bacilliform virus 1*, genus *Clavavirus*	Hyper thermophilic archaea from the genus *Aeropyrum*
	Corticoviridae	Non-enveloped, isometric	Circular ds DNA	Pseudoalteromonas virus PM2 (AF155037), species *Pseudoalteromonas virus PM2*, genus *Corticovirus*	Bacteria, gram-negative *Pseudoalteromonas* strains
	Cystoviridae	Enveloped, spherical	Segmented ds RNA	Pseudomonas phage phi6 (Segment S, M12921; Segment M, M17462; Segment L, M17461), species *Pseudomonas virus phi6*, genus *Cystovirus*	Gram-negative bacteria, mostly *Pseudomonas* species

TABLE 6.3 *(Continued)*

Order	Family	Morphology	Nucleic acid	Genus	Host specific
	Fuselloviridae	Non-enveloped, lemon shaped	Circular ds DNA		
	Globuloviridae	Enveloped, isometric	Linear ds DNA	*Pyrobaculum spherical virus,* Thermoproteus tenax spherical virus 1	hyperthermophilic archaeal genera *Pyrobaculum* and *Thermoproteus*
	Guttaviridae	Non-enveloped, ovoid	Circular ds DNA	Aeropyrum pernix ovoid virus 1 (HE580237), species *Aeropyrum pernix ovoid virus 1,* genus *Betaguttavirus*	Hyperthermophilic archaea, phylum Crenarchaeota
	Inoviridae	Non-enveloped, filamentous	Circular ds DNA	*Enterobacteria phage AE2*, M13 *Pseudomonas phage Pf1,* *Vibrio phage 493,* *Xanthomonas phage Cf16*	*Enterobacteria, Pseudomonas, Xanthomonas, Vibrio sp.*
	Leviviridae	Non-enveloped, isometric	Linear ss RNA	Enterobacteria phage GA, Enterobacteria phage MS2	*Enterobacteria, Caulobacter, Pseudomonas, Acinetobacter and* other gram negative bacteria
	Microviridae	Non-enveloped, isometric	Circular ss DNA	Enterobacteria phage $\Phi \times 174$	*Enterobacteria*
	Plasmaviridae	Enveloped, pleomorphic	Circular ds DNA	Acholeplasma virus L2 (L13696), species *Acholeplasma virus L2,* genus *Plasmavirus*	*Acholeplasma* species

TABLE 6.3 *(Continued)*

Order	Family	Morphology	Nucleic acid	Genus	Host specific
	Pleolipoviridae	Enveloped, pleomorphic	Circular ds DNA, Circular ss DNA, or linear ds DNA	Halorubrum pleomorphic virus 1 (FJ685651), species *Halorubrum virus HRPV1*, genus *Alphapleolipovirus*	Archaea, euryarchaeal *Halorubrum*, *Haloarcula* or *Halogeometricum* strains
	Portogloboviridae	Enveloped, isometric	Circular ds DNA		
	Sphaerolipoviridae	Enveloped, isometric	Linear ds DNA		
	Spiraviridae	Non-enveloped, rod shaped	Circular ss DNA		
	Tectiviridae	Non-enveloped, isometric	Linear ds DNA	*Bacillus phage AP50,* *Enterobacteria phage PRD1,* *Thermus phage P37-14*	*Enterobacteria, Bacillus*
	Tristomaviridae	Enveloped, rod shaped	Linear ds DNA	Pyrobaculum filamentous virus 1 (KU307456), species *Pyrobaculum filamentous virus 1*, genus *Alphatristromavirus*	Hyperthermophilic archaea of the order Thermoproteales
	Turriviridae	Enveloped, isometric	Linear ds DNA		
	Bicaudaviridae	Non-enveloped, lemon shaped	Circular ds DNA	Acidianus two-tailed virus (AJ888457), species *Acidianus two-tailed virus*, genus *Bicaudavirus*	Hyperthermophilic archaea from the genus *Acidianus*

Source: Adapted from Ref. [24]

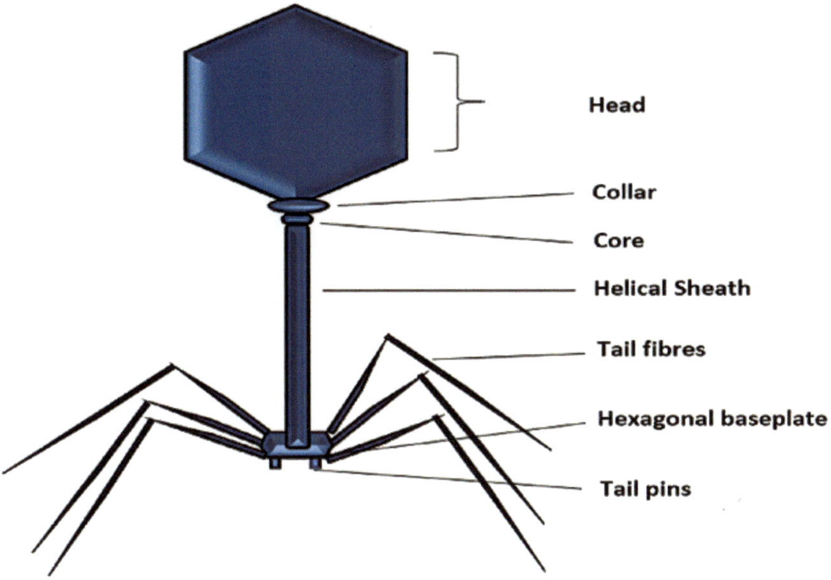

FIGURE 6.2 Structure of bacteriophages.

Source: Adapted from Ref. [21]

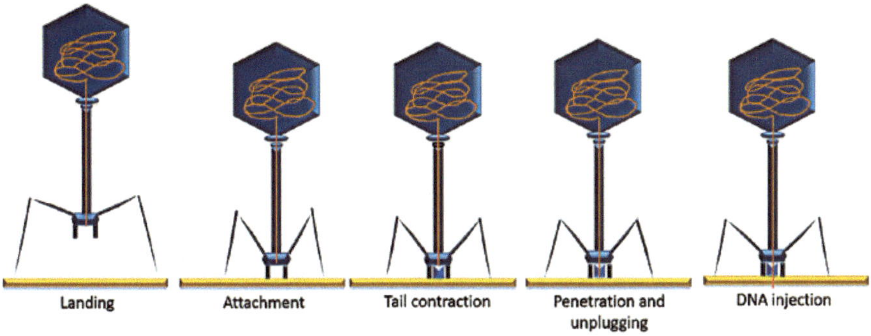

FIGURE 6.3 Adsorption of bacteriophage to the bacterial cell.

Source: Adapted from Ref. [14]

Phages bind to various receptors on bacterial cells, including proteins on the bacterium's outer surface, lipopolysaccharides (LPS), pili, and lipoproteins. This is a reversible method. The permanent attachment of phage to a bacterium is mediated by one or more components of the base plate.[9]

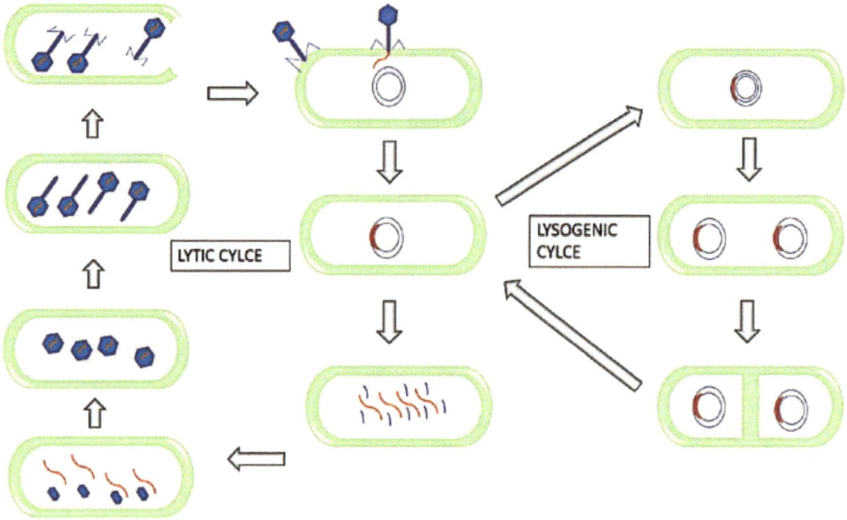

FIGURE 6.4 Life cycle of bacteriophages to establish the infection of the bacterial cell.
Source: Adapted from Ref. [14]

The sheath (for those phages that have one) contracts as the phage binds irreversibly to the bacterium, and the hollow tail fibre is forced through the bacterial envelope. Enzymes in certain phages digest different parts of the bacterial envelope. Nucleic acid from the head reaches the bacterial cell through the hollow tail. The rest of the phage remains as a "ghost" on the bacterium's surface. A non-susceptible bacterium can be contaminated artificially by injecting phage DNA via a mechanism called transfection. Phages can be either lytic (virulent) or lysogenic (temperate) depending on their life cycle (Fig. 6.4). Unlike lytic phages, which kill the cells they invade, temperate phages cause a long-term infection without killing.[14]

6.4.1.1 LYTIC CYCLE

Lytic or virulent phages replicate in bacteria and destroy the cell at the end of their life cycle by lysis. The phage cycle is said to be in eclipse time shortly after the nucleic acid is injected. There are no contagious phage particles within or outside the bacterial cell during the eclipse process. The eclipse process is the time between the phage nucleic acid entering the bacterial cell and the mature phage being released from the infected cell. The synthesis of phage components and their assembly into mature phage particles is the subject of this process. The phage nucleic acid hijacks the host's biosynthetic machinery, resulting in the production of phage-specific mRNAs and proteins (Fig. 6.4).[15,19]

Early phage proteins can degrade the host chromosome in some cases. The structural proteins (head, tail) that make up the phage, as well as the proteins needed for bacterial cell lysis, are synthesized separately. The tail is then attached to the head after the nucleic acid is packaged within the head. Maturation is the method of assembling phage components into a mature infective phage particle. The bacteria begin to lyse as the phage lysis protein accumulates, and intracellular phages are released into the medium during the lysis and release process. Phage enzymes are thought to weaken the bacteria's cell walls.

6.4.1.2 LYSOGENIC CYCLE

Lysogenic or temperate phages may either multiply or enter a dormant state in the cell through the lytic cycle. In the majority of cases, the phage DNA integrates into the host chromosome and is replicated alongside the host chromosome and is passed on to the daughter cells (Fig. 6.4). Prophage refers to the integrated state of phage DNA. Lysogeny is the term for this process, and the bacteria that nurture prophages are known as lysogenic bacteria. Since the prophage contains genes, it may give bacteria new abilities. When a cell is lysogenized, the phage's extra genes are periodically expressed in the cell. These genes have the ability to modify the bacterial cell's properties. This is referred to as phage conversion or lysogenic conversion.

6.5 SAFETY CONCERNS

Food safety agencies and industries are focusing their efforts on the most promising interventions and demonstrating their transformational impacts on food safety. Risk assessments ~~on~~ of foodborne disease prevention using emerging technologies are needed more than that technology is required for food safety. The incorporation of bacteriophages within food packaging or related products requires further research to understand and overcome the problems associated with their use. When high initial load is used in order to check the degradation of food by bacteria during the processing, storage and transportation, what happens to the remaining load after all bacteria have been removed from the system. It has been shown that treatments currently applied to foods, such as pasteurization, can reduce the titer of viable bacteriophages within the sample, thus raising doubts as to their ability to deliver an antibacterial effect.[8]

Bacteriophage survival in the food chain, and within the cold chain used by the food industry, may pose a health risk. It is little known about the co-evolution of bacteria and bacteriophages in live tissues. How this co-evolution will affect absorption rate from the gastrointestinal tract or how it would react with body tissues if they leached out from food packaging into the food are some unanswered questions. Dissemination of bacteriophages within a human or animal body, is unfamiliar at present. Interactions that might occur in the human body need more study interest as this will decide the fate of this particular emerging technology.

6.6 SUMMARY

Active food packaging with bacteriophages is a novel emerging technology. As people's understanding of eco-friendly natural products rises, such packaging shows great promise for use in the food industry. Research activities for the incorporation of bacteriophages are being studied extensively. This chapter documents the knowledge of bacteriophages incorporated in active packaging and uses of bacteriophages in the food industry other than in packaging. More detailed research is needed to assess the real potential and availability of this technology as the interaction of bacteriophages in the human body and in animal bodies is still unexplored. This technique can be an excellent alternative to optimizing and improving the quality of food safety measures.

KEYWORDS

- **active food packaging**
- **antimicrobial food packaging**
- **bacteriophages**
- **food safety**
- **target organisms**

REFERENCES

1. Ackermann, H. W. Bacteriophage Observations and Evolution. *Res. Microbiol.* **2003,** *154* (4), 245–251.
2. Alves, D.; Cerqueira, M. A.; Pastrana, L. M.; Sillankorva, S. Entrapment of a Phage Cocktail and Cinnamaldehyde on Sodium Alginate Emulsion-Based Films to Fight Food Contamination by *Escherichia Coli* and *Salmonella Enteritidis*. *Food Res. Int.* **2020,** *128*, E-article ID 108791.
3. Anany, H.; Brovko, L. Y.; Arabi, T. E.; Grifith, M. W. Handbook of Natural, and Undefined: *Bacteriophages as Antimicrobials in Food Products: History, Biology and Application*. books.google.com; 2014; pp 69–88.
4. Boggione, D. M. G.; Batalha, L. S.; Gontijo, M. T. P.; Lopez, M. E. S.; Teixeira, A. V. N. C.; Santos, I. J. B.; Mendonça, R. C. S. Evaluation of Microencapsulation of the UFV-AREG1 Bacteriophage in Alginate-Ca Microcapsules Using Microfluidic Devices. *Colloids Surf. B: Biointerfaces* **2017,** *1* (158), 182–189.
5. Chibeu, A.; Agius, L.; Gao, A.; Sabour, P. M.; Kropinski, A. M.; Balamurugan, S. Efficacy of Bacteriophage Listex™p100 Combined With Chemical Antimicrobials in Reducing Listeria Monocytogenes in Cooked Turkey and Roast Beef. *Int. J. Food Microbiol.* **2013,** *167* (2), 208–214.
6. Choi, I.; Yoo, D. S.; Chang, Y.; Kim, S. Y.; Han, J. Polycaprolactone Film Functionalized with Bacteriophage T4 Promotes Antibacterial Activity of Food Packaging Toward Escherichia Coli. *Food Chem.*, **2021,** *346,* E-article ID 128883
7. Choińska-Pulit, A.; Mituła, P.; Śliwka, P.; Łaba, W.; Skaradzińska, A. Bacteriophage Encapsulation: Trends and Potential Applications. *Trends Food Sci. Tech.* **2015,** *45* (2), 212–221.
8. Cooper, I. R. A Review of Current Methods Using Bacteriophages in Live Animals, Food and Animal Products Intended for Human Consumption. *J. Microbiol. Methods* **2016,** *130,* 38–47.
9. Eiserling F. A. Bacteriophage Structure. In *Comprehensive Virology Volume 13: Structure and Assembly. Comprehensive Virology*; Fraenkel-Conrat H., Wagner R. R. Eds; Springer, Boston, MA, 1979; https://doi.org/10.1007/978-1-4684-3453-8_8
10. Gong, C.; Jiang, X. Application of Bacteriophages to Reduce *Salmonella* Attachment and Biofilms on Hard Surfaces. *Poult. Sci.* **2017,** *96* (6), 1838–1848.
11. Gouvêa, D. M.; Mendonça, R. C. S.; Soto, M. L.; Cruz, R. S. Acetate Cellulose Film With Bacteriophages for Potential Antimicrobial Use in Food Packaging. *LWT - Food Sci. Technol.* **2015,** *63* (1), 85–91.
12. Hoelzer, K.; Moreno Switt, A. I.; Wiedmann, M.; Boor, K. J. Emerging Needs and Opportunities in Foodborne Disease Detection and Prevention: From Tools to People. *Food Microbiol.* **2018,** *75*, 65–71.
13. Huang, C.; Shi, J.; Ma, W.; Li, Z.; Wang, J.; Li, J.;Wang, X. Isolation, Characterization, and Application of a Novel Specific Salmonella Bacteriophage in Different Food Matrices. *Food Res. Int.* **2018,** *111*, 631–641.
14. Hudson, J. A.; Billington, C.; Carey-Smith, G.; Greening, G. Bacteriophages as Biocontrol Agents in Food. *J. Food Prot.* **2005,** *68* (2), 426–437.
15. James, C. E.; Stanley, K. N.; Allison, H. E.; Flint, H. J.; Stewart, C. S.; Sharp, R. J.; Saunders, J. R.; McCarthy, A. J. Lytic and Lysogenic Infection of Diverse Escherichia

Coli and Shigella Strains with a Verocytotoxigenic Bacteriophage. *Appl. Environ. Microbiol.* **2001,** *67* (9), 4335–4337.

16. King, A. M. Q.; Lefkowitz, E. J.; Mushegian, A. R.; Adams, M. J.; Dutilh, B. E.; Gorbalenya, A. E.; Harrach, B.; Harrison, R. L.; Junglen, S.; Knowles, N. J.; Kropinski, A. M.; Krupovic, M.; Kuhn, J. H.; Nibert, M. L.; Rubino, L.; Sabanadzovic, S.; Sanfaçon, H.; Siddell, S. G.; Simmonds, P.; Varsani, A.; Zerbini, F. M.; Davison, A. J. Changes to Taxonomy and the International Code of Virus Classification and Nomenclature Ratified by the International Committee on Taxonomy of Viruses. *Arch. Virol.* **2018,** *163* (9), 2601–2631.
17. Liana, A. E.; Marquis, C. P.; Gunawan, C.; Justin Gooding, J.; Amal, R. Antimicrobial Activity of T4 Bacteriophage Conjugated Indium Tin Oxide Surfaces. *J. Colloid Sci.* **2018,** *514*, 227–233.
18. Lone, A.; Anany, H.; Hakeem, M.; Aguis, L.; Avdjian, A. C.; Bouget, M.; Atashi, A.; Brovko, L.; Rochefort, D.; Griffiths, M. W. Development of Prototypes of Bioactive Packaging Materials Based on Immobilized Bacteriophages for Control of Growth of Bacterial Pathogens in Foods. *Int. J. Food Microbiol.* **2016,** *18* (217), 49–58.
19. Maurice, C. F.; Bouvier, C.; de Wit, R.; Bouvier, T. Linking the Lytic and Lysogenic Bacteriophage Cycles to Environmental Conditions, Host Physiology and their Variability in Coastal Lagoons. *Environ. Microbiol.* **2013,** *15* (9), 2463–2475.
20. Meyer, A.; Greene, M.; Kimmelshue, C.; Cademartiri, R. Stabilization of T4 Bacteriophage at Acidic And Basic pH by Adsorption on Paper. *Colloids Surf. B: Biointerfaces* **2017,** *160,* 169–176.
21. Vonasek, E.; Le, P.; Nitin, N. Encapsulation of Bacteriophages in Whey Protein Films for Extended Storage and Release. *Food Hydrocoll.* **2014,** *37*, 7–13.
22. Wurtz, M. Bacteriophage Structure. *Electron Microsc. Rev.* **1992,** *5* (2), 283–309.
23. Xu, J.; Zhao, C.; Chau, Y.; Lee, Y. K. The Synergy of Chemical Immobilization and Electrical Orientation of T4 Bacteriophage on a Micro Electrochemical Sensor for Low-Level Viable Bacteria Detection via Differential Pulse Voltammetry. *Biosens. Bioelectron.* **2020,** E-article ID 11914.
24. Adams, M.J., Lefkowitz, E.J., King, A.M.Q. et al. Changes to taxonomy and the International Code of Virus Classification and Nomenclature ratified by the International Committee on Taxonomy of Viruses (2017). Arch Virol 162, 2505–2538 (2017). https://doi.org/10.1007/s00705-017-3358-5

PART II

NANOTECHNOLOGY-BASED IMPROVEMENTS IN FOOD PACKAGING

CHAPTER 7

SCOPE OF NANOTECHNOLOGY IN FOOD PACKAGING

RAMANDEEP KAUR and KAMALJIT KAUR

ABSTRACT

Nanotechnology contributes to novel techniques for food-packaging systems with benefits; it improved barrier properties, enhanced shelf-life and stability, reduced food spoilage, and improves health. Nowadays, monitoring of food during storage is the main focus of researchers related to nanotechnology research. Nanotechnology can improve traditional packaging systems by introducing ease of handling/transportation, protection against contaminants, alteration in the permeability of the package, and improvement in barrier, mechanical, and heat-resistant properties. However, there are some safety issues, which need attention. The quantity of nanoparticles that will be incorporated into the packaging films and leached into the food products is unclear. Moreover, what are the effects of exposure to various nanomaterials on our health is not known. Hence, this chapter elaborates on the benefits and applications of nanopackaging, types of nanomaterials, and their contribution to the monitoring of food security and safety issues.

7.1 INTRODUCTION

In recent years, consumers demand safe, wholesome, healthy, and ready-to-eat foods, which also follow strict food safety regulations. Packaging plays a vital role during the transportation and storage of the product while maintaining freshness and nutritional quality throughout the chain. Nanotechnology is emerging as a vibrant field with wide applications in the packaging of several food products. Nanotechnology comprises the fabrication,

Advances in Sustainable Food Packaging Technology. Megh R. Goyal, Preeti Birwal, Anand K. Pandey, & Ritesh B. Watharkar, (Eds.)

manipulation, and characterization of different materials under the range of 1–100 nm.[68] Biomaterials can be reformed at the nanoscale into nanofibers, nanoclays, nanoparticles, nanoplates, etc.[51]

The history of food nanotechnology started with the pasteurization process, which was developed by Louis Pasteur to kill the disease-causing microbes of 1000 nm. Various applications of nanotechnology in the food industry are flavor enhancement, development of specific delivery systems, shelf-life improvement, encapsulation, tracing, tracking, and brand protection. The leading role of food packaging is to minimize the wastage of food and to assist the consumer in the right purchase of food with the help of different techniques.[15]

In the era of globalization, consumers prefer foods that require less cooking time and are healthy and safe too. This makes the situation highly challenging for the food industry. These factors have also led to the improvement of packaging technologies. The important role of food packaging is to contain the product, protect it, and sell the product so that the product is delivered to the consumer with the best attribute. By using a suitable and protective packaging system in the food industry, the shelf stability of the food can be increased. This can be achieved by a smart food-packaging system incorporated with nanosensors that provide real-time information about the food package. Nanopackaging is helpful in releasing flavors, antioxidants, antimicrobials, nutraceuticals, and enzymes to enhance the shelf stability of food products.[68]

The application of nanotechnology ranges from designing, manufacturing, and processing the filling of polymers with nanoparticles.[51] In food packaging, the primary use of nanotechnology is to modify the numerous characteristics of traditional packaging materials, including easiness of handling and transportation, convenience to use, protection/preservation against microbial contaminants, communication of the product, alteration of the permeability of package, enhancement in barrier forming potential, mechanical and heat-resistance properties, development of antimicrobial surfaces, and nanobiodegradable packaging materials.[8]

This chapter focuses on the concept of nanotechnology for the use of nanopackaging material in the food industry.

7.2 BENEFITS OF NANOTECHNOLOGY IN FOOD PACKAGING

Food nanopackaging can control the moisture and temperature parameters and can improve the elasticity and gas barrier properties. Nanotechnology is being used to manufacture intelligent and active packaging systems for enhancing the shelf-life of foods so that they could be transported to far-off places. Active packaging provides protection and preservation by incorporating the active element into a food-packaging material, which can interact with external and internal factors, thus enhancing the quality, food safety, and shelf stability of the food product.[39] Intelligent packaging assures better functionality of packages related to marketing and communication by incorporating the indicators in the food package. It delivers a dynamic response about the food quality and behaves as a protector in opposition to fraudulent replication.[48]

Smart packaging systems with nanocomposites/antimicrobials/nanosensors are manufactured in a manner to detect the spoilage of food and to release the nanoforms of antimicrobials to enhance the shelf-life, so that product can be kept for a longer period in supermarkets before its sale.[34] Packaging material containing nanocomposites is helpful in examining the internal and external environment or food conditions, pellets, and containers over the whole chain of food supply. Nanosensors can detect gases released from food and can give an alert indication to customers by varying the food color. Further, films of silicate nanoparticles can reduce the flow of oxygen and moisture into the package and out of it to maintain the freshness of food. They can inhibit the growth of mold in the refrigerators. Nanotechnology is helpful in reducing packaging waste. It can identify the microbial pathogens in food products in 2–7 days and also can be used for the detection of toxins, pesticides, and spoilage.[39]

The significant use of nanotechnology is also in the production of biodegradable materials. These films can degrade rapidly under natural conditions. Biodegradable polymers can be categorized into three categories, such as (1) polymers extracted directly from the biomass (proteins, polysaccharides, polynucleotides, polypeptides); (2) polymers made by chemical synthesis of mixed biomass and bio-based monomers, such as poly-lactic acid or bio-polyester; and (3) polymers produced by microorganisms or genetically modified bacteria, such as xanthan, poly-hydroxy butyrate, pullan, and bacterial cellulose.[56]

7.3 PROSPECTS OF NANOTECHNOLOGY IN FOOD PACKAGING

The functional additives, which are added to food packages, include a number of nanomaterials, such as nanoclays, nanozinc oxide, nanotitanium dioxide, titanium nitride, silver, and gold nanoparticles. Nanotechnology has budding applications to enhance food safety in the food sector, to improve packaging, and resulting nutrition and processing.[30] The important applications of nanotechnology are as follows:

- Application as antimicrobial ingredients for food packaging.
- Enhancement of the shelf stability of the food.
- Safety of food.
- Preparation of flavoring, coloring, and nutritional additives.

Nanomaterials (nanosensors) are also used for the detection of food spoilage.[47] Nanotechnology-based food packaging can be categorized into the following three categories.

7.3.1 IMPROVED PACKAGING

The improved packaging system is helpful in providing better temperature resistance, gas transmission resistance, and humidity resistance to packaging material when nanoparticles are mixed with polymer chains. The United States Food and Drug Administration (FDA) approved the use of nanocomposites for food materials. Nylon nanocomposites are good barriers to oxygen and carbon dioxide flow. In food packaging, nylon nanocomposites are being used to preserve freshness and prevent off-flavor development in packaged foods. Multilayer polyethylene terephthalate (PET) bottles are used as nanocomposites for beer and alcoholic beverages.[24]

Graphene nanoplate-based nanocomposite materials provide improved food-packaging applications which have heat resistance and barrier properties. Carbon nanotubes and nanofibers have electrical and mechanical properties, but they are not used in food packaging due to their high cost and difficulty in processing dispersions.[5] The starch-based biodegradable polymers with better gas barrier and mechanical properties were developed by the addition of bentonite nanoparticles. Bentonite is incorporated into polymers to prevent gas permeation.[6] Nanoclays can be embedded and incorporated to improve the barrier attribute of packaging material. The layered silicates comprising two-dimensional layers are commonly used in nanocomposites. These are 1 nm thick and several microns long that depends upon the kind of silicate. The tortuosity of the path for the penetration of the molecule increases with the incorporation of nanoclays and also provides excellent barrier properties.[49] The extensive penetration of the polymer is enhanced with exfoliated nanocomposites, randomly dispersed polymers, and de-laminated layers of clay.[41] Exfoliated nanocomposites possess excellent properties owing to the interaction between the clay and polymer.[3]

Montmorrillonite, the most commonly used kind of clay filler, has two tetrahedral silica layers (e.g., hydrated clay layered with aluminium-silicate having an edge-shared octahedral aluminum hydroxide sheet). The compensation of the imbalanced negative charges on the surface is done with exchangeable cations such as Ca^{2+} and Na^{+}. Weak electrostatic forces link the parallel layers of ions.[74] Montmorrillonite is considered having a negative charge on the surface. It is a vital parameter, that can explain the spacing present in between the layer of equilibrium. The charge of the layer varies as it is different for different layers and should be estimated as an average over the whole surface of crystal. It is an efficient support filler, because of its large aspect ratio (50–1000) and wide surface area.[75]

7.3.2 ACTIVE PACKAGING

In food preservation, active packaging plays an important role besides acting as an inert barrier. Active packaging system releases desirable compounds such as antioxidant and antimicrobial compounds, which results in enhanced shelf stability of food. This packaging system also includes antimicrobials, oxygen scavengers, and enzyme immobilization. It is also beneficial in controlled-release packaging, where nanocomposites are utilized in delivery systems, which improves the migration of functional additives, such as probiotics, vitamins, and minerals into the food.[27]

Active packaging contains the use of metal and metal oxide nanoparticles such as silver-based nanofillers, copper oxide, zinc oxide, and titanium dioxide, owing to their antimicrobial and antioxidant properties, and are used in the form of nanocomposites for food packaging. Nanomaterials interact directly with food and its environment, which further enhanced retention of the nutrients. The nanomaterials such as carbon nanotubes, nanomagnesium oxide, nanotitanium dioxide, and nanocopper oxide act as antimicrobials. Silver nanoparticles are widely used in food packaging as antibacterial agents. Titanium oxide nanoparticles can be used in food packaging to kill food spoilage bacteria. In addition, titanium oxide and zinc oxide nanoparticles could be used in packaging materials to keep the package hygienic.[36]

7.3.3 INTELLIGENT (SMART) PACKAGING

The main objective of intelligent or smart packaging is to analyze microbial and biochemical changes in the products. In a smart packaging system, the specific pathogens and gases which evolve due to food spoilage can be easily detected. This technique has been used as a tracking device for food safety. Major food industries like MonoPrix, British Airways, and Nestle are using chemical sensors to sense the change in color.[43]

In smart packaging of foods, nanoparticles are incorporated, which act as reactive particles inside the package, are used for sensing changes in food due to the action of microorganisms or biochemical changes, and provide information about the packaged product. The nanosensors are sensitive to external changes and help in the communication and identification of the product to upkeep its quality and safety. The smart food-packaging system comprises oxygen and spoilage indicators, identification of products, and traceability. With the help of nanosensors, the shelf stability of the product can be determined. For instance, enzymes incorporated with gold nanoparticles are used for microbes detection, gas sensors describe the condition of the food, and nanofibrils (perylene-based fluorophores) tell about meat and fish spoilage by sensing amines. Moreover, nanobarcodes are used for security and tagging. Therefore, smart sensors are valuable for the customers for identification of food quality and for producers to encourage fast distribution and authentication of the food. Nanocantilevers are rapid biosensors, such as micromechanical oscillators, which are used for the detection of bacterial growth of *Escherichia coli*.[45]

Radiofrequency identification (RFID) tags are known as an advanced form of intelligent packaging, which can recognize and trace a product with a unique tag that emits radio waves. RFID is further categorized into four categories: active, semi-active, passive, and semi-passive, depending on the communication. These devices coupled with pallets, containers, boxes, and articles thus can be identified and recognized. RFID tags can also read away from the line of sight and several meters, having a reading range of 91 m. It also contained a battery, which enables it to communicate. There is no power supply in passive tags; thus, these are unable to respond unless the RFID reader emission is activated. Sufficient power produces for integrated circuits. On produced power, in the label, the energy reflects the reader. The transmission range can reach up to 6 m.[67]

RFID systems are also categorized based on the frequency range: low frequency (from 125 to 134.2 kHz), high frequency (13.56 MHz) ultrahigh-frequency (from 868 to 956 MHz), and active/microwave frequency (2.45 GHz). RFID technologies are grouped among the systems named automatic identification or Auto-ID. For the implementation, RFID is still an expensive alternative having some obstacles, that is, low power, narrow bandwidth, and short range. RFID prints labels directly on plastic or paper instead of silicon, but, investments in the components such as software, controllers, networks, cables, connectors, readers, antennas, tags, sensors, and implementation processes are costly. Although printed circuits cover a large area; it has low resolution compared to traditional circuits.[67]

7.3.3.1 TIME AND TEMPERATURE INDICATORS

During long-distance transportation of food material chill storage is required, which is a critical control point, therefore to maintain the quality and safety of chilled and frozen foods time–temperature indicators (TTIs) should be used. The attractive features of TTIs are that they are efficient, affordable, cheap, and simple. Therefore, nowadays, TTIs are used to monitor and translate the food quality for consumers.[60]

7.3.3.2 INTEGRITY INDICATORS

Gas indicators help to know the gas composition present in the package, and it helps in controlling the toxic composition of gases produced due to the decomposition of food inside the food container. In the presence of gas indicators, color changes take place due to enzymatic or chemical reactions. Indicators are always in direct contact with the surrounding environment of the food. Different types of nanobased indicators are used in food products. The existence of oxygen may show that the package was sealed wrongly, leaking, or maybe tampered. The detection of carbon monoxide in conventional and modified atmosphere packaging has gained significant attention in the food industry.[46]

7.3.3.3 FRESHNESS INDICATORS

These are one type of label that directly tell or indicate product quality. These indicators help in the detection of the very first change inside the package, such as gas composition and pH. The change, which takes place within is shown by indicators and then, transformed into a response, generally a color response, which is simply measured and related to the freshness of food.[10]

7.3.3.4 ACTIVE TAGS AND TRACEABILITY

Usually, RFID is used in active packaging. Active tags are electronic information-based systems in which, radiofrequency is used to identify and trace the object that transfers data from a tag attached to the object. Furthermore, RFID is an enhancement in the earlier barcodes and manual tracking systems. In addition, it is robust and works under extreme pressure and temperature conditions and also has a wide reading range, which can read away from 100 m.

Nanotechnology can also enable the packaging of sensors to integrate cheap tags of RFID. These nanoenabled tags of RFID are flexible, smaller, and can easily be printed on thin labels. This increases the versatility of tags and also reduces their cost during the process of production.[1,2]

TABLE 7.1 Nanotechnology and Food Systems.

Applications	Properties	Reference
Delivery system	• Act as a vehicle for functional compounds from destruction and control release of compounds.	[78]
Food additive	• Utilized to modify nutritional quality, functionality, flavor, and texture of products. • Detection of microbes and pathogens. • Change the taste and appearance of the food by using nanodeveloped colors.	[58]
Nanoemulsions	• Using nanoemulsions in food products can make possible the utilization of less fat with no compromise in creaminess and provide a healthier option to consumers • Products of this type such as ice cream, spreads, and low-fat nanostructured mayonnaise.	[13]
Nanoencapsulation	• Delivers the fat-soluble compounds. • Prevent the degradation during processing or in gastrointestinal tract. • Have site-specific release and other food compatibility. • More residence time and more absorption. • Retain the bioactive compounds (lipids, proteins, antioxidants, carbohydrates, and vitamins).	[14]
Nanoceuticals	• As bioactive proteins to impart health benefits. • Used to enhance nutraceuticals (vitamin D, probiotics, omega-3 fatty acid, minerals, and carotenoids) stability during processing, storage, and distribution.	[14, 57]

7.4 DIFFERENT TYPES OF NANOMATERIALS (FIG. 7.1)

7.4.1 NANOCOATING

Nanocoating of edible substances can be used as thin films in food placed in between the components of food to provide a barrier against gas and water vapor transfer. The edible coatings are made with different products such as polysaccharides (cellulose derivatives, alginates, pectins, starches, chitosan, and other polysaccharides) and lipid compounds (waxes, animal fat, acylglycerols, vegetable fat, and fatty acids). The fat-based films have exceptional moisture barrier properties and are used as coating agents for confectionery products to add gloss. Waxes are mostly used for fruit and vegetable coating to slow down respiration and moisture evaporation.[11]

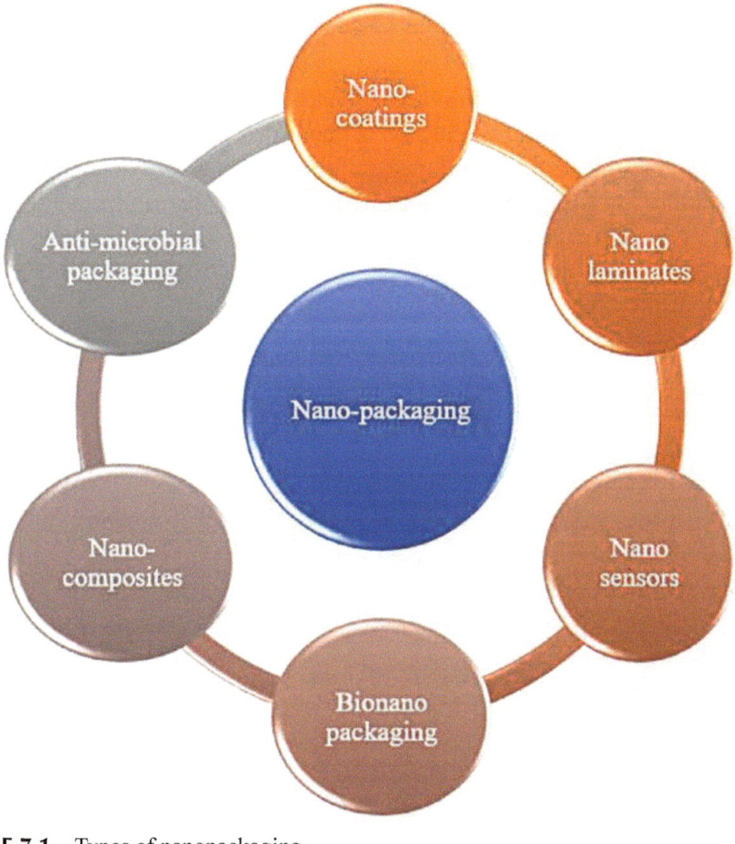

FIGURE 7.1 Types of nanopackaging.

Conversely, few studies are available on the addition of nanoparticles in coating films to enhance their physical properties. The clay montmorillonite has been added to the pectins to lower the diffusion of oxygen. Similarly, a nanocoating developed by montmorillonite and gelatin has been used for the enhancement of the physical properties.[80]

The inorganic nanofillers are also available which improved the retention of color, texture, acids, sugars flavor, and enhanced appearance and strength while storage and shipping and thus reduce spoilage. Nanoparticles have been also used as additives and antimicrobial carriers. Further, they can be used for the stabilization of the additives and properly manage their diffusion into the different regions of the food, that is, bulk vs surface of the food system. This is also for the long-term storage of foods and imparts specific desirable features, such as flavor to the food system.[22]

In food packaging, numerous kinds of nanocoatings applications are known such as inside and outside nanocoating packages, and multilayer laminated packaging films sandwiched with the nanocoating. Nanocoatings have excellent barrier properties and are used as edible films and coatings on extensive foods types, which act as gas, lipid, and moisture barriers. The vacuum-based aluminum coatings with organosilanes are used on plastic films, food surfaces, and containers of beverages. The nanosurface biocides, exchange oxide evaporated films, and polyvinylidene fluoride coated films, also contribute to the moisture and oxygen barrier and preservation of aroma, good printability, improved storage life, and reduced cost of laminated machinability and production.[11]

They are environment friendly, without dioxins and are made through incineration, and provide good optical and mechanical properties. Nanosurface biocide includes metals MgO, TiO_2, ZnO, Ag, and $AgNO_3$, which are attained by nanomaterial incorporation, and have antimicrobial properties on the surface package. The food is maintained under hygienic conditions, which is provided in reusable food containers in freezers and refrigerators, such as boxes and crates. The well-established technologies of coating are immobilization utilizing non-covalent migratory, covalent non-migratory, deposition in a layer-by-layer manner, and chemistry of photo grafting that are embedded for release in a controlled manner.[76]

7.4.2 NANOLAMINATES

A nanolaminate comprises two or more layers of materials that have nanometer dimensions and chemically or physically bond to each other. It uses the deposition technique in a layer-by-layer manner where the surfaces which are charged are coated by the interfacial films containing a layer of nanodimensions of different materials. Nanolaminates have a number of advantages in the production of edible films and coatings over conventional methods and may therefore have extensive applications in the food industry.[78]

Several kinds of adsorbing substances have been utilized to develop several layers of laminates, such as colloidal particles (droplets, vesicles, micelles), charged lipids (surfactants, phospholipids), and natural polyelectrolytes (proteins, polysaccharides). There is also the possibility to add active functional compounds including enzymes, antioxidants, anti-browning, anti-browning, antimicrobials, added colors, and flavors into films.[18]

These functional compounds enhance the quality and shelf stability of coated foods. The basic characteristics of film-forming materials define the functional properties of laminated films. The nanolaminated coatings are also prepared by using edible compounds (lipids, carbohydrates, proteins) with easy processing operations such as washing and dipping. The thickness, composition, properties, and structure of the multi-layered laminate made can be controlled in several ways, such as by changing the adsorbing compounds types in the dipping solutions, the type, and a number of dipping steps used, the way that the material is to introduce to the several dipping solutions and environmental conditions used (as temperature, dielectric constant, ionic strength, pH, etc.).[59]

7.4.3 NANOSENSORS

Nanosensors are devices that are used for the detection of the physical amount of the substance and convert it into observable and readable signals. They are also utilized to control the inner environment of the food materials and their properties, which are sensed by the sensors. Many studies reported nanobiosensors that have been previously developed and commercialized to detect product tampering, chemical contaminants, spoilage, pathogens, and track products during food processing.[52]

There are several techniques for nanosensors development, that involves molecular self-assembly, top–down lithography, and bottom–up assembly approaches. Presently, nanosensors devices can be categorized into several categories: nanoelectromechanical systems (NEMS), nanosystems: cantilevers, nanowires, nanosensors, nanoprobes, nanoparticle-based sensors, and nanostructured materials, for example, porous silicon. Table 7.2 summarizes

TABLE 7.2 Sensors: Types, Properties, and Applications.

Type of sensor	Manufacturing	Properties	Applications	Reference
Chemical and biosensors	Developed a miniaturized gas ionization detector based on CNTs.	Gas sensors based on nanotubes.	Nanotube molecular wire-based chemical sensors detect gaseous molecules like NH_3 and NO_2 in the environment.	[50]
Deployable nanosensors	"Chemical lab-on-a-chip" detector.	Security of homeland and during war times.	One sensor is the Sniffer STAR (a nanotechnology-based chemical sensor that can be integrated into a microunmanned aerial vehicle).	[31]
Surface plasmon resonance sensors which are localized and propagating.	Surface plasmon resonance (SPR) biosensors, optical biosensors.	Detect biological and chemical entities.	A method to increase the wavelength shift detected from localized surface plasmon resonance (LSPR) bioassays using antibodies labeled with gold nanoparticles for biotechnology and environmental fields.	[29]
Physical sensor	Mounted by a single particle on the carbon nanotube end (CNT) and then an electrical charge is applied to it.	Electrical and mechanical properties of carbon nanotubes.	Used to determine the mass of individual biomolecules.	[38]
Sensors based on Bulk nanostructured materials	Catalytic behavior of nanoparticles.	The high surface area makes nanoparticles appropriate for immobilizing biomolecules, polymers, and biomaterials that generate composite materials with tunable surface properties.	Applications in electrochemical and biosensors.	[42]

TABLE 7.2 *(Continued*

Type of sensor	Manufacturing	Properties	Applications	Reference
Sensors based on CNTs	The electrical conductance of semiconducting Single-walled carbon nanotubes (SWNTs) (band gaps ~0.5 eV) can vary by orders of magnitude under electro-static gating or doping. This is the basis for the nanotube field effect transistors.		Detection of molecules in a gas mixture was demonstrated with the multiplexed nanotube sensors.	[27]

the major types of nanosensors, their properties, and applications. In food analysis, based on applications nanosensors can be divided into optical nanosensors, electrochemical nanosensors, nanoparticle-based nanosensors, and APTA-sensors.[40]

Aptasensors are biosensors, which comprise target-recognition elements—aptamers and nanomaterials—which act as signal enhancer or transducers. Aptamers are peptide molecules or single-stranded nucleic acids of natural and synthetic origin having a size less than 25 kDa. These are very specific and selective towards their target compound (viruses, toxins, proteins, ions) owing to their accurate and definite three-dimensional structures. Due to its selection and production through SELEX- Systemic Evolution of Ligands by Exponential enrichment (in vitro combinatorial molecular technique), aptamers are knowns as synthetic antibodies.

The aptamers have dissociation constants in the picomolar or nanomolar range. These are broadly used in the fabrication of aptasensors as recognition compounds. There are numerous kinds of nanomaterials, which can be used in aptasensors such as magnetic nanoparticles, carbon nanoparticles, semiconductor nanoparticles, nanoclusters, and metal nanoparticles. The various types of transducing systems have been engaged in aptasensors for food assessment. The aptasensors uses nanoparticle. On the basis of the detection systems, aptamers can be categorized into electrochemical and optical systems.[68]

Nanoporous silicon comprises a complex silicon threads network having a thickness of 2–5 nm and shows 500 m^2/cm^3 of internal surface. The tremendously tiny pores provide strong luminescence to the material at room temperature. The unique property of the emission is the wavelength of the emitted light which depends on the material porosity. This property leads to the novel design types of gas sensors, which can be monitored by observing visually a color change.[28]

In the past few years, the development of a multitude of solid-state, robust, and simple sensors has been reported. The operation is based on the transduction of the bound analyte to a computable signal at the sensor's active surface, resulting in the alteration of temperature, resistance, and capacitance of the active element.[44] In chemical sense, the recognition at the molecular level can be achieved by specific interaction similar to the interaction of the key–lock model, as they are identified from microstructured man-made devices and natural biological systems. A complete understanding is required about the chemical compositions and electronic and optical properties of the involved materials with particular prominence on their interfaces so that the specific detection of ions and molecules can be performed reliably. An increase in interest has been noticed in organic structure, molecular, bioelectronic devices, and chemical sensing with potential future applications. The latest research efforts focus on appropriate choices of prototype materials to be microfabricated, consistent technologies in the making and modification of materials, new experimental approaches, and adequate theories of interfaces to produce micro-structured devices up to atomic scale.[26]

7.4.4 BIONANO PACKAGING

Bionano packaging is also known as biodegradable packaging films that are used to control the migration of moisture and gases inside the food product and to preserve the quality, nutritional, and sensory properties inside the food products.[72] Biobased packaging is supposed to be more eco-friendly as compared to conventional packaging. Bio-based packaging, similar to other packaging materials, provides a barrier over the food product and protects them from the growth of unwanted microorganisms thereby enhancing shelf-life. The bionano films are able to decompose with the action of living organisms and their end products are completely natural which include, carbon dioxide, water, and biomass. This packaging uses a minimal amount of fossil fuels to develop the package and also uses renewable sources, after that covering disposal energy can be done with incineration.[19]

Generally, biodegradable plastics are utilized in the formation of biobased packaging. These require a minimum of one step for the degradation process with the naturally found metabolism of the organism. With suitable conditions of moisture, oxygen, and temperature accessibility, this biodegradation leads to the disintegration of the plastics without the release of any toxic and environmentally dangerous compounds. Different types of biodegradable films are used. There are various organic nanomaterials mainly categorized into three categories as discussed above: nanomaterials of plastic polymer, protein-based nanomaterials, and polysaccharide-based nanomaterials, that provide a matrix of biopolymers for nanocomposite materials.[12]

Polyvinyl acetate (PVA) is made by the process of hydrolysis and degraded easily in the water through microorganisms. It has broadly been included in other compounds which are polymer-based to improve the mechanical properties which are related to its compatible structure and hydrophilic properties (such as mechanical high hydrophilicity, biocompatibility, solvent resistance, and mechanical performance).[23] Chitosan and PVA hydrogels containing lignin nanoparticles were developed by a freezing–thaw procedure.[79]

Polyhydroxyalkanoates (PHAs) have also received great consideration as biodegradable and biocompatible thermoplastics for developing food-packaging materials. The main compound is poly(3-hydroxybutyrate) of the PHAs, which is partially crystalline having high melting temperature and rigidity. With the insertion of 3-hydroxyvalerate, the copolymer is prepared which has low crystallinity, hence enhancing the handling properties of the PHAs packaging film.[23]

The blends of poly lactic acid and PHA at different weight ratios of these compounds (0/100, 25/75, 50/50, 75/25, 100/0) by using a melt compounding process were developed. The mechanical properties of polylactic acid have been improved and the degradation process has been accelerated by adding clays to the polylactic acid.[7] The mixtures of polylactic acid and organophilic clay in chloroform have been used which results in the enhancement of crystallization capacity and Young's modulus.[53] Conversely, the glass transition temperature rises only to some extent with raising clay content. This may be because of the microcomposites structure instead of the nanocomposite structure. The melt extrusion was used to prepare polylactic acid/layered silicate nanocomposites and possess outstanding enhancement of material properties in both melt and solid states related to the without clay.[71]

Starch is a plant carbohydrate used as a renewable source, because of its availability from several plants and huge production naturally and low cost.[25] There are numerous possibilities for utilizing starch as a packaging material.[37] However, starch does not form films without any chemical modification due to a lack of appropriate mechanical properties. The thermoplastic material can be manufactured by treating the starch in an extruder with both mechanical and thermal energy. The plasticizers are incorporated in the production of thermoplastic starches, as they decrease the intramolecular hydrogen bonds and maintain the product stability. Corn is an ultimate source of starch to make bioplastics, however, new research studies are focusing on the utilization of the starches from wheat, rice, potato, oat, soy, and barley sources for bioplastics.

The graphene oxide-filled starch and lignin nanocomposite was developed by Aqlil et al.[4] This study showed that grapheme had a high effect on the mechanical properties and also lessens the water vapor permeability and moisture uptake of the prepared film. The film of starch having carbon nanotubes of multiple walls, without or with hydroxylation was prepared. It was found that the hydrophobic properties of the film were highly enhanced with the addition of a nanotube. The potato starch nanocomposite films were also prepared and mechanical, thermal, and barrier properties were evaluated.[54] The results display that the temperature of the glass transition, the film tensile strength, and the melting point of the film were enhanced. The transmittance of UV-A, UV-B, and UV-C along with visible light waves and permeability for water vapor were decreased.

7.4.5 NANOCOMPOSITES

Polymer nanocomposite contains multiphase solid material with nanoscale fillers having at least one dimension at one phase less than 100 nm which is present within a polymer matrix.[70] Due to the dispersion of nanometer-sized particles, in the polymer matrix, these compounds show highly improved optical, thermal, and mechanical physicochemical properties, as compared to the pure polymer and conventional compounds having very less filler loading, around 5% and below. Many research studies have proved the positive impact of nanocomposite polymers on the barrier properties after filling with nanofillers. The nanocomposite packaging has a future prospective as an advanced and improved technology of food packaging, that preserves or reserves the food quality features and improves the shelf-life of packed products of food.[20,61] Polymer nanocomposites mainly comprise the polymer matrix, plasticizers, compatibilizers, and nanofillers.[61]

The transmission electron microscope (TEM) and scanning electron microscopy (SEM) are used to study the surface morphology of nanocomposites polymers. By using TEM, internal structure, spatial distribution, and dispersion of nanofillers inside the matrix of polymer were evaluated. Furthermore, X-ray spectroscopy (XRD) is used to study the chemical compositions of the packaging films. The crystallite size, crystallinity, and composite structures include intercalated, tactoid, and exfoliated structures of nanocomposite of polymers and these clay composite of polymers can be evaluated by XRD. The Fourier transform infrared spectroscopy (FTIR) studies the functional group and chemical changes by the interaction of the polymer with nanofillers. It indicates the vibration peaks of a specific chemical group at a particular frequency. Moreover, numerous techniques for analytics have been utilized to illustrate the films of polymer nanocomposites such as atomic force microscopy (AFM) technique, thermogravimetric analysis (TGA), and stricture analysis by nuclear magnetic resonance (NMR).[61,62,66]

The proof of the presence of nanofillers inside the polymer matrix has to be revealed for the significant improvement in the nanocomposites performance properties for various applications in food packaging, which includes packaging of cheese, boil-in-bag foods, food grains, processed meat products, confectionary, bakery products, beer bottles, carbonated beverages, fruit juices, and dairy products. There are several reports existing on the potential nanocomposite application in the industry of food packaging. However, they mostly deal with the preparation of nanocomposites with thermoset and thermoplastic polymers. Additionally, the use of nanocomposite used as a food-packaging material is limited due to the comparatively more cost of production and low feasibility for production at a mass level. Moreover, concerns aroused about the safety of food products from the government and the consumer related to the utilization of nanomaterials as material for packaging.[73]

7.4.6 ANTIMICROBIAL PACKAGING

Antimicrobial packaging of nanocomposites is the most effective system of smart packaging which provides a prolonged shelf-life to the food by decreasing and inhibiting spoilage and pathogenic microorganisms which contaminate the foods.[65] Polymer antimicrobial food-packaging material is particularly beneficial owing to the high surface-to-volume ratio of nanofillers and it also improves the surface reactivity of the nanosized antimicrobial compounds as compared to bulk equivalent, and making them capable to kill or to inactivate the microorganisms.[62] Packaging films having antimicrobial properties are made by mixing the antimicrobial substance into the polymeric matrix through the polymer processing. Polymer nanocomposite materials have been used as microbial agents for antimicrobial activity as agents,[32] antimicrobial food, and carriers,[9] and growth inhibitors in food-packaging films. Metal and its oxide nanoparticles show high antimicrobial potential. These are mainly used in antimicrobial food packaging due to their large area of surface and specificity.

The antibacterial properties of nanoparticles made up of metal and their oxide including zinc, titanium, copper, and silver nanoparticles have received significant attention in the field of food packaging as well as in biotechnological and biomedical applications.[65] Among these, nanocomposite of biopolymers such as agar,[62] chitosan,[63] and gelatin[35] and silver nanoparticles (AgNPs) showed strong antimicrobial activity against the Gram-positive and Gram-negative bacteria. Because of the robust antimicrobial activity of nanoparticles of copper, they have gained effective attention for their application in the packaging of food. Nanomaterials that are copper-based own insignificant sensitivity to tissues of human but are highly sensitive to microbes in comparison to silver.

Shankar et al.[66] made nanocomposites of agar polymer and other six nanoparticle types, which displayed increased antimicrobial activity for the Gram-negative and positive pathogens of food. Oxides of metal including MgO, ZnO, and TiO_2 have been utilized for preparing the films for antimicrobial packaging due to their outstanding antimicrobial activity along with higher stability. The films of polymer-based nanocomposite made from nanoclay which is modified organically have been established for their significantly high antimicrobial activity.[77] Rhim et al.,[63] prepared films of nanocomposites of chitosan/clay with nanoclays of two types, that possess antimicrobial activity to kill microorganisms that are pathogenic. It was noticed that the film of nanocomposite prepared with MMT which is organically modified (Cloisite® 30B) expresses antimicrobial activity to kill bacteria which are Gram-positive bacteria like Listeria monocytogenes and Staphylococcus aureus while the MMT which occurs naturally do not have any antimicrobial activity. Although, chitosan on its own has antimicrobial activity against numerous varieties of microbes, and can also have applications as insect repellent, gas permeable, suitable microwave heating, anti-sticking, antifogging, light absorbing, color providing, and antioxidant releasing; films for packaging prepared by nanocomposites of polymers are anticipated to gain increased attention in the near future.[55]

7.5 CONTRIBUTION OF NANOTECHNOLOGY FOR MONITORING OF FOOD SECURITY

Nanotechnology includes the design, study, synthesis, creation, manipulation, and application of devices, materials, and functional systems by controlling and exploiting properties and phenomena of matter on a very small scale, typically between the length of 1 and 100 nm.[8] The newly developing packaging technologies will focus on the progress and development of nanomaterials and nanoparticles; these may include, nanotubes, nanofibers, nanoparticles, fullerenes, nanosheets, and nanocylinders. The unique electronic and optical properties of nanomaterials permit the growth of electronic devices of a new generation, for example, nanotransistors to build future nanomemory and nanoprocessors, nanosensors, and nanobattery.

Nanotechnology is a powerful interdisciplinary tool for developing intelligent packaging systems.[29,32] The technological advancement and integration of nanosensors, sensors, and indicators are essential for developing intelligent packaging. A nanosensor/sensor estimates only a few aspects, while measurement and display are integrated by an indicator. The sensors and nanosensors must be linked to a single device for the receptor's signal transduction, whereas an indicator provides directly the semiquantitative or qualitative information of quality related to the visible change.[1]

Nanotechnology promotes nanosensor application in food packaging for the control over the quality, at different steps of the logistic process, and to confirm that the consumer gets a quality product. Nanotechnology using intelligent packaging can help in authentication, locating, and tracking features of products to avoid adulteration, falsification, and product diversity for product planned for a specific market.[5,8] Some concerns for nanotechnology food consumers include the uncertainty of nanoparticles' behavior in the body and their toxic effects. For this, it is important to establish regulations and protocols related to the implication of intelligent packaging and food security. The use of nanotechnology depends on geographical location and the market as consumers residing in China show more willingness towards new technologies, while Switzerland consumers are less responsive to technology.[1]

7.6 SAFETY ISSUES IN NANOPACKAGING

Consumer safety is necessary, and so is the assessment of the migration of compounds of packaging in food and to estimate of the probable hazards for measurement of complete risk. Although, to date, few studies have reported the impact of nanopackaging on human health and the potential interaction of nanomaterial with food components,[33] the migration limit of nanocompounds (10 mg) which can move into foodstuff from packaging material in contact (Commission Regulation (EU) No. 10/ 2011) is 10 mg in Europe. For 1 kg of food package; this equals 60-mg substance migration per kg of food.

The movement of Ag and Cu from nanocomposites, which is used for antimicrobial properties in food packaging, has been also studied.[17] This study concluded that nanofiller percentage, particle size, contact time, and temperature were the most critical parameters in driving migration. A model to study the movement of particles from packaging to food has been developed and was an effective interpreter of the migration level of nanosilver and of nanocopper to some extent into food. Further, development and confirmation could be advantageous to the industry by decreasing the cost and time associated with studies based on migration.

Silver migration from three kinds of nanocomposites into foodstuff, has been studied.[21] The results presented that silver migrated into food and the highest level of migration was found in the acidic food. Furthermore, heating also caused an increase in migration, especially heating in microwaves persuades significant migration compared to classical ovens. It was proposed that silver migration could happen via two mechanisms: the silver nanoparticles' detachment from composites or the oxidative silver ions dissolution. Principles for regulation of migration limits of plastics of different types into the matrix of food are usually applied for nanoparticles. Although the variation in the physiochemical characteristics of compounds of nanosize, which leads to different limits of safety for migrations.[68]

7.7 SUMMARY

Nanotechnology has great potential to modify the properties of conventional packaging. It is an innovative invention in the field of food packaging with the enhancement of its properties. Additionally, nanotechnology can be utilized to provide active improved, and smart properties to the systems of food packaging, such as improved moisture and gas barrier properties, antimicrobial properties, oxygen scavenging ability, and communication about the condition of food inside the package. Therefore, nanomaterial not only protects foods against environmental factors but also incorporates desirable properties into the packaging system so as to enhance the shelf stability of foods and preservation of nutrients. Smart packaging, such as nanosensors, could have been used to make the consumers aware of the point of time when the food is not safe for consumption. Sensors can alert before the food spoils or can update consumers on the exact quality of the food inside packages.

However, there are some safety concerns about the use of nanotechnology for food materials. The properties and safety of source material are well known. Conversely, nanosized materials show different properties from those they are made up of. As less literature is available regarding the occurrence of the migrations of nanomaterial from packaging to food products. Hence, significant research is still required to assess the potential toxicity of nanotechnology products, as well as the environmental safety of their use.

KEYWORDS

- **active packaging**
- **intelligent packaging**
- **nano-packaging**
- **safety issues**
- **smart packaging**

REFERENCES

1. Abad, E.; Palacio, F.; Nuin, M.; De Z'arate, A. G.; Juarros, A.; Gomez, J.; Marco, S. RFID Smart Tag for Traceability and Cold Chain Monitoring of Foods: Demonstration in an Intercontinental Fresh Fish Logistic Chain. *J. Food Eng.* **2009,** *93,* 394–399.
2. Abad, E.; Zampolli, S.; Marco, S. Flexible Tag MicroLab Development: Gas Sensors Integration in RFID Flexible Tags for Food Logistic. *Sens. Actuators B Chem.* **2007,** *127,* 2–7.
3. Adame, D.; Beall, G. W. Direct Measurement of the Constrained Polymer Region in Polyamide/ Clay Nanocomposites and the Implications for Gas Diffusion. *Appl. Clay Sci.* **2009**, *42,* 545–552.
4. Aqlil, M.; Moussemba Nzenguet, A.; Essamlali, Y.; Snik, A.; Larzek, M.; Zahouily, M. Graphene Oxide Filledlignin/Starch Polymer Bionanocomposite: Structural, Physical, and Mechanical Studies. *J. Agric. Food Chem.* **2017,** *65,* 1057–1058.
5. Arora, A.; Padua, G. W. Review: Nanocomposites in Food Packaging. *J. Food Sci.* **2010,** *75,* 43–50.
6. Avella, M.; De Vlieger, J. J.; Errico, M. E.; Fischer, S.; Vacca, P.; Volpe, M. G. Biodegradable Starch/Clay Nanocomposite Films for Food Packaging Applications. *Food Chem.* **2005,** *93,* 467–474.
7. Bandyopadhyay, S.; Chen, R.; Giannelis, E. P. Biodegradable Organic-Inorganic Hybrids Basedon Poly. *Poly. Mater. Sci. Eng.* **1999,** *81,* 159–160.
8. Berekaa, M. M. Nanotechnology in Food Industry; Advances in Food Processing, Packaging and Food Safety. *Int. J. Curr. Microbiol. Appl. Sci.* **2015,** *4,* 345–357.
9. Bi, L. L.; Yang, G.; Narsimhan, A. K.; Bhunia, Y.; Yao. Designing Carbohydrate Nanoparticles for Prolonged Efficacy of Antimicrobial Peptide. *J. Control. Release* **2011,** *150,* 150–156.
10. Biji, K. B.; Ravishankar, C. N.; Mohan, C. O.; Gopal, T. K. S. Smart Packaging Systems for Food Applications: A Review. *J. Food Sci. Technol.* **2015,** *52,* 6125–6135.
11. Cagri, A.; Ustunol, Z.; Ryser, E. T. Antimicrobial Edible Films and Coatings. *J. Food Protocol* **2004,** *67,* 833–848.
12. Chandra, R.; Rustgi, R. Biodegradable Polymers. *Prog. Polym. Sci.* **1998,** *23,* 1273–1335.
13. Chaudhry, Q.; Scotter, M.; Blackburn, J. Applications of Nanotechnologies for the Food Sector. *Food Addit. Contam.* **2008,** *25,* 241–258.

14. Chen, H.; Weiss, J.; Shahidi, F. Nanotechnology in Nutraceuticals and Functional Food Technology. *Food Technol.* **2006,** *60,* 30–36.
15. Chrisey, L.; Lee, G.; O'Ferrall, C. Covalent Attachment of Synthetic DNA to Self-Assembled Monolayer Films. *J. Nucleic Acid Res.* **1996,** *24,* 3031–3039.
16. Cioffi, N.; Torsi, L.; Ditaranto, N.; Tantillo, G.; Ghibelli, L.; Sabbatini, L.; Traversa, E. Copper Nanoparticle/Polymer Composites with Antifungal and Bacteriostatic Properties. *Chem. Mater.* **2005,** *17,* 5255–5262.
17. Cushen, M.; Kerry, J.; Morris, M.; Cruz-Romero, M.; Cummins, E. Evaluation and Simulation of Silver and Copper Nanoparticle Migration from Polyethylene Nanocomposites to Food and an Associated Exposure Assessment. *J. Agric. Food Chem.* **2014,** *62,* 1403–1411.
18. Dasgupta, N.; Ranjan, S.; Patra, D.; Srivastava, P.; Kumar, A.; Ramalingam, C. Bovine Serum Albumin Interacts with Silver Nanoparticles with A "Side-On" or "End On" Conformation. *Chem. Biol. Interact.* **2016,** *253,* 100–111.
19. Del Nobile, M. A.; Conte, A.; Buonocore, G. G.; Incoronato, A. L.; Massaro, A.; Panza, O. Active Packaging by Extrusion Processing of Recyclable and Biodegradable Polymers. *J. Food Eng.* **2009,** *93,* 1–6.
20. Duncan, T. V. Applications of Nanotechnology in Food Packaging and Food Safety: Barrier Materials, Antimicrobials and Sensors. *J. Colloid Interface Sci.* **2011,** *363,* 1–24.
21. Echegoyen, Y.; Nerín, C. Nanoparticle Release from Nano-Silver Antimicrobial Food Containers. *Food Chem. Toxicol.* **2013,** *62,* 16–22.
22. Gaaz, T. S.; Sulong, A. B.; Akhtar, M. N.; Kadhum, A. A.; Mohamad, A. B.; Al-Amiery, A. A. Properties and applications of Polyvinyl Alcohol, Halloysite Nanotubes and Their Nanocomposites. *Molecules* **2015,** *20,* 22833–22847.
23. Gfeller, K. Y.; Nugaeva, N.; Hegner, M. Micromechanical Oscillators as Rapid Biosensor for the Detection of Active Growth of *Escherichia coli. Biosens. Bioelectron.* **2005,** *21,* 528–533.
24. Gonera, A.; Cornillon, P. Gelatinization of Starch/Gum/Sugar Systems Studied by Using DSC, NMR, and CSLM. *Starch* **2002,** *54* (11), 508–516.
25. Göpel, W. Chemical Sensing, Molecular Electronics and Nanotechnology: Interface Technologies Down to the Molecular Scale. *Sens. Actuators B Chem.* **1991,** *4,* 7–21.
26. Gravel, Bikker, J. F.; DeKruif, C. G. Unique Milk Protein Based Nanotubes: Food and Nanotechnology Meet. *Trends Food Sci. Technol.* **2006,** *17,* 196–203.
27. Hagen, A.; Hertel, T. Quantitative Analysis of Optical Spectra from Individual Single-Wall Carbon Nanotubes. *Nano Lett.* **2003,** *3,* 383–388.
28. Hall, W. P.; Ngatia, S. N.; Van Duyne, R. P. LSPR Biosensor Signal Enhancement Using Nanoparticle Antibody Conjugates. *J. Phys. Chem. C* **2011,** *115,* 1410–1414.
29. He, X.; Hwang, H. M. Nanotechnology in Food Science: Functionality, Applicability, and Safety Assessment. *J. Food Drug Anal.* **2016,** *24,* 671–681.
30. Himizu, Y.; Egashira, M. Basic Aspects and Challenges of Semiconductor Gas Sensors. *J. Mater. Res. Soc. Bull.* **1999,** *24,* 18–24.
31. Hong, S. I.; Rhim, J. W. Antimicrobial Activity of Organically Modified Nano-Clays. *J. Nanosci. Nanotechnol.* **2008,** *8,* 5818–5824.
32. Jain, A.; Ranjan, S.; Dasgupta, N.; Ramalingam, C. Nanomaterials in Food and Agriculture: An Overview on Their Safety Concerns and Regulatory Issues. *Crit. Rev. Food Sci. Nutr.* **2018,** *58,* 297–317.

33. Jim, D. Guest Commentary: Nanotechnology: Its Impact on Food Safety. *J. Environ. Health* **2008,** *70,* 47–50.
34. Kanmani, P.; Rhim, J. W. Physical, Mechanical and Antimicrobial Properties of Gelatin Based Active Nanocomposite Films Containing AgNPs and Nanoclay. *Food Hydrocoll.* **2014,** *35,* 644–652.
35. Kavitha, P.; Manjunath, M.; Huey-Min, H. Nanostructures: Current Uses and Future Applications in Food Science. *J. Food Drug Anal.* **2017,** *25,* 245–253.
36. Kim, M.; Pometto, A. L. Food Packaging Potential of Some Novel Degradable Starch-Polyethylene Plastics. *J. Food Protect.* **1994,** *57,* 1007–1012.
37. Kong, J.; Franklin, N. R.; Zhou, C.; Chapline, M. G.; Peng, S.; Cho, K.; Dai, H. Nanotube Molecular Wires as Chemical Sensors. *J Sci.* **1999,** *287*, 622–625.
38. Lim, L. T. Active and Intelligent Packaging Materials. Chapter 4.52; In *Comprehensive Biotechnology*; Moo-Young, M., Ed.; Elsevier: Amsterdam, 2011.
39. Liu, W. J.; Yang, H. L.; Wang, Z.; Dong, L. S.; Liu, J. J. Effect of Nucleating Agents on the Crystallization of Poly (3-Hydroxybutyrate-co-3-hydroxyvalerate). *J. Appl. Polym. Sci.* **2002,** *86*, 2145–2152.
40. Luduena, L. N.; Alvarez, V. A.; Vazquez, A. Processing and Microstructure of PCL/Clay Nanocomposites. *Mater. Sci. Eng.* **2007,** *460,* 121–129.
41. Ma, Y. Z.; Stenger, J.; Zimmermann, J.; Bachilo, S. M.; Smalley, R. E.; Weisman, R. B.; Fleming, G. R. Ultrafast Carrier Dynamics in Single-Walled Carbon Nanotubes Probed by Femtosecond Spectroscopy. *J. Chem. Phys.* **2004,** *120*, 3368–3373.
42. Majid, I.; Nayik, G. A.; Dar, M. S.; Nanda, V. Novel Food Packaging Technologies: Innovations and Future Prospective. *J. Saudi Soc. Agric. Sci.* **2016,** *17* (4), 454–462.
43. Mandelis, A.; Christofides, C.; Winefordner J. D. Eds.; In *Physics, Chemistry and Technology of Solid-State Gas Sensor Devices*; 1st Ed.; Wiley Interscience: New York, 1993; p 352.
44. Maria S. Possibilities of Nanotechnology in Food Processing and Packaging Applications. Unpublished Presentation at *VTT Workshop on Converging Technologies for Food: Nanotech-, Bioinfo-, Cognitive- Sciences*, Dec 20, 2005, Brussels.
45. Matindoust, S.; Baghaei-Nejad, M.; Abadi, M. H. S.; Zou, Z.; Zheng, L. R. Food Quality and Safety Monitoring Using Gas Sensor Array in Intelligent Packaging. *Sens. Rev.* **2016,** *36,* 169–183.
46. McClements, D. J.; Xiao, H. Potential Biological Fate of Ingested Nanoemulsions: Influence of Particle Characteristics. *Food Funct.* **2012,** *3*, 202–220.
47. Mihindukulasuriya, S. D. F.; Lim, L. T. Nanotechnology Development in Food Packaging: A Review. *Trends Food Sci. Technol.* **2014,** *40,* 149–167.
48. Mirzadeh, A.; Kokabi, M. The Effect of Composition and Draw-Down Ratio on Morphology and Oxygen Permeability of Polypropylene Nano-Composite Blown Films. *Eur. Polym. J.* **2007,** *43,* 3757–3765.
49. Modi, A.; Koratkar, N.; Lass, E.; E.; Wei, B.; Ajayan, P. M. Miniaturized Gas Ionization Sensors Using Carbon Nanotubes. *J. Nature* **2003,** *424*, 171–174.
50. Momin, J. K.; Joshi, B. H. Nanotechnology in Foods. Chapter 1; In *Nanotechnologies in Food and Agriculture*; Rai, M., Ribeiro, C., Mattoso, L., Duran N., Eds.; Springer International Publishing: Switzerland, 2015; pp 3–24.
51. Nachay, K. Analyzing Nanotechnology. *Food Technol.* **2007,** *61,* 34–36.

52. Ogata, N.; Jimenez, G.; Kawai, H.; Ogihara, T. Structure and Thermal/Mechanical Properties of poly (L-Lactide)-Clay Blend. *J. Polym. Sci. Polym. Phys.* **1997,** *35,* 389–396.
53. Oleyaei, S. A.; Zahedi, Y.; Ghanbarzadeh, B.; Moayedi, A. A. Modification of Physicochemical and Thermal Properties of Starch Films by Incorporation of TiO_2 Nanoparticles. *Int. J. Biol. Macromol.* **2016,** *89,* 256–264.
54. Ozdemir, M.; Floros, J. D. Active Food Packaging Technology. *Crit. Rev. Food Sci. Nutr.* **2004,** *44,* 185–193.
55. Pal, M. Nanotechnology: A New Approach in Food Packaging. *J. Food Microbiol. Saf. Hyg.* **2017**, *2*, 8–9.
56. Qureshi, M. A.; Karthikeyan, P.; Ahmed, D.; Uprit, A.; Kumar, U. Application of Nanotechnology in Food and Dairy Processing: An Overview. *Pak. J. Food Sci.* **2012,** *22,* 23–31.
57. Radha, K.; Thomas, A.; Sathian, C. T. Application of Nano Technology in Dairy Industry: Prospects and Challenges—A Review. *Indian J. Dairy Sci.* **2014,** *67,* 367–374.
58. Ranjan, S.; Dasgupta, N.; Srivastava, P.; Ramalingam, C. Spectroscopic Study on Interaction Between Bovine Serum Albumin and Titanium Dioxide Nanoparticle Synthesized from Microwave-Assisted Hybrid Chemical Approach. *J. Photochem. Photobiol. B Biol.* **2016,** *161,* 472–481.
59. Ranjan, S.; Dasgupta, N.; Chakraborty, A. R.; Samuel, S. M.; Ramalingam, C.; Shanker, R. Nanoscience and Nanotechnologies in Food Industries: Opportunities and Research Trends. *J. Nanopart. Res.* **2014,** *16,* 1–23.
60. Rhim, J. W.; Ng, P. K. W. Natural Biopolymer-Based Nanocomposite Films for Packaging Applications. *Crit. Rev. Food Sci. Nutr.* **2007,** *47,* 411–433.
61. Rhim, J. W.; Wang, L. F.; Hong, S. I. Preparation and Characterization of Agar/Silver Nanoparticles Composite Films with Antimicrobial Activity. *Food Hydrocoll.* **2013,** *33,* 327–335.
62. Rhim, J. W.; Hong, S. I.; Park, H. M.; Ng, P. K. Preparation and Characterization of Chitosan-Based Nanocomposite Films with Antimicrobial Activity. *J. Agric. Food Chem.* **2006,** *54,* 5814–5822.
63. Schedin, F.; Geim, A. K.; Morozov, S. V.; Hill, E. V.; Blake, P.; Katsnelson, M. I.; Novoselov, K. S. Detection of Individual Gas Molecules Adsorbed on Graphene. *Nat. Mat.* **2007,** *6,* 652–655.
64. Shankar, S.; Chorachoo, J.; Jaiswal, L.; Voravuthikunchai, S. P. Effect of Reducing Agent Concentrations and Temperature on Characteristics and Antimicrobial Activity of Silver Nanoparticles. *Mater. Lett.* **2014,** *137,* 160–163.
65. Shankar, S. X.; Teng, X.; Rhim, J. W. Properties and Characterization of Agar/Cu np bionanocomposite Films Prepared with Different Copper Salts and Reducing Agents, *Carbohydr. Polym.* **2014,** *114*, 484–492.
66. Sharma, D.; Panwar, K. Radio Frequency Identification (RFID). *Int. J. Res.* **2014,** *1,* 125–129.
67. Sharma, C.; Dhiman, R.; Rokana, N.; Panwar, H. Nanotechnology: An Untapped Resource for Food Packaging. *Front. Microbiol.* **2017,** *8,* 1735–1746.
68. Sharma, R.; Raghavan, K. V.; Thakur, M. S.; Raghvarao, K. S. M. S. Recent Advances in Nanoparticle Based Aptasensors for Food Contaminants. *Biosens. Bioelectron.* **2015,** *74,* 612–627.

69. Sinha Ray, S.; Bousmina, M. Biodegradable Polymers and Their Layered Silicate Nanocomposites: In Greening the 21st Century Materials World. *Prog. Mater. Sci.* **2005,** *50,* 962–1079.
70. Sinha, R. S.; Maiti, P.; Okamoto, M.; Yamada, K.; Ueda, K. New Polylactide/Layered Silicate Nanocomposites. Preparation, Characterization and Properties. *Macromolecules* **2002,** *35,* 3104–3110.
71. Siracusa, V.; Rocculi, P.; Romani, S.; Dalla, R. M. Biodegradable Polymers For Food Packaging: A Review. *Trends Food Sci. Technol.* **2008,** *19,* 634–643.
72. Smolander, M.; Chaudhry, Q. Nanotechnologies in Food Packaging. Chapter 6; In *Nanotechnologies in Food*; Chaudhry, Q., Castle, L., Watkins R., Eds.; Royal Society of Chemistry Publishing: Cambridge, UK, 2010; pp 86–101.
73. Tan, W.; Zhang, Y.; Szeto, Y. S.; Liao, L. A Novel Method to Prepare Chitosan/Montmorillonite Nanocomposites in the Presence of Hydroxyl-Aluminum Oligomeric Cations. *Compos. Sci. Technol.* **2008,** *68,* 2917–2921.
74. Uyama, H.; Kuwabara, M.; Tsujimoto, T.; Nakano, M.; Usuki, A.; Kobayashi, S. Green Nanocomposites from Renewable Resources: Plant Oil–Clay Hybrid Materials. *Chem. Mater.* **2003,** *15,* 2492–2494.
75. Vasile, C. Polymeric Nanocomposites and Nanocoatings for Food Packaging: A Review. *Materials* **2018,** *11,* 1834–1840.
76. Wang, X, Du, Y.; Yang, J.; Wang, X.; Shi, X.; Hu, Y. Preparation, Characterization and Antimicrobial Activity of Chitosan/Layered Silicate Nanocomposites. *Polymer* **2006,** *47,* 6738–6744.
77. Weiss, J.; Takhistov, P.; McClements, D. J. Functional Materials in Food Nanotechnology. *J. Food Sci.* **2006,** *71,* 107–116.
78. Yang, W.; Fortunati, E.; Bertoglio, F.; Owczarek, J. S.; Bruni, G.; Kozanecki, M.; Kenny, J. M.; Torre, L.; Visai, L.; Puglia, D. Polyvinyl Alcohol/Chitosan Hydrogels with Enhanced Antioxidant and Antibacterial Properties Induced by Lignin Nanoparticles. *Carbohydr. Polym.* **2018,** *181,* 275–284.
79. Zheng, J. P.; Li, P.; Ma, Y. L.; Yao, K. D. Gelatin Montmorillonite Hybrid Nanocomposite, I: Preparation and Properties. *J. Appl. Polym. Sci.* **2002,** *86,* 1189–1194.

CHAPTER 8

SCOPE OF NANOTECHNOLOGY IN SMART PACKAGING SYSTEMS

GAURAV YADAV and ANAND KUMAR PANDEY

ABSTRACT

The physicochemical and biological properties of packaging material have concerns related to quality, transport, shelf-life, spillage, and contamination of food. The increasing demand and variety of food and consumer awareness encourage to improve the packaging technologies. Nanotechnology, the study of materials at a nano-meter scale, provides great scope to develop versatile material capable of monitoring and maintaining the quality and freshness of the food to detect microbial growth and to increase the shelf-life. Nanotechnology-based smart packaging for foods employs materials, which monitor and provide information about a wide range of changes occurring in the food products like temperature, pH, moisture, microbial growth, and their surrounding environment when developed into intelligent systems like sensors, indicators, and Radiofrequency Identification (RFID) systems. These assist and increase the ease of product selection by consumers. Nanomaterials, including nano-clays, synthetic, and natural polymer-based nano-composite films or coatings, and metallic nanoparticles, seem promising to develop such stimuli responsive smart packaging, proving the effective role of nanotechnology. Though effective results have been reported, yet limitations related to efficiency, effectivity, specificity, and toxicity of nanomaterials demand exhaustive research to develop economically feasible packaging.

Advances in Sustainable Food Packaging Technology. Megh R. Goyal, Preeti Birwal, Anand K. Pandey, & Ritesh B. Watharkar, (Eds.)

8.1 INTRODUCTION

Interaction with the environment degrades the food affecting its quality and durability and thus makes it unfit for consumption.[112,160] Packaging of food is an important practice to maintain its quality, storage, shipping, and delivery.[39] The packaging acts as a barrier between food and the environment and thus increases the durability of food by maintaining its quality.[21] It also plays a role in the marketing of the food product as various types of information related to food like food content, food weight, volume, nutritional value, calorie content, color, etc., can be presented on the packaging material.[39]

Modifications in packaging like dispensable caps, sprays, enclosable openings, and other such features make products more usable and convenient for customers.[39] Every year a huge amount of produced food gets wasted because of microbial contamination, poor transportation facilities, or environmental factors. An increase in global demand for industrial processing of food, ease in the transport of food products, and reduction in preparation time of fresh foods is encouraging the food and beverage packaging industry to explore more advanced packaging solutions.[65]

Earlier, protection and preservation of food were considered as the principal roles of food packaging, while in the present scenario, other factors related to packaging viz, transportation ease, traceability, tamper indication, and sustainability are equally important.[65,112] Nanotechnology-based interventions in packaging industries overcome various limitations of current packaging practices and pave the way for meeting the global demand.[122,126] Unlike traditional packaging practices that act as a passive barrier, nanotechnology-based food packaging is actively involved in maintaining the quality of food, its shelf-life, and other related attributes.[21]

The amalgamation of nanotechnology in packaging industry of food has upgraded the traditional packaging.[47,114] In general, three different types of packaging systems have been developed by utilizing nanotechnology, namely: active packaging, improved packaging, and intelligent/smart packaging.[2]

Studies are being conducted to develop nano paper composed of hybrid material like silver nanoparticles embedded in bacterial cellulose, which will be useful in the detection of analytes in the gas phase above the food when packed in such packaging.[11] Bilayer films incorporated with ZnO check the fish spoilage.[12] Smart labels are becoming an integral part of smart packaging systems and provide immense information about food quality in real time. Carbon nanoparticles are being tested to be used in inkjet-printed functional labels with photoluminescence inks. They are responsive to the internal gas environment of the packaging. Research studies suggest that ZnO and SiO_2 nanoparticles can be used as ink component for functional label printing.[13,14]

All these advanced nanotechnologies possess high potential to be exploited at a wide scale in packaging industries. The condition of food can be tracked in real time from the production to consumption in the entire supply chain is an exceptional quality of smart packaging. They also provide more authentic information about the condition of the product. Smart packaging offers a much better experience of purchase to customers than traditional packaging. The product directly communicates with the customers and helps them in making a better choice for themselves.

This chapter provides deep insights into different nano-based packaging technologies to improve and maintain the quality of food to solely satisfy the consumer's needs and demands. Different packaging nanomaterials and nano-systems developed to date have been detailed in this chapter with a focus on the most updated technologies of smart packaging systems to provide in-depth knowledge and further encourage the scientific community in this concern.

8.2 IMPROVED PACKAGING SYSTEMS

The principal properties of food packaging material, that is, barrier, mechanical, and thermal properties, are enhanced by the introduction of nanoparticles in its polymer chain.[2,8,78,92,170] Barrier properties vary with the polymers used to develop the packaging material, for example, polyethylene terephthalate (PET) in comparison to low-density polyethylene (LDPE) display good barrier properties against oxygen on the other hand, HDPE which is high-density polyethylene shows improved resistance against water vapor compared to PET.[28,111] Such barrier properties are result of various factors that include branching degree, polar nature, presence of hydrogen bonding, amount of cross-linking, crystallinity, etc.[22,111]

Traditional packaging materials, when fused with nanoparticles, form nanocomposite, which displays excellent mechanical performance and high resistant characteristics with anti-microbial effect against a wide spectrum of microbes. Polymer nanocomposites (PNCs) have been introduced as a novel packaging material in which an inert, nanoscale filler is dispersed throughout a polymer matrix. Clay, silica nanoparticles, and graphene are the most commonly used filler materials, along with carbon nanotubes (CNTs), starch nanocrystals, and chitosan nanoparticles.[26,105,114,147]

The addition of nanofillers in polymer creates a tortuous path altering the diffusion rate of the molecules.[89] This increase in tortuosity explains the increase in barrier properties of polymer nanocomposites. For effective improvement, the distribution of fillers must be uniform all over the polymer matrix, that is not easy to achieve.[111] The proportion of the largest dimension to the smallest of the filler, that is, the aspect ratio of filler material, plays a significant role in its associated properties. Fillers with high aspect ratios possess a more specific surface area with associated high reinforcing properties.[15,114] Several studies have reported that PNCs show improved strength,[67,120] flame resistance,[120,140] and thermal conductivity.[98,120]

8.2.1 CLAY AND SILICA NANOCOMPOSITES

Clay and silicates are the most extensively used fillers due to high cost-effectiveness, associated ease of processing, and sufficient availability. They are considered remarkably stable, and are apparently non-toxic in nature, and thus possess great potential to be used for the purpose. The silicate layers in nanocomposites are usually employed in their two-dimensional structure, with thickness ranging from 1 nm to few microns. The combination of silicates and polymers increases the tortuosity of the diffusive path for an infiltrate molecule and hence provides excellent barrier properties.[16,79,114] The morphology of nanoparticles and their dispersion pattern in packaging polymers are leading aspects that determine the bulk properties of the PNC materials.[111] Typically, micro composites are present in the tactoid structure, that is, the layered silicate is overall covered by polymer, but interactions between polymers and layered silicates may lead to formation of exfoliated or intercalated nanocomposites. Intercalated nanocomposites are structures made up of multiple layers with layers of polymer present alternatively layers lying few nanometers apart.[94,114,132] In exfoliated nanocomposites, random dispersion of clay layers penetrates the polymer extensively. Barrier properties of monodispersed exfoliated systems are better than intercalated or interposed and tactoid systems.[111]

Montmorillonite is the most widely studied clay filler. It consists of an octahedral aluminum hydroxide sheet in between two tetrahedral layers of silica[80,114,132] linked by weak forces that are electrostatic in nature. Exchangeable cations (typically Na^+ and Ca^{2+}) compensate for the non-uniformity of negative charges on surface, where linkage between the parallel layers is due to electrostatic weak forces.[114,123] Montmorillonite has a high surface area and large aspect ratio, making it an effective reinforcement filler.

Several polymers and biopolymers have been made for the development of packaging material by introducing clay layer, which improves their mechanical strength, offers resistance to gases, and reduces water vapor permeability.[35,40,64,73,114] Clay nanoparticles affect the glass transition and thermal degradation temperatures.[93,114] Thus, their use in packaging material offers improved packaging, which enhances the shelf-life of food, provides shutter proof, is light in weight, and heat resistant.

8.2.2 NANO-COATINGS

Coatings are thin films of edible material placed between food components that serve as moisture, lipid, and gas barriers.[7,39,60] Nanocoatings are <1–100 nm in width, and these ultra-thin layers of nanocoatings neither modify the surface topography nor they fill in defects or make a smooth surface as paint does. They impart particular chemical or physical function/(s) to the surface, as gas-barrier coatings improve corrosion resistance and enhance insulating or conductive properties.[84,129] The addition of additives in small quantities to coating material modify a large variety of properties, like, flow behavior, surface tension, gloss, structure, UV, and weather resistance.

Nanomaterial-containing coatings, in comparison to conventional coatings, offer high-performance materials and better processing properties (e.g., high elasticity, fast-drying, high water vapor permeability).[129] Components of edible coating can be categorized as: lipids and hydrocolloids (water soluble polysaccharides). Cellulose derivatives, pectin, starches, alginates, and chitosan are the suitable polysaccharides, while waxes, acylglycerols, and fatty acids are suitable lipids.[23,60,173] Lipid films exhibit excellent barrier properties for moisture control. The incorporation of nanoparticles in these coating films improves their physical properties.[6,129] However, edible coatings has been investigated for utilization on a wide range of foods, consisting of vegetables, fruits, meats, cheese, chocolate, bakery products, candies, and French fries, only a few studies have been conducted.[11,82,104,129] The addition of clay montmorillonite into pectins is known to lower the diffusion of oxygen. Similarly, nanocomposites prepared by gelatin and montmorillonite show considerable improvement in their physical properties.[129,184]

8.2.3 NANOLAMINATES

Nanolaminates are multiple layers of nanomaterial that are bonded to each other, physically or chemically.[17,60] They offer great advantages over conventional technologies in the preparation of edible coatings and films, thus find important application in the food and dairy industry.[60,132] The layers can be created using different adsorbing substances, including natural polyelectrolytes like proteins and polysaccharides, charged lipids like phospholipids and surfactants, and colloidal particles like micelles, vesicles, and droplets. The addition of these agents increases the durability and characteristics of the coated foods. Like nanocoating, these nanolaminates can also be prepared entirely from edible materials, including proteins, polysaccharides, and lipids.[60]

8.2.4 CARBON NANOTUBES

Carbon nanotubes (CNTs), hollow cylindrical nanotubes made of carbon atoms, are usually available in two forms, as single wall nanotube (SWNT) and multi-walled nanotubes (MWNTs). SWNTs are around one atom thick, whereas MWNTs are made up of some tubes arranged in concentric manner having very high elastic modulus and aspect ratios.[81,114,185] Many studies have reported that the incorporation of CNTs into several polymers like polyvinyl alcohol, polyamide, and polypropylene has improvised their tensile strength.[96,114,177] CNTs display high antibacterial properties, ascribed to their capability of direct penetration through microbial cells.[60,53] Chemoresistive detectors have been developed using SWCNs with cobalt mesoaryl porphyrin complexes that can detect amines generated during spoilage of meat.[70] Though CNTs have a wide range of applications, their high cost, processing, and dispersion related aspects limit their incorporation in nanocomposites.[5,81,114,180]

8.3 ACTIVE PACKAGING

European Regulation (EU) 450/2009/EC defines "active material and article" as "components which would discharge or absorb substances purposely from or inside the packed food or the surrounding environment of the food."[10,20,78,92,114] The conventional system of food packaging acts as passive barriers, protecting food by providing a barrier against environmental contaminations,[20] while in an active packaging system, the product, package, and the environment interact in a positive way, resulting in enhanced food conservation and extended shelf-life.[8,161] Nanomaterials utilized in the development of active packaging have direct interaction with the environment or the food, thus providing an improved defense to product. These active agents can work as absorbers or emitters of compounds that play an important role in food preservation.[20,179]

8.3.1 ANTIMICROBIAL PACKAGING

Undesirable microorganisms can be present in a packaged food or packaging materials themselves. Antimicrobial packing acts to reduce, inhibit, or retard the growth of these microorganisms.[4,5,8] Substances having antimicrobial properties can be incorporated into or coated onto food packaging materials to control the growth of undesirable microorganisms.[8,62] To increase the shelf-life, these materials act by extending the lag phase and reducing the growth phase of microorganisms.[8,41] The major potential food applications of such antimicrobial agents include meat, poultry, bread, cheese, fruits, and vegetables.

Ethanol and carbon dioxide, along with chlorine dioxide and sulfur dioxide, are released in volatile antimicrobial packaging systems which do not come-in direct contact with the food. The barrier properties of this volatile antimicrobial system are very high to avoid the loss through permeation.[4,8] Various oxide(s) of nanoparticles such as zinc oxide (ZnO), titanium dioxide (TiO_2), magnesium oxide (MgO), and silicon oxide (SiO_2) act as UV blockers and photo-catalytic disinfecting agents.[30,114] The antimicrobial activity of nanoparticles might be due to one of the following mechanisms: direct interaction with the microbial cells (interrupting the transfer of transmembrane electrons, cell envelope destruction), oxidization of the cell components, and secondary product formation, leading to cell damage.[66,114]

8.3.1.1 SILVER NANOPARTICLES

These are among the most studied nanoparticles. Their antimicrobial potential is well established, and they are active not only against bacterial strains but also inhibit various fungal and viral strains.[59,114] Silver targets the metabolism of bacteria by interacting and binding to proteins, DNA, and enzymes, thus results in bacteriostatic effects.[13,114] It also disrupts and destabilizes both the cytoplasmic and the outer membranes.[83,114] Silver nanoparticles inhibit enzymes of respiratory chain and also stimulates reactive oxygen species (ROS) production.[24,114] Ethylene, responsible for the ripening of fruits and vegetables, is absorbed and decomposed by silver nanoparticles leading to extended shelf-life. Stable nature of AgNP-based nanocomposites makes it a preferred choice for use, and the slow silver ions release in the stored foods leads to persistence of antimicrobial activity.

8.3.1.2 TITANIUM DIOXIDE NANOPARTICLES

Titanium dioxide nanoparticles (TiO_2-NPs), are another type of most explored nanomaterials that exhibit a high degree of thermostability and have the ability to modify the properties of biodegradable food packaging films. Being cheap, non-toxic, and photostable, these nanoparticles have emerged as a superior photo-catalyst material.[47,137] Their surface energy increases with an increase in size, which is an important factor in polymer/filler interaction.[87,114] TiO_2-NPs show very high surface reactivity. The antibacterial properties of TiO_2-NPs are well known but are confined to the exposure of UV irradiation.[74,114,116] When irradiated with UV light, they produce numerous ROS which can kill microorganisms. Further, in the form of nano-additives, TiO_2 is reported to improve the mechanical properties of polymer nanocomposites.[47,106,146]

8.3.1.3 ZINC OXIDE

Metal oxide nanoparticles are known to show higher stability even under extreme conditions. Their ability to exhibit antimicrobial activity at low concentrations and non-toxic nature for humans makes them an attractive counterpart to be used in packaging material. ZnO-NPs show diverse morphology and are strong antimicrobial agents showing robust inhibition against the growth of broad-spectrum bacterial species.[25,47] Their antibacterial activity increases with a further decrease in particle size. Also, ZnO-NPs play role of UV-protective agents for area of food packaging.[49,114,139,176] Unlike TiO_2, which requires UV light for stimulation, ZnO-NPs get stimulated with visible light. Direct contact with microbial cell wall results in the destruction of bacterial cell integrity, the liberation of antimicrobial ions, and the generation of ROS.[112,114]

8.3.1.4 OTHER NON-METAL OXIDE AND METAL OXIDE NANOPARTICLES

Apart from the above mentioned metal oxides, many other non-metal and metal oxide nanomaterials, such as MgO-NPs, Fe_3O_4-NPs, CuO, and SiO_2-NPs, also possess increased potential to be used in packaging materials.[17,47,69,113,121]

Copper oxide nanoparticles (CuO-NPs) are known for their antimicrobial activity and are capable of checking the growth of viruses, bacteria, and fungi. As they have enormous area of surface, they interact with the membrane of cell and display enhanced antimicrobial effect.[47,3] The antibacterial effect of these nanoparticles has been exceptionally utilized in the water purification and food packaging materials field.

The integration of SiO_2-NPs in packaging also enhances the barrier and mechanical properties of different matrices of polymers.[114,134] The amalgamation of $nSiO_2$ with starch matrix increases the tensile properties while decreasing the water absorption by the starch.[114,136] In a past study by Jia et al., vinyl silica nanoparticles were copolymerized with vinyl acetate and were able to generate polyvinyl alcohol and SiO_2 nanocomposites. This nanocomposite, in comparison to pure polyvinyl alcohol, had better thermal and mechanical properties. The SiO_2-NPs properties can be explained by their high area of surface, average particle size, less toxicity, poor thermal conductivity, and extremely high insulation.[48,114]

MgO is a renewable, colorless, and crystalline mineral that can be produced on a large scale economically. Their use in food is considered safe by Food and Drug Administration.[47,121]

8.3.2 OXYGEN SCAVENGERS

Oxygen scavenging is the most widely used active packaging technology for foods. Access to O_2 in the food package promotes various deteriorations like browning, rancidity, development of oxidative microflora, a decline in vitamins, foul odors, bad flavors, unwanted and unpleasant color changes, leading to reduced shelf-life of the food product. Directly or indirectly, oxygen (O_2) is responsible for the deterioration of many types of foods. For example, the browning of fruits and the rancidity of vegetable oils are a result of direct oxidation reactions. Food spoilage by aerobic microorganisms is an indirect effect of O_2. To confine the rate of such deteriorating reactions in food, control of oxygen levels in food packages is a must.[8,44,60,114] The use of modified atmosphere packaging (MAP) or vacuum packaging for oxygen-sensitive food also does not remove oxygen completely. In order to circumvent the detrimental effects of O_2, it should be eliminated or reduced to acceptable levels in accordance with the food. The addition of O_2 scavengers in food packages maintain extremely low levels of O_2, minimizing the quality change in oxygen-sensitive food, thus enhancing the shelf-life of food.[8,54,60,130] The use of gases like N_2 and CO_2 creates an inert atmosphere which also reduces the O_2 content in the headspace.[95,114]

8.3.3 MOISTURE REGULATORS

The presence of moisture is a major cause of food spoilage. In the packaging of fresh and unprocessed fruits and vegetables, when one part of the package is cooler than the other areas, respiration followed by condensation occurs. Soluble nutrients from food leach into the water leading to microbial growth and food spoilage hence results into low consumer appeal. Moisture content causes softening of dry, crispy products, caking of hygroscopic products like milk powder, instant coffee powder, sweets, etc. Moisture regulators decrease the water activity, which suppresses the microbial growth in the product.[8] They are active non-migratory packaging which controls the excess moisture. Possible reasons for the high content of water into the package can be water trapped during packaging and released due to change in temperature, respiration of fresh products, low or poor vapor permeability of the package, etc.[10,20,32]

Moisture absorbers use the hygroscopic substance to reduce water activity, providing an environment less suitable for the growth of mold, yeast, and bacteria. Moisture scavengers, if reduce humidity in the headspace, are defined as "relative humidity controllers" and "moisture removers" if they absorb liquids exuded from food.[20,141] Drip absorbent sheets are two-layered structures of microporous non-woven plastic like polyethylene or polypropylene and an active adsorbent substance like silica gel-based adsorbents, superabsorbent polymer, etc.[8,103,107] For dry food, desiccants like silica gel, natural clays, molecular sieves, calcium oxide are used, and inside the carton, solid polymeric humectants are layered with tiny permeable bags or packs of inorganic salts that can retain a high amount of water as the humidity increases. Magnesium oxide, barium oxide, and calcium oxide are metal oxides that are also used as moisture absorbers. They irreversibly react with water providing the corresponding oxides.[20,72]

Great care is to be taken while designing moisture absorbers for high water content products to avoid undesired food dehydration.[108] The material for moisture absorption can be spread into the matrix of polymer or in between the two plastic film layers which constitute the package. Organic absorbers (such as fructose sorbitol[10] and cellulose, and modified starch[91]) have also been reported.

8.3.4 *ETHYLENE SCAVENGERS*

Ethylene, a natural plant growth hormone, accelerates the respiration of fruits and vegetables, inducing fruit ripening, softening, and senescence. It is an odorless and colorless gas, purely unsaturated hydrocarbon, which is effective even at low concentrations. As it promotes ripening, it has a detrimental impact on the shelf-life of many fruits and vegetables. Being a hormone, it plays an important role in the respiration rate, growth cycle, somatic embryogenesis, root growth, seed germination, and plant development.[1,8,20,109] Ethylene absorbers and scavengers, which physically absorb and hold ethylene molecules, are used to remove ethylene. Potassium permanganate is immobilized on an inert material such as alumina or silica gel and is most widely used as an ethylene scavenger.[33,131] Embedding in nanoparticles enhances its scavenging ability. Being toxic in nature, it cannot be integrated with a surface that comes in contact with food, so it is available as sachets for packages that can be placed in product holdings.[31] The ethylene gas gets converted into ethylene glycol, carbon dioxide, and water by oxidation.[127] Nanoparticles, due to their high surface activity and reactivity, oxidize ethylene by photocatalytic reaction into water and carbon dioxide.[109,174]

8.4 INTELLIGENT OR SMART PACKAGING

The material and articles capable of monitoring the state of packaged food and/or the surroundings of the food are termed as intelligent materials and articles. By providing the user with information about food and its environment, smart packaging systems elevate the communication property of packaging.[10,92,114,162] They are capable of recognizing any change in characteristics of the packaged food.[8,99,102] Nanomaterials are used to detect any biochemical or microbial change, spoilage-associated change, or presence of chemical contaminants in the food.[92] The nano-sensors (nanomaterial-based sensors) respond to external stimuli change in order to identify and assure any change in quality and safety of food and then communicate it to provide information. Smart packaging informs the manufacture, retailer, and consumer about the condition of the product in real-time.[68,86]

Three types of technologies are primarily used in intelligent packaging systems, that is, indicators, sensors, and data carriers.[20,34] The main function of indicators and sensors is to provide information related to product quality, while data carriers are mainly related to the management of supply chain logistics. These indicators or sensors can be placed on the primary packaging or inside or outside of the secondary packaging or that of the tertiary packaging.[20,42,175]

8.4.1 INDICATORS

Indicators act as information transmitters for the consumer regarding absence or of some specific substance. They either provide indication for a reaction which is occurring between two or more components or monitors the amount or concentration of any particular substance. Such information is translated to signals, usually in the form of immediate visual change.[20,138]

8.4.1.1 TIME-TEMPERATURE INDICATOR

Temperature plays role as initial factor which is of utmost importance and evaluate the shelf-life of product, that is, food. To maintain the quality of a food product especially perishable food, a major challenge is to protect it from the unwanted fluctuations of temperature. Improper management of the temperature can lead to wastage of huge amount of food during its production, and also during the distribution. To meet these needs, time-temperature indicators (TTI) provided valuable support,[20,38,157] According to EC/450/2009, TTIs communicate the information regarding whether a particular critical temperature has surpassed over the time and/or evaluate the least amount of time, which is spent by the product above that temperature (time-temperature history). TTI labels provide a visual temperature history of a product during distribution and storage. TTIs are cost-effective and user-friendly devices that monitor, record, and translate the overall effect of temperature history on food quality.[36,37,124,158] TTIs help manufacturers in monitoring the supplied food stuff, allow retailers to ensure that the food products have been stored within the appropriate temperature range, and assist the buyers in getting information regarding the quality of the food product they are buying.[76,97,114,183]

There are three basic types of TTIs available in the market: (1) critical temperature indicators, (2) partial history indicators, and (3) full history indicators.[117] Critical indicators indicate the attainment of a particular temperature. Indicators of partial history of temperature display the history of time–temperature only when the temperature goes above a certain set limit.[19,34,124] Indicators of full temperature history provide continuous monitoring of changes in temperature with time.[20,90] Change in color of an indicator or migration of a temperature-dependent dye through a porous material communicates the change in food quality.[114] On the basis of working principles, TTI systems are classified as chemical, physicochemical, or biological systems, with their integrated time- and temperature-dependent change being demonstrated as an irreversible, visible color change, a mechanical change or a motion toward a color change.[124,125] TTIs are small, self-adhering labels placed over the containers or individual packages to increase visibility and interpretability.[115]

MonitorMark™ from 3M™ is available in two versions, one is a threshold indicator for monitoring distribution, and the other is a smart label for consumer information. The former is a critical temperature indicator, and it yields a response only after exceeding a determined temperature. It contains a blue dye and a substance having a selected melting point. The wick and reservoir are separated by a film strip, which is removed at the activation stage. On exceeding the critical temperature, the substance begins to melt. A blue coloring starts to appear as the substance diffuses through the porous wick. The critical temperature indicators are available in a range from −15 to 26°C. The smart label for consumer information is a partial-history indicator. When exposed to higher storage temperature than recommended or when the product reaches the end time of its shelf-life, it shows a change in color. The working principle is similar to the threshold indicator, based on the melting and diffusion of the blue dye.[36,45,51,61]

TimeStrips are smart labels monitoring elapsed time on a perishable product. It observes and records the total usage time of the product from its opening till closing. They are single-use customer-activated smart labels. They can evaluate time elapsed from minutes up to a year in the freezer, refrigerator, at normal ambient, or even at elevated temperatures. Inside the timestrip, a food-grade liquid diffuses through a special porous membrane in a consistent and repeatable way. By squeezing a start button, the liquid moves into direct contact with the membrane, thus activating the timestrip. On the top surface of the timestrip, the markers have been printed, which communicate the important time since activation. Timestrip adhered to a package or a product gets activated either automatically when the consumer opens the food package, or can manually be activated, when supplied as an external label, by a consumer, when they use the product for the first time.[36,46,61]

8.4.1.2 FRESHNESS INDICATOR

Freshness indicators are a class of intelligent devices that monitor the quality of food, thus providing actual information about the quality of food products during storage, transport, and display. Shelf-life longer than that established or exposure to harmful conditions mainly leads to freshness decay. Unlike temperature indicators, freshness indicators give the information in a direct manner about the product quality.[20,52,120] They analyze the food deterioration chemical reaction between metabolites of microbes and unify indicators thus provide a visual information about food quality.[8,114] The freshness indicators incorporated are sensitive to the changes in the metabolites concentration like concentrations of carbon dioxide, ethanol, organic acids, volatile compounds of nitrogen, glucose, or derivatives of sulfur generated during spoilage of food product.[29,76,114] Freshness indicator can be printed on the packaging film or can be incorporated inside the packaging film.

8.4.2 SENSORS

A sensor provides a quantifiable signal in response to a change in chemical, biological, or physical property, which is proportional to the measurement.[20,34] Sensors are designed to be used for the measurement of pH, humidity, temperature, and light exposure, etc.[20,128] Sensors in general contain four main components: (1) Receptor, which is the sensing part and has a selective coating which detects the composition, presence, or concentration of a particular chemical analyte via adsorption over the surface[128]; (2) Transducer, it detects the change in receptor and converts it into an output signal. Transducer is called "active" if it requires an external power supply for measurement and "passive" if not[118]; (3) Electronic unit that processes the signal; (4) A signal display unit.[34]

8.4.2.1 CHEMICAL SENSORS

Chemical sensors have two elements, receptor and transducer. The receptor is a selective chemical coating having ability to detect the presence, composition, concentration, and activity of a particular chemical or gas molecule (H_2, NO_2, CO, H_2S, CO_2, etc.) which is primarily responsible for spoilage, through gas adsorption. The receptor observes the presence of a particular chemical and converts it into a signal by transducer.[118,128] Transducers are either active or passive. Carbon-based nanomaterials, applied in chemical sensors, have a high surface area and exhibit excellent electrical and mechanical properties.[128] They are used for the detection of pathogens, chemical contaminants, and spoilage.[71,86]

8.4.2.2 BIOSENSORS

Biosensors are much like chemical sensors, except that the receptor in the biosensor is made of organic or biological material. These biological materials can be DNA, RNA, enzymes, antibodies, antigens, microbes, hormones, etc.[118,128,138] The transducers can be optical or electrochemical.[63] Electrochemical transducers are most widely used. These bioreceptors detect the target analyte, and the transducer converts biochemical signals into measurable electronic responses.[8,138] Biosensors are used to identify, measure, and record allergens and analytes like amino acids, sugars, alcohols, pathogens, lipids, etc.[20,128] They targets the produced metabolites of biochemical reactions happening in the process of food degradation.[20,118]

8.4.2.3 GAS SENSOR

Gas sensors detect the presence of a gaseous analyte in the package. It includes oxygen sensors, carbon dioxide sensors, water vapor sensors, ethanol sensors, etc.[20,54,57] These systems are based on the objective of luminescence quenching or changes in absorbance which occur by directly contacting with analyte. Optochemical sensors sense the gas analyte such as hydrogen sulfide, carbon dioxide, and volatile amines to detect the quality of the product.[133] There are three types of optochemical sensing methods including fluorescence-based system utilizing an indicator which is pH sensitive, colorimetric sensing which is absorption based, and approach of energy transfer which uses phase fluorometry for detection.[8,77,88]

8.4.2.4 EDIBLE SENSORS

These are the most interesting class of sensors. Capable of detecting food deterioration, these sensors are made from natural and biodegradable material. They don't have any adverse or hazardous effects on humans. Recently, a sensor made up of a pectin matrix has been developed which contains extract of red cabbage for colorimetric indication. Pectin, a polysaccharide found in nature, is commercially extracted from citrus fruit and apples. The red cabbage extract contains a significant quantity of anthocyanins. Anthocyanins, when exposed to pH change, have the ability to change color. The presence of amines can also be detected with anthocyanins.[20,21,182] Other films which are edible are made up of gellan gum, gelatin, and extract of red radish anthocyanin have been developed that are gas sensitive and display change in color from reddish orange to yellow, with pH ranging from 2 to 12.[20,52,168]

8.4.3 DATA CARRIERS

Data carrier devices are a different class of intelligent packaging than indicators and sensors. They do not provide information regarding the quality of the food product. Instead, they are used for automated traceability, theft prevention, or protection against counterfeit.[20,85] Food safety improves with better traceability and helps in making a better market for consumers.[20,52] The management and transport of food packages have become more convenient by the use of data carriers. Barcode labels and radio frequency identification are the utmost essential data carrier devices in the sector of food packaging and are mostly placed on the tertiary packaging (e.g., pallets, containers, etc.).[20,34]

Typically, RFID system is composed of two elements, a tag and a reader. The tag consists of two modules, a microchip used for processing and storage of information and a small antenna used for transmitting and receiving information. Reader obtains information by receiving the radio frequency.[20,58,110] Presently RFID, the technology uses two types of tags: active and passive tags. Active tags are named so based on the fact they have their own power supply (battery). Their transmission distance is in the range of 20–100 m. Communication with a reader is possible at any point of time. They are expensive and larger in size than passive tags. Passive tags have no direct power source associated with them. Incoming radio waves from the reader provide energy to transmit information. They have a short transmission range from a few centimeters to 10 m. They are activated when present in the range of RFID readers. They are less expensive and smaller in size than active tags.[20,34,128]

The use of RFID improves the identification and traceability of the product. In a supply chain, the speed and efficiency of distribution increase with the help of RFID as multiple items are monitored at every stage. It is crucial in current modern chains of supply where lots of raw or packed food material is collected from different regions for processing at a particular site, and is then distributed in different regions to consumers.[61,119] Because of their ability to identify, categorize, and manage the flow, RFIDs have been successfully applied to traceability control and supply chain management.[8,50,110] RFID is an advanced technology that provides supply chain visibility, enables fast, automated processes such as exception management and information sharing.[8,122]

8.5 SCOPE OF NANOTECHNOLOGY IN SMART PACKAGING

Various studies have been conducted to make and modify nanotechnology-based films/strips. These added nanoparticles are sensitive to the slightest change. Degradability has also been a major issue related to packaging, for which natural polymers are being used in place of petroleum-based polymers.[10,136] This change has also led the way to incorporate nanoparticles in the packaging, which has increased the properties and characteristics of packaging to many folds.

Acetobacter xylinum produces cellulose, using D-Glucose as a carbon source. It is termed as bacterial cellulose (BC). It is a renewable polymer[10,136,137] with fiber thickness in the range of 30–100 nm.[138,139] High water retention capability and high crystallinity are properties exhibited by BC that intrigue researchers to use it over other types of cellulose.[140,141] Novel films based on BC are being synthesized using polypyrrole (PPy) and bacterial cellulose PPy-Zinc oxide nanocomposites (BC-PPy-ZnO).[10] Polypyrrole is a biodegradable conducting polymer with very high conductivity in comparison to other conducting polymers. Also, its polymerization is fast and quite simple and is thermally very stable.[10,142] Zinc oxide being biocompatible, is usually used in sensors, electronic devices, antioxidant, and antibacterial material. Having a bandgap of 3.47eV, ZnO is a good semiconductor metal oxide. As BC films are porous and renewable in nature, ZnO can be easily amalgamated with BC.[49,143–145] XRD and FIR analysis of the results from the experiments done for the formation of these novel films showed that new interactions were formed in BC, PPy, and PPy-ZnO. In the case of ZnO and cellulose interaction, it was seen that ZnO particles bonded physically with the cellulose. Nano-sized cellulose fiber, nano-sized PPy, and ZnO and PPy-ZnO were confirmed on cellulose fiber surface by SEM analysis. It was also observed that film conductivity had a good relationship with food parameters like food odor, temperature, pressure, and storage time. Hence, BC-PPy and BC-PPy-ZnO films could potentially be used in food packaging and will contribute in the estimation of food spoilage and time of storage.[10]

Spoilage is a big issue in seafood like fish as enzymatic activity, and microbial contamination make them highly perishable. Because of microbial activity, amino acids, and proteins are degraded into volatile amines like trimethylamines, dimethylamines, and ammonia (NH_3). They are also known as Total Volatile Basic Nitrogen (TVB-N), broadly used as fish spoilage indicator.[143–145] New research is being focused on developing smart packaging films/strips that can trace the various factors related to food quality. For tracing fish spoilage, pH-sensitive calorimetric films are being explored the most, as they show a change in color when exposed to volatile amines generated because of degradation. These pH-sensitive films are composed of solid support and a dye.[12,148,149,178] Anthocyanin extracted from mulberry fruits is non-toxic with pH-sensitive properties was evaluated to develop such calorimetric films.[24,29,150,151,155,181] Anthocyanin also shows color stability to UV light and temperature.

ZnO shows good UV absorptivity and is also considered generally recognized as safe (GRAS). Gelatin, a film-forming agent, forms the upper layer of the bilayer calorimetric film. It is incorporated with ZnO nanoparticles which improve the light blocking property. For lower layer formation, Gellan gum, an additive which is an extracellular polysaccharide secreted by *Sphingomonas paucimobilis*, with anthocyanin, was used.[152–154] Results of the study suggested that bilayer calorimetric films were well prepared and properly functional. The SEM analysis confirmed the fabrication of the bilayer. The addition of ZnO improved the illumination stability of the bilayer significantly without interfering with the NH_3-sensing ability of film. The bilayer film did show proper color change to indicate spoilage. In terms of cost and function, the development of bilayer is very much successful, but safety issues need exhaustive evaluation before incorporating in the packaging of food products.[148]

Carbon nanoparticles find many applications in smart packaging because of their biocompatibility, availability, water solubility, and chemical inertness. Apart from these, carbon nanoparticles show luminescence bands when subjected to the visible region of the spectrum, which is quite useful in a very interesting application, that is, to be used as a composition of ink for printing functional labels for smart packaging. The carbon nanoparticle reacts with the changes in the composition of food which is visible as a change in luminescence intensity.[11,13] Putting carbon nanoparticles on packaging material through printing has come out as a very appealing way of monitoring food composition. Inkjet printing is already in use for the printing of manufacturing date, expiry date, and other product-related information at the time of packaging. Applying carbon nanoparticles-based ink to printing through inkjet printers is compatible, feasible, and as convenient as normal printing and saves the cost of setting up a separate system for nanoparticle-based ink printing. These labels can also be used for security-based printing purposes. Other than carbon nanoparticles, ZnO nanocomposites and Ag nanoparticles are also suitable for inkjet printing.[11,13,156]

8.6 POTENTIAL RISKS OF NANOTECHNOLOGY

The field of nanotechnology science is attracting great attention from researchers. Significant development has been done, and much more is yet to be done. Its application in the development of food products and food packaging is also increasing and becoming more evident. Being small in size and having high reactivity makes a good chance that they can be hazardous for the body if consumed.[114]

Possible ways for nanoparticles to enter the body are through inhalation, ingestion, or cutaneous exposure.[75,114] Once nanoparticles enter into the biological environment, their interaction with biomolecules is inevitable.[27] Studies have revealed that titania and silver nanoparticles may enter the blood circulation and, being insoluble, get accumulated in organs.[12,56,104] The liver and spleen are mainly responsible for the distribution of nanoparticles as they mediate their passage to the blood circulation from the intestine.[18] Tarantini et al., showed that silica nanoparticles (NPs) could be toxic as they induce oxidative stress, which is responsible for the induction of apoptosis and DNA damage. They exposed Caco-2, an intestinal epithelial cell line, to silica NPs. At a dosage of 32 μg/mL, there was a decrease in cell viability and an increase in ROS production in the intestinal epithelial cell line.[163,159] While, in contrast, Moos et al., revealed null toxicity of NPs of silica in the epithelial intestinal cells. Minimal induction of toxicity was displayed up to a 100 μg/cm^2 dosage.[164] Schubbe et al. also reported no cytotoxicity up to a dosage of 200 μg/mL of NPs.[165]

TiO_2 NPs are known to induce oxidative stress resulting in arrest of cell cycle, damage of DNA, or dysfunction of mitochondria.[166] Some studies also suggest that these NPs can disrupt the structure of normal microvilli present in cells of intestinal epithelium affecting the normal function of the cells, mainly absorption of nutrients.[167] However, Chalew and Schwab treated intestinal epithelial cell lines, Caco-2 and SW480, with 100 μg/mL of NPs and found no toxicity-related evidence.[18] However, cases of lung tumor because of inhalation of very high doses (10 mg/m^3) of TiO_2 are also present.

Silver NPs are known to be genotoxic, cytotoxic, and even carcinogenic. They form free radicals and lead to oxidative damage to cells and tissues.[94] Exposure to Ag NPs increases ROS production leading to mitochondrial injury and DNA damage in human lung fibroblasts and glioblastoma cells. Similar toxicities have also been reported in intestinal epithelial cells.[169]

Several nanoparticles find their way through the bloodstream affecting the blood vessel lining and its function. It may lead to a blood clot and could also cause cardiovascular diseases. Also, the possibility of migration of NPs to the fetus in pregnant women and crossing blood–brain barrier cannot be ignored.[18] Hence, further studies are of utmost importance to evaluate the migration ability of these nanoparticles in the body. Avella et al., reported a tiny amount of nanoparticle migration from nanocomposites to food, but they were within the prescribed limit of EC.[7]

The chemical and physical properties of polymer and food involved extremely affect the nanoparticle(s) migration to the matrix of food, moreover some parameters which in turn control the migration are molecular weight, particle size, diffusivity and solubility, of a specific substance in the polymer.[114]

8.7 FUTURE PROSPECTS OF NANOTECHNOLOGY

Nanotechnology-based packaging systems have pave the way for new and better packaging practice but concerns like its migration followed by the toxic effect on the body put forth a challenge on how to commercially use this technology. The existing study about their toxicity and migration are not in full agreement with each other and are not conclusive.[171,172] Henceforth, research is required to evaluate such instances. By the time, with the help of proper regulation, we can keep their concentration in check.

As per the regulations about "Plastic Food Contact Materials" made by European Union (EU) 10/2011, only those type of nanoparticles that have been denoted in Annex-I specifications of EU report are authorized and can be used in plastic packaging for food. Silica, carbon black, and titanium nitride are initially listed, and thus authorized, nanoparticle.[26,111,135] Though, the specific migration limit (SML) and SML-total (SML (T)) have not been specified in the regulation. For titanium nitride nanoparticles, a concentration of up to 20 mg/kg is allowed. For carbon black nanoparticles, the maximum use level allowed is 2.5 wt%, without the conditions for the SML (T) or SML in the regulation. Evan for silica nanoparticles, there is no specified value for SML (T) or SML (T).[26,65,101]

The available scientific data about the toxicity or migration of NPs is still at the infancy stage. The nanomaterial toxicity in packaging of food is mediated by the dosage and physiochemical characteristics. Systematic characterization and assessment are required for the safe application of nanotechnology in food packaging. More research and data are needed before NPs can be tagged as toxic or safe.[43,114] For nanomaterial detection and characterization, only a few techniques are available that have sufficiently high sensitivity and selectivity. For nanomaterial analysis inside a matrix of polymer, X-ray Diffraction (XRD), transmission electron microscopy (TEM), and X-ray fluorescence (XRF), are widely used. For detection of the nanomaterial in the food simulants, asymmetric flow field-flow fractionation (AF4) technique in association with multi-angle light scattering (MALS) is utilized.[9,35] For chemical evaluation, AF4 is associated with the techniques having specificity for elements, including inductively coupled plasma-mass spectrometry (ICP-MS).[14] ICP-MS basic form does not distinguish between nano-sized element metal and ions of metal.[32] As more study and data are required to comprehend the commercial use of nanoparticles, we should look to develop a better test that can measure the values more accurately and help us set a more definite use limit. The nanoparticles must be evaluated on a case-to-case study basis thus considering standardized conditions of test, timing of test, test temperature, and medium matter for analysis, of a particular nanoparticle. An individual toxicological assay is always vital. Also, more focus needs to be put on in in vivo and in vitro studies of metabolism of nanostructures.[75]

8.8 SUMMARY

Packaging of food protects the food from the deteriorating effect of the environment. Traditional food packaging is passive as it is not directly involved in the maintenance of the quality of food. An increase in global demand of food is pushing the food industry to look for better food packaging options. The use of nanomaterials and nanotechnology in food packaging has been seen as a more advanced and better alternative to the traditional packaging system. Smart packaging is the use of nanosensors and indicators to monitor the quality of food and its environment. The user gets real-time data about the food package. This helps the customer in better decision making. Though use of nanomaterial is revolutionizing the food packaging industry, yet it has health-related concerns also. Various studies have shown accumulation of nanoparticles in our body.

KEYWORDS

- **food quality**
- **food safety**
- **nanocomposites**
- **nanotechnology**
- **sensing technologies**
- **smart packaging**

REFERENCES

1. Abeles, F. B.; Morgan, P. W.; Saltveit, M. E. Eds.; In *Ethylene in Plant Biology*, 2nd Ed.; Academic Press Inc: San Diego, California, 1992; pp 413.
2. Ahvenainen, R. Active and Intelligent Packaging: An Introduction. Chapter 2; In *Novel Food Packaging Techniques*; Ahvenainen, R., Eds.; Woodhead Publishing Ltd.: Cambridge, UK, 2003; pp 5–21.
3. Almasi, H.; Jafarzadeh, P.; Mehryar, L. Fabrication of Novel Nanohybrids by Impregnation of CuO Nanoparticles into Bacterial Cellulose and Chitosan Nanofibers: Characterization, Antimicrobial and Release properties. *Carbohydr. Polym.* **2018,** *186*, 273–281.
4. Amin, K. A. M.; Panhuis, M. Polyelectrolyte Complex Materials from Chitosan and Gellan Gum. *Carbohydra. Polym.* **2011,** *86* (1), 352–358.
5. Appendini, P.; Hotchkiss, J. H. Review of Antimicrobial Food Packaging. *Innov. Food Sci. Emerg. Technol.* **2002,** *3* (2), 113–126.
6. Arora, A.; Padua, G. W. Nanocomposites in Food Packaging. *J. Food Sci.* **2010,** *75* (1), R43–R49.
7. AshaRani, P. V.; Low Kah Mun, G.; Hande, M. P.; Valiyaveettil, S. Cytotoxicity and Genotoxicity of Silver Nanoparticles in Human Cells. *ACS Nano* **2009,** *3* (2), 279–290.
8. Avella, M.; De Vlieger, J. J.; Errico, M. E.; Fischer, S.; Vacca, P.; Volpe, M. G. Biodegradable Starch/Clay Nanocomposite Films for Food Packaging Applications. *Food Chem.* **2005,** *93* (3), 467–474.
9. Avena-Bustillos, R. J.; Krochta, J. M.; Saltveit, M. E. Water Vapor Resistance of Red Delicious Apples and Celery Sticks Coated with Edible Caseinate-Acetylated Monoglyceride Films. *J. Food Sci.* **1997,** *62* (2), 351–354.
10. Baalousha, M.; Kammer, F. V.; Motelica-Heino, M.; Hilal, H. S.; Le Coustumer, P. Size Fractionation and Characterization of Natural Colloids by Flow-Field Flow Fractionation Coupled to Multi-Angle Laser Light Scattering. *J. Chromatogr. A* **2006,** *1104* (1-2), 272–281.

11. Bąkowska, A.; Kucharska, A. Z.; Oszmiański, J. The Effects of Heating, UV Irradiation, and Storage on Stability of the Anthocyanin–Polyphenol Co-pigment Complex. *Food Chem.* **2003,** *81* (3), 349–355.
12. Baldwin, E. A.; Nisperos, M. O.; Chen, X.; Hagenmaier, R. D. Improving Storage Life of Cut Apple and Potato with Edible Coating. *Postharvest Biol. Technol.* **1996,** *9* (2), 151–163.
13. Biji, K. B.; Ravishankar, C. N.; Mohan, C. O.; Gopal, T. S. Smart Packaging Systems for Food Applications: A Review. *J. Food Sci. Technol.* **2015,** *52* (10), 6125–6135.
14. Bovi, G. G.; Caleb, O. J.; Klaus, E.; Tintchev, F.; Rauh, C.; Mahajan, P. V. Moisture Absorption Kinetics of Fruit Pad for Packaging of Fresh Strawberry. *J. Food Eng.* **2018,** *223*, 248–254.
15. Bovi, G. G.; Caleb, O. J.; Linke, M.; Rauh, C.; Mahajan, P. V. Transpiration and Moisture Evolution in Packaged Fresh Horticultural Produce and the Role of Integrated Mathematical Models: A Review. *Biosyst. Eng.* **2016,** *150*, 24–39.
16. Bustos, A. R. M.; Encinar, J. R.; Sanz-Medel, A. Mass Spectrometry for the Characterization of Nanoparticles. *Anal. Bioanal. Chem.* **2013,** *405* (17), 5637–5643.
17. Cagri, A.; Ustunol, Z.; Ryser, E. T. Antimicrobial Edible Films and Coatings. *J. Food Protect.* **2004,** *67* (4), 833–848.
18. Cai, X.; Du, X.; Cui, D.; Wang, X.; Yang, Z.; Zhu, G. Improvement of Stability of Blueberry Anthocyanins by Carboxymethyl Starch/ Xanthan Gum Combinations Microencapsulation. *Food Hydrocoll.* **2019,** *91*, 238–245.
19. Carrero-Sanchez, J. C.; Elias, A. L.; Mancilla, R.; Arrellin, G.; Terrones, H.; Laclette, J. P.; Terrones, M. J. N. L. Biocompatibility and Toxicological Studies of Carbon Nanotubes Doped with Nitrogen. *Nano Lett.* **2006,** *6* (8), 1609–1616.
20. Cavaliere, E.; De Cesari, S.; Landini, G.; Riccobono, E.; Pallecchi, L.; Rossolini, G. M.; Gavioli, L. Highly Bactericidal Ag Nanoparticle Films Obtained by Cluster Beam Deposition. *Nanomed. Nanotechnol. Biol. Med.* **2015,** *11* (6), 1417–1423.
21. Chalew, T. E. A.; Schwab, K. J. Toxicity of Commercially Available Engineered Nanoparticles to Caco-2 and SW480 Human Intestinal Epithelial Cells. *Cell Biol. Toxicol.* **2013,** *29* (2), 101–116.
22. Chen, Z.; Lu, C. Humidity Sensors: A Review of Materials and Mechanisms. *Sensor Lett.* **2005,** *3* (4), 274–295.
23. Choi, I.; Lee, J. Y.; Lacroix, M.; Han, J. Intelligent pH Indicator Film Composed of Agar/ Potato Starch and Anthocyanin Extracts from Purple Sweet Potato. *Food Chem.* **2017,** *218*, 122–128.
24. Choi, M. Y.; Choi, D.; Jin, M. J.; Kim, I.; Kim, S. H.; Choi, J. Y.; Lee, S. Y.; Kim, J. M.; Kim, S. W. Mechanically Powered Transparent Flexible Charge-Generating Nanodevices with Piezoelectric ZnO Nanorods. *Adv. Mater.* **2009,** *21* (2), 2185–2189.
25. Ciabocco, M.; Cancemi, P.; Saladino, M. L.; Caponetti, E.; Alduina, R.; Berrettoni, M. Synthesis and Antibacterial Activity of Iron-Hexacyanocobaltate Nanoparticles. *JBIC J. Biol. Inorg. Chem.* **2018,** *23*, 385–398.
26. Czaja, W.; Romanovicz, D.; Malcolm Brown, R. Structural Investigations of Microbial Cellulose Produced in Stationary and Agitated Culture. *Cellulose* **2004,** *11* (3), 403–411.
27. Dalmas, F.; Cavaillé, J. Y.; Gauthier, C.; Chazeau, L.; Dendievel, R. Viscoelastic Behavior and Electrical Properties of Flexible Nanofiber Filled Polymer Nanocomposites. Influence of processing conditions. *Compos. Sci. Technol.* **2007,** *67* (5), 829–839.

28. De Azeredo, H. M. Nanocomposites for Food Packaging Applications. *Food Res. Int.* **2009,** *42* (9), 1240–1253.
29. Decher, G.; Schlenoff, J. B. Eds.; In *Multilayer Thin Films: Sequential Assembly of Nanocomposite Materials*; Wiley-VCH: Weinheim, 2003; pp 510.
30. Dimitrijevic, M.; Karabasil, N.; Boskovic, M.; Teodorovic, V.; Vasilev, D.; Djordjevic, V.; Kilibarda, N.; Cobanovic, N. Safety Aspects of Nanotechnology Applications in Food Packaging. *Procedia Food Sci.* **2015,** *5*, 57–60.
31. Dobrucka, R.; Cierpiszewski, R. Active and Intelligent Packaging Food-Research and Development - A Review. *Polish J. Food Nutr. Sci.* **2014,** *64* (1), 7–15.
32. Drago, E.; Campardelli, R.; Pettinato, M.; Perego, P. Innovations in Smart Packaging Concepts for Food: An Extensive Review. *Foods* **2020,** *9 (11)*, 628.
33. Dudnyk, I.; Janeček, E. R.; Vaucher-Joset, J.; Stellacci, F. Edible Sensors for Meat and Seafood Freshness. *Sens. Actuators B Chem.* **2018,** *259*, 1108–1112.
34. Duncan, T. V. Applications of Nanotechnology in Food Packaging and Food Safety: Barrier Materials, Antimicrobials and Sensors. *J. Colloid. Interface Sci.* **2011,** *363* (1), 1–24.
35. El-Ghaouth, A.; Arul, J.; Ponnampalam, R.; Boulet, M. Use of Chitosan Coating to Reduce Water Loss and Maintain Quality of Cucumber and Bell Pepper Fruits. *J. Food Process. Preserv.* **1991,** *15* (5), 359–368.
36. Emamifar, A.; Kadivar, M.; Shahedi, M.; Soleimanian-Zad, S. Effect of Nanocomposite Packaging Containing Ag and ZnO on Inactivation of Lactobacillus plantarum in Orange Juice. *Food Control* **2011,** *22* (3-4), 408–413.
37. Esmailzadeh, H.; Sangpour, P.; Shahraz, F.; Hejazi, J.; Khaksar, R. Effect of Nanocomposite Packaging Containing ZnO on Growth of Bacillus subtilis and Enterobacter aerogenes. *Mater. Sci. Eng. C* **2016,** *58*, 1058–1063.
38. European Commission Regulation No. 450/2009 of 29 May 2009 on Active and Intelligent Materials and Articles Intended to Come into Contact with Food. *Official Journal of European Union,* 2009; pp 9. https://eur-lex.europa.eu/LexUriServ/LexUriServ.do?uri=OJ:L:2009:135:0003:0011:EN:PDF (Accessed Apr 11, 2021).
39. Fabricius, A. L.; Duester, L.; Meermann, B.; Ternes, T. A. ICP-MS-Based Characterization of Inorganic Nanoparticles: Sample Preparation and Off-line Fractionation Strategies. *Anal. Bioanal. Chem.* **2014,** **406** (2), 467–479.
40. Farhoodi, M. Nanocomposite Materials for Food Packaging Applications: Characterization and Safety Evaluation. *Food Eng. Rev.* **2016,** *8* (1), 35–51.
41. Fraunhofer, W.; Winter, G.; Coester, C. Asymmetrical Flow Field-Flow Fractionation and Multiangle Light Scattering for Analysis of Gelatin Nanoparticle Drug Carrier Systems. *Anal. Chem.* **2004,** *76* (7), 1909–1920.
42. Fuertes, G.; Soto, I.; Carrasco, R.; Vargas, M.; Sabattin, J.; Lagos, C. Intelligent Packaging Systems: Sensors and Nanosensors to Monitor Food Quality and Safety. *J. Sens.* **2016,** *2016* (4046061), 1–8.
43. Fujishima, A.; Rao, T. N.; Tryk, D. A. Titanium Dioxide Photocatalysis. *J. Photochem. Photobiol. Phytochem. Rev.* **2000,** *1*, 1–21.
44. Gaikwad, K. K.; Lee, Y. S. Current Scenario of Gas Scavenging Systems Used in Active Packaging-A Review. *Korean J. Packag. Sci. Technol.* **2017,** *23*, 109–117.
45. Gaikwad, K. K.; Singh, S.; Ajji, A. Moisture Absorbers for Food Packaging Applications. *Environ. Chem. Lett.* **2019,** *17*, 609–628.

46. Gaikwad, K. K.; Singh, S.; Negi, Y. S. Ethylene Scavengers for Active Packaging of Fresh Food Produce. *Environ. Chem. Lett.* **2020,** *18* (2), 269–284.
47. Gelin, K.; Bodin, A.; Gatenholm, P.; Mihranyan, A.; Edwards, K.; Strømme, M. Characterization of Water in Bacterial Cellulose using Dielectric Spectroscopy and Electron Microscopy. *Polymer* **2007**, *48* (26), 7623–7631.
48. Ghaani, M.; Cozzolino, C. A.; Castelli, G.; Farris, S. An Overview of the Intelligent Packaging Technologies in the Food Sector. *Trends Food Sci. Technol.* **2016,** *51*, 1–11.
49. Ghanbarzadeh, B.; Oleyaei, S. A.; Almasi, H. Nano-structured Materials Utilized in Biopolymer-Based Plastics for Food Packaging Applications. *Crit. Rev. Food Sci. Nutr.* **2015,** *55*, 1699–1723.
50. Ghoshal G. Recent Trends in Active, Smart, and Intelligent Packaging for Food Products. *Food Packag. Preserv.* **2018,** *9*, 343–374.
51. Giannakourou, M. C.; Koutsoumanis, K.; Nychas, G. J. E.; Taoukis, P. S. Field Evaluation of the Application of Time Temperature Integrators for Monitoring Fish Quality in the Chill Chain. *Int. J. Food Microbiol.* **2005,** *102* (3), 323–336.
52. Göransson, M.; Nilsson, F. J. Temperature Performance and Food Shelf-life Accuracy in Cold Food Supply Chains: Insights from Multiple Field Studies. *Food Contr.* **2018,** *86*, 332–341.
53. Guilbert, S.; Cuq, B.; Gontard, N. Recent Innovations in Edible and/or Biodegradable Packaging Materials. *Food Addit. Contam.* **1997,** *14* (6), 741–751.
54. Gutiérrez, T. J.; Ponce, A. G.; Alvarez, V. A. Nano Clays from Natural and Modified Montmorillonite with and Without Added Blueberry Extract for Active and Intelligent Food Nanopackaging Materials. *Mater. Chem. Phys.* **2017,** *194*, 283–292.
55. Han, J. H. Antimicrobial Food Packaging. In *Novel Food Packaging Techniques*, 2003; vol *8*, pp 50–70.
56. Han, J. H.; Ho, C. H. L.; Rodrigues, E. T. Intelligent Packaging. Chapter 9; In *Innovations in Food Packaging*; Han, J. H., Ed.; Elsevier: Amsterdam, The Netherlands, 2005; pp 138–155.
57. Hao, X.; Ma, J.; Zhang, D.; Yang, T.; Ma, H.; Yang, Y.; Cheng, C.; Huang, J. Thickness Dependence of Structural, Optical and Electrical Properties of ZnO: Al Films Prepared on Flexible Substrates. *Appl. Surf. Sci.* **2001,** *183* (1-2), 137–142.
58. He, X.; Hwang, H. M. Nanotechnology in Food Science: Functionality, Applicability, and Safety Assessment. *J. Food Drug Anal.* **2016,** *24* (4), 671–681.
59. Hogan, S. A.; Kerry, J. P. Smart Packaging of Meat and Poultry Products. Chapter 3; In *Smart Packaging Technologies for Fast Moving Consumer Goods*; Kerry, J., Butler, P., Eds.; John Wiley & Sons Ltd: West Sussex, England, 2008; pp 33–59.
60. Hosseini, S. F.; Gómez-Guillén, M. C. A State-of-the-Art Review on the Elaboration of Fish Gelatin as Bioactive Packaging: Special Emphasis on Nanotechnology-Based Approaches. *Trends Food Sci. Technol.* **2018,** *79*, 125–135.
61. Hrytsenko, O.; Shvalagin, V.; Grodziuk, G.; Kompanets, M. The Use of Carbon Nanoparticles for Inkjet-Printed Functional Labels for Smart Packaging. *J. Nanomater.* **2018,** *2018* (6485654), 1–10.
62. Hrytsenko, O.; Shvalagin, V.; Grodziuk, G.; Granchak, V. Influence of Parameters of Screen Printing on Photoluminescence Properties of Nanophotonic Labels for Smart Packaging. *J. Nanotechnol.* **2017,** *2017* (7125682), 1–12.
63. http://solutions.3m.com/wps/portal/3M/en_US/Microbiology/FoodSafety/product-information /product-catalog (accessed Apr 11, 2021).

64. https://timestrip.com/industries/food/ (accessed Apr 11, 2021).
65. Huang, S.; Xiong, Y.; Zou, Y.; Dong, Q.; Ding, F.; Liu, X.; Li, H. A Novel Colorimetric Indicator Based on Agar Incorporated with *Arnebia euchroma* Root Extracts for Monitoring Fish Freshness. *Food Hydrocoll.* **2019,** *90*, 198–205.
66. Huang, Y.; Mei, L.; Chen, X.; Wang, Q. Recent Developments in Food Packaging Based on Nanomaterials. *Nanomaterials* **2018,** *8* (10), 830.
67. Jia, X.; Li, Y.; Cheng, Q.; Zhang, S.; Zhang, B.; Preparation and Properties of Poly (Vinyl Alcohol)/ Silica Nanocomposites Derived from Copolymerization of Vinyl Silica Nanoparticles and Vinyl Acetate. *Eur. Polym. J.* **2007,** *43* (4), 1123–1131.
68. Jones, N.; Ray, B.; Ranjit, K. T.; Manna, A. C. Antibacterial Activity of ZnO Nanoparticle Suspensions on a Broad Spectrum of Microorganisms. *FEMS Microbiol. Lett.* **2008,** *279* (1), 71–76.
69. Jones, P.; Clarke-Hill, C.; Shears, P.; Comfort, D.; Hillier, D. Radio Frequency Identification (RFID) in the UK: Opportunities and Challenges. *Int. J. Retail Distrib. Manag.* **2004,** *32* (3), 164–171.
70. Joseph, T.; Morrison, M. *Nano-Forum Report*: *Nanotechnology in Agriculture and Food*, 2006; p 15. https://www.nanowerk.com/nanotechnology/reports/reportpdf/report61.pdf (accessed Apr 11, 2021).
71. Jung, M.; Kim, J.; Koo, H.; Lee, W.; Subramanian, V.; Cho, G. Roll-to-Roll Gravure with Nanomaterials for Printing Smart Packaging. *J. Nanosci. Nanotechnol.* **2014,** *14* (2), 1303–1317.
72. Kalpana, S.; Priyadarshini, S. R.; Leena, M. M.; Moses, J. A.; Anandharamakrishnan, C. Intelligent Packaging: Trends and Applications in Food Systems. *Trends Food Sci. Technol.* **2019,** *93*, 145–157.
73. Kang, S.; Pinault, M.; Pfefferle, L. D.; Elimelech, M. Single-Walled Carbon Nanotubes Exhibit Strong Antimicrobial Activity. *Langmuir* **2007,** *23* (17), 8670–8673.
74. Kerry, J. P.; O'grady, M. N.; Hogan, S. A. Past, Current and Potential Utilization of Active and Intelligent Packaging Systems for Meat and Muscle-Based Products: A Review. *Meat Sci.* **2006,** *74* (1), 113–130.
75. Kim, J. Y.; Han, S. I.; Hong, S. Effect of Modified Carbon Nanotube on the Properties of Aromatic Polyester Nanocomposites. *Polymer* **2008,** *49* (15), 3335–3345.
76. Kim, Y. S.; Kim, J. S.; Cho, H. S.; Rha, D. S.; Kim, J. M.; Park, J. D.; Choi, B. S.; Lim, R.; Chang, H. K.; Chung, Y. H.; Kwon, I. H. Twenty-Eight-Day Oral Toxicity, Genotoxicity, and Gender-Related Tissue Distribution of Silver Nanoparticles in Sprague-Dawley Rats. *Inhal. Toxicol.* **2008,** *20* (6), 575–583.
77. Klemm, D.; Heublein, B.; Fink, H. P.; Bohn, A. Cellulose: Fascinating Biopolymer and Sustainable Raw Material. *Angewandte Chemie Int. Ed.* **2005,** *44* (22), 3358–3393.
78. Ko, H. U.; Mun, S.; Min, S. K.; Kim, G. W.; Kim, J. Fabrication of Cellulose ZnO Hybrid Nanocomposite and Its Strain Sensing Behavior. *Materials* **2014,** *7* (10), 7000–7009.
79. Kress-Rogers, E.; and Brimelow C. J. B. Eds.; In *Instrumentation and Sensors for the Food Industry*, 2nd Ed.; CRC Press: Boca Raton, 1998; pp 817.
80. Kumar, P.; Reinitz, H. W.; Simunovic, J.; Sandeep, K. P.; Franzon, P. D. Overview of RFID Technology and Its Applications in the Food Industry. *J. Food Sci.* **2009,** *74* (8), R101–R106.
81. Kumar, R.; Münstedt, H. Silver Ion Release from Antimicrobial Polyamide/Silver Composites. *Biomaterials* **2005,** *26* (14), 2081–2088.
82. Kuswandi, B. Environmentally Friendly Food Nano-Packaging. *Environ. Chem. Lett.* **2017,** *15* (2), 205–221.

83. Kuswandi, B.; Wicaksono, Y.; Abdullah, A.; Heng, L. Y.; Ahmad, M. Smart Packaging: Sensors for Monitoring of Food Quality and Safety. In *Sensing and Instrumentation for Food Quality and Safety*, 2011; vol *5* (3), pp 137–146.
84. Labuza, T. P.; Breene, W. M. Applications of "Active Packaging" for Improvement of Shelf-life and Nutritional Quality of Fresh and Extended Shelf-Life Foods. *J. Food Process. Preserv.* **1989,** *13* (1), 1–69.
85. Lamba, A.; Garg, V. Recent Innovations in Food Packaging: A Review. *Int. J. Food Sci. Nutr.* **2019,** *4*, 123–129.
86. Lee, H. S.; Im, S. J.; Kim, J. H.; Kim, H. J.; Kim, J. P.; Min, B. R. Polyamide Thin-Film Nanofiltration Membranes Containing TiO2 Nanoparticles. *Desalination* **2008,** *219* (1-3), 48–56.
87. Lee, J.; Lee, D.; Lim, D.; Yang, K. Structural, Electrical and Optical Properties of ZnO: Al Films Deposited on Flexible Organic Substrates for Solar Cell Applications. *Thin Solid Films* **2007,** *515* (15), 6094–6098.
88. Legislation, European Commission: https://ec.europa.eu/food/safety/chemical_safety/food_contact_materials/legislation_en (accessed June 19, 2021).
89. Li, Q.; Mahendra, S.; Lyon, D. Y.; Brunet, L.; Liga, M. V.; Li, D.; Alvarez, P. J. Antimicrobial Nanomaterials for Water Disinfection and Microbial Control: Potential Applications and Implications. *Water Res.* **2008,** *42* (18), 4591–4602.
90. Li, Y. C.; Schulz, J.; Grunlan, J. C. Polyelectrolyte/Nanosilicate Thin-Film Assemblies: Influence of pH on Growth, Mechanical Behavior, and Flammability. *ACS Appl. Mater. Interfaces* **2009,** *1*, 2338–2347.
91. Liao, F.; Chen, C.; Subramanian, V. Organic TFTs as Gas Sensors for Electronic Nose Applications. *Sens. Actuators B Chem.* **2005,** *107* (2), 849–855.
92. Liu, S.; Li, X.; Chen, L.; Li, L.; Li, B.; Zhu, J.; Liang, X. Investigating the H2O/O2 Selective Permeability from A View of Multi-Scale Structure of Starch/Sio2 Nanocomposites. *Carbohydr. Polym.* **2017,** *173*, 143–149.
93. Liu, S. F.; Petty, A. R.; Sazama, G. T.; Swager, T. M. Single-Walled Carbon Nanotube/Metalloporphyrin Composites for the Chemi-resistive Detection of Amines and Meat Spoilage. *Angewandte Chemie Int. Ed.* **2015,** *54* (22), 6554–6557.
94. Liu, Y.; Chakrabartty, S.; Alocilja, E. C. Fundamental Building Blocks for Molecular Biowire Based Forward Error-Correcting Biosensors. *Nanotechnology* **2007,** *18* (42), 424017.
95. López-Rubio, A.; Almenar, E.; Hernandez-Muñoz, P.; Lagarón, J. M.; Catalá, R.; Gavara, R. Overview of Active Polymer-Based Packaging Technologies for Food Applications. *Food Rev. Int.* **2004,** *20*, 357–387.
96. Lotti, C.; Isaac, C. S.; Branciforti, M. C.; Alves, R. M.; Liberman, S.; Bretas, R. E. Rheological, Mechanical and Transport Properties of Blown Films of High-Density Polyethylene Nanocomposites. *Eur. Polym. J.* **2008,** *44* (5), 1346–1357.
97. Macwan, D. P.; Dave, P. N.; Chaturvedi, S. A Review on Nano-TiO_2 Sol–Gel Type Syntheses and Its Applications. *J. Mater. Sci.* **2011,** *46* (11), 3669–3686.
98. Maisanaba, S.; Pichardo, S.; Puerto, M.; Gutiérrez-Praena, D.; Camean, A. M.; Jos, A. Toxicological Evaluation of Clay Minerals and Derived Nanocomposites: A Review. *Environ. Res.* **2015,** *138*, 233–254.
99. McCracken, C.; Dutta, P. K.; Waldman, W. J. Critical Assessment of Toxicological Effects of Ingested Nanoparticles. *Environ. Sci. Nano* **2016,** *3* (2), 256–282.

100. Mihindukulasuriya, S. D. F.; Lim, L. T. Nanotechnology Development in Food Packaging: A Review. *Trends Food Sci. Technol.* **2014,** *40* (2), 149–167.
101. Mills, A.; Chang, Q.; McMurray, N. Equilibrium Studies on Colorimetric Plastic Film Sensors for Carbon Dioxide. *Anal. Chem.* **1992**, *64* (13), 1383–1389.
102. Miltz, J.; Passy, N.; Mannheim, C. H. Trends and Applications of Active Packaging Systems. In *Food and Packaging Materials-Chemical Interaction*; Jagersta, M., Ohlsson, M., Eds.; The Royal Society of Chemistry: London, England, 1995; pp 201–210.
103. Mirzadeh, A.; Kokabi, M. The Effect of Composition and Draw-Down Ratio on Morphology and Oxygen Permeability of Polypropylene Nanocomposite Blown Films. *Eur. Polym. J.* **2007,** *43* (9), 3757–3765.
104. Mittal, V. Polymer Layered Silicate Nanocomposites: A Review. *Materials* **2009,** *2* (3), 992–1057.
105. Moghadam, A. D.; Omrani, E.; Menezes, P. L.; Rohatgi, P. K. Mechanical and Tribological Properties of Self-Lubricating Metal Matrix Nanocomposites Reinforced by Carbon Nanotubes (CNTs) and Graphene–A Review. *Compos. B Eng.* **2015,** *77*, 402–420.
106. Moos, P. J.; Olszewski, K.; Honeggar, M.; Cassidy, P.; Leachman, S.; Woessner, D.; Cutler, N. S.; Veranth, J. M. Responses of Human Cells to Zno Nanoparticles: A Gene Transcription Study. *Metallomics* **2011,** *3* (11), 1199–1211.
107. Morillon, V.; Debeaufort, F.; Blond, G.; Capelle, M.; Voilley, A. Factors Affecting the Moisture Permeability of Lipid-Based Edible Films: A Review. *Crit. Rev. Food Sci. Nutr.* **2002,** *42* (1), 67–89.
108. Morones, J. R.; Elechiguerra, J. L.; Camacho, A., Holt, K., Kouri, J. B., Ramírez, J. T. Yacaman, M. J. The Bactericidal Effect of Silver Nanoparticles. *Nanotechnology* **2005,** *16* (10), 2346.
109. Müller, K.; Bugnicourt, E.; Latorre, M.; Jorda, M.; Echegoyen Sanz, Y.; Lagaron, J. M.; Miesbauer, O.; Bianchin, A.; Hankin, S.; Bölz, U.; Perez, G. Review on The Processing and Properties of Polymer Nanocomposites and Nanocoatings and Their Applications in the Packaging, Automotive and Solar Energy Fields. *Nanomaterials* **2017,** *7* (4), 74.
110. Müller, P.; Schmid, M. Intelligent Packaging in the Food Sector: A Brief Overview. *Foods* **2019,** *8* (1), 16.
111. Nachay, K. Analyzing Nanotechnology. *Food Technol.* **2007,** *61* (1), 34–36.
112. Naicker, P. K.; Cummings, P. T.; Zhang, H.; Banfield, J. F. Characterization of Titanium Dioxide Nanoparticles Using Molecular Dynamics Simulations. *J. Phys. Chem. B* **2005,** *109* (32), 15243–15249.
113. Neurauter, G.; Klimant, I.; Wolfbeis, O. S. Microsecond Lifetime-Based Optical Carbon Dioxide Sensor Using Luminescence Resonance Energy Transfer. *Anal. Chim. Acta* **1999,** *382* (1-2), 67–75.
114. Nielsen, L. E. Models for the Permeability of Filled Polymer Systems. *J. Macromol. Sci. Chem.* **1967,** *1* (5), 929–942.
115. O'Grady, M. N.; Kerry, J. P. Smart Packaging Technologies and Their Application in Conventional Meat Packaging Systems. Chapter 19; In *Meat Biotechnology*; Toldra, F., Ed.; Springer: London, UK, 2008; pp 425–451.
116. Ozdemir, M.; Floros, J. D. Active Food Packaging Technologies. *Crit. Rev. Food Sci. Nutr.* **2004,** *44*, 185–193.
117. Pal, M. Nanotechnology: A New Approach in Food Packaging. *J. Food Microbiol. Safety Hygiene* **2017,** *2* (02), 8–9.

118. Pang, J.; Liu, X.; Zhang, X.; Wu, Y.; Sun, R. Fabrication of Cellulose Film with Enhanced Mechanical Properties in Ionic Liquid 1-Allyl-3-Methylimidaxolium Chloride (AmimCl). *Materials* **2013,** *6* (4), 1270–1284.
119. Petersson, L.; Oksman, K. Biopolymer Based Nanocomposites: Comparing Layered Silicates and Microcrystalline Cellulose as Nano reinforcement. *Compos. Sci. Technol.* **2006,** *66* (13), 2187–2196.
120. Pirsa, S.; Shamusi, T.; Kia, E.M. Smart Films Based on Bacterial Cellulose Nanofibers Modified by Conductive Polypyrrole and Zinc Oxide Nanoparticles. *J. Appl. Polym. Sci.* **2018,** *135* (34), E-article 46617.
121. Pourjavaher, S.; Almasi, H.; Meshkini, S.; Pirsa, S.; Parandi, E. Development of a Colorimetric pH Indicator Based on Bacterial Cellulose Nanofibers and Red Cabbage (*Brassica oleraceae*) Extract. *Carbohydr. Polym.* **2017,** *156*, 193–201.
122. Pradhan, N.; Singh, S.; Ojha, N.; Shrivastava, A.; Barla, A.; Rai, V.; Bose, S. Facets of Nanotechnology as Seen in Food Processing, Packaging, and Preservation Industry. *BioMed Res. Int.* **2015,** *2015* (365672), 1–18.
123. Prasad, P.; Kochhar, A. Active Packaging in Food Industry: A Review. *J. Environ. Sci. Toxicol. Food Technol.* **2014,** *8* (5), 1–7.
124. Prashantha, K.; Soulestin, J.; Lacrampe, M. F.; Krawczak, P.; Dupin, G.; Claes, M. Masterbatch-Based Multi-Walled Carbon Nanotube Filled Polypropylene Nanocomposites: Assessment of Rheological and Mechanical Properties. *Compos. Sci. Technol.* **2009,** *69* (11-12), 1756–1763.
125. Quero, F.; Padilla, C.; Campos, V.; Luengo, J.; Caballero, L.; Melo, F.; Li, Q.; Eichhorn, S. J.; Enrione, J. Stress Transfer and Matrix-Cohesive Fracture Mechanism in Microfibrillated Cellulose-Gelatin Nanocomposite Films. *Carbohydr. Polym.* **2018,** *195*, 89–98.
126. Ranjan, S.; Dasgupta, N.; Chakraborty, A. R.; Samuel, S. M.; Ramalingam, C.; Shanker, R.; Kumar, A. Nanoscience and Nanotechnologies in Food Industries: Opportunities and Research Trends. *J. Nanopart. Res.* **2014,** *16* (6), 1–23.
127. Ray, S. S.; Okamoto, M. Polymer/Layered Silicate Nanocomposites: A Review from Preparation to Processing. *Prog. Polym. Sci.* **2003,** *28*, 1539–1641.
128. Realini, C. E.; Marcos, B. Active and Intelligent Packaging Systems for a Modern Society. *Meat Sci.* **2014,** *98* (3), 404–419.
129. Regulation, H. A. T., Commission Implementing Regulation (EU) No 406/2011. *Official Journal of the European Union*, 2011; vol *54*, pp 11–12
130. Reig, C. S.; Lopez, A. D.; Ramos, M. H.; Ballester, V. A. C. Nanomaterials: A Map for Their Selection in Food Packaging Applications. *Packag. Technol. Sci.* **2014,** *27* (11), 839–866.
131. Restuccia, D.; Spizzirri, U. G.; Parisi, O. I.; Cirillo, G.; Curcio, M.; Iemma, F.; Puoci, F.; Vinci, G.; Picci, N. New EU Regulation Aspects and Global Market of Active and Intelligent Packaging for Food Industry Applications. *Food Contr.* **2010,** *21* (11), 1425–1435.
132. Rhim, J. W. Increase in Water Vapor Barrier Property of Biopolymer-Based Edible Films and Coatings by Compositing with Lipid Materials. *Food Sci. Biotechnol.* **2004,** *13* (4), 528–535.
133. Rhim, J. W.; Park, H. M.; Ha, C. S. Bio-nanocomposites for Food Packaging Applications. *Prog. Polym. Sci.* **2013,** *38* (10-11), 1629–1652.

134. Robertson, G. L. Ed.; In *The Stability and Shelf – Life of Food*; Woodhead Publishing: London, 2016; pp 77–106.
135. Roilo, D.; Maestri, C. A.; Scarpa, M.; Bettotti, P.; Checchetto, R. Gas Barrier and Optical Properties of Cellulose Nanofiber Coatings with Dispersed TiO_2 Nanoparticles. *Surf. Coat. Technol.* **2018,** *343*, 131–137.
136. Rooney, M. L. Introduction to Active Food Packaging Technologies. Chapter 5; In *Innovations in Food Packaging*; Han, J. H., Ed.; Elsevier Ltd, 2005**;** pp. 63–79.
137. Rux, G.; Mahajan, P. V.; Linke, M.; Pant, A.; Sängerlaub, S.; Caleb, O. J.; Geyer, M. Humidity-Regulating Trays: Moisture Absorption Kinetics and Applications for Fresh Produce Packaging. *Food Bioproc. Technol.* **2016,** *9*, 709–716.
138. Sadeghi, K.; Lee, Y.; Seo, J. Ethylene Scavenging Systems in Packaging of Fresh Produce: A Review. *Food Rev. Int.* **2019,** *3*, 1–22.
139. Sarac, A.; Absi, N.; Dauzère-Pérès, S. Literature Review on the Impact of RFID Technologies on Supply Chain Management. *Int. J. Prod. Econ.* **2010,** *128* (1), 77–95.
140. Sarfraz, J.; Gulin-Sarfraz, T.; Nilsen-Nygaard, J.; Pettersen, M. K. Nanocomposites for Food Packaging Applications: An Overview. *Nanomaterials* **2021,** *11* (1), 10–19.
141. Sawai, J., Quantitative Evaluation of Antibacterial Activities of Metallic Oxide Powders (ZnO, MgO and CaO_2) by Conductimetric Assay. *J. Microbiol. Meth.* **2003,** *54* (2), 177–182.
142. Schübbe, S.; Schumann, C.; Cavelius, C.; Koch, M.; Müller, T.; Kraegeloh, A. Size-Dependent Localization and Quantitative Evaluation of the Intracellular Migration of Silica Nanoparticles in Caco-2 Cells. *Chem. Mater.* **2012,** *24* (5), 914–923.
143. Shariatinia, Z.; Fazli, M. Mechanical Properties and Antibacterial Activities of Novel Nanobiocomposite Films of Chitosan and Starch. *Food Hydrocoll.* **2015,** *46*, 112–124.
144. Sharma, C.; Dhiman, R.; Rokana, N.; Panwar, H. Nanotechnology: An Untapped Resource for Food Packaging. *Front. Microbiol.* **2017**, *8*, 1735–1742.
145. Shetty, J. M. Time Temperature Indicators for Monitoring Environment Parameters During Transport and Storage of Perishables: Review. *Environ. Conserv. J.* **2018,** *19*, 101–106.
146. Shi, H.; Magaye, R.; Castranova, V.; Zhao, J. Titanium Dioxide Nanoparticles: A Review of Current Toxicological Data. *Part. Fiber Toxicol.* **2013,** *10* (1), 1–33.
147. Shi, L.; Zhao, Y.; Zhang, X.; Su, H.; Tan, T. Antibacterial and Anti-Mildew Behavior of Chitosan/Nano-TiO2 Composite Emulsion. *Korean J. Chem. Eng.* **2008,** *25* (6), 1434–1438.
148. Singh, R. P. Scientific Principles of Shelf-life Evaluation. In *Shelf-life Evaluation of Foods*; Man, C. M. D., Jones, A., Eds.; Springer: Boston, MA, 1994; pp 3–26.
149. Siracusa, V.; Lotti, N. Intelligent Packaging to Improve Shelf-life. Chapter 8; In *Food Quality and Shelf-life*; Galanakis, C. M., Ed.; Academic Press: Vienna, Austria, 2019; pp 261–279.
150. Störmer, A.; Bott, J.; Kemmer, D.; Franz, R.; Critical Review of the Migration Potential of Nanoparticles in Food Contact Plastics. *Trends Food Sci. Technol.* **2017,** *63*, 39–50.
151. Subramanian, V.; Fréchet, J. M.; Chang, P. C.; Huang, D. C.; Lee, J. B.; Molesa, S. E.; Murphy, A. R.; Redinger, D. R.; Volkman, S. K. Progress Toward Development of All-Printed RFID tags: Materials, Processes, and Devices. *Proc. IEEE Food Sci.* **2005,** *93* (7), 1330–1338.
152. Suppakul, P. Intelligent Packaging. Chapter 38; In *Handbook of Frozen Food Processing and Packaging*; Sun, D.W., Ed.; CRC Press: Boca Raton, USA, 2012; pp 835–858.

153. Svensson, A.; Nicklasson, E.; Harrah, T.; Panilaitis, B.; Kaplan, D. L.; Brittberg, M.; Gatenholm, P. Bacterial Cellulose as a Potential Scaffold for Tissue Engineering of Cartilage. *Biomaterials* **2005,** *26* (4), 419–431.
154. Swaroop, C.; Shukla, M. Nano-magnesium Oxide Reinforced Polylactic Acid Biofilms for Food Packaging Applications. *Int. J. Biol. Macromol.* **2018,** *113*, 729–736.
155. Tajima, M. Strategic Value of RFID in Supply Chain Management. *J. Purch. Supply Manag.* **2007**, *13* (4), 261–273.
156. Tan, W.; Zhang, Y.; Szeto, Y. S.; Liao, L. A Novel Method to Prepare Chitosan/Montmorillonite Nanocomposites in the Presence of Hydroxy-Aluminum Oligomeric Cations. *Compos. Sci. Technol.* **2008,** *68* (14), 2917–2921.
157. Taoukis, P. S.; Labuza, T. P. Time-Temperature Indicators (TTIs). Chapter 6; In *Novel Food Packaging Techniques*; Ahvenainen, R., Ed.; Woodhead Publishing Ltd.: Cambridge, UK, 2003; pp 103–126.
158. Taoukis, P. S.; Labuza, T. P. Applicability of Time-Temperature Indicators as Shelf-life Monitors of Food Products. *J. Food Sci.* **1989,** *54* (4), 783–788.
159. Tarantini, A.; Lanceleur, R.; Mourot, A.; Lavault, M. T.; Casterou, G.; Jarry, G.; Hogeveen, K.; Fessard, V. Toxicity, Genotoxicity and Proinflammatory Effects of Amorphous Nanosilica in the Human Intestinal Caco-2 Cell Line. *Toxicol. In Vitro* **2015,** *29* (2), 398–407.
160. Thakur, V. K; Gupta, R. K. Recent Progress on Ferroelectric Polymer-Based Nanocomposites for High Energy Density Capacitors: Synthesis, Dielectric Properties, and Future Aspects. *Chem. Rev.* **2016,** *116* (7), 4260–4317.
161. Utto, W.; Mawson, J.; Bronlund, J. E.; Wong, K. K. Y. Active Packaging Technologies for Horticultural Produce. *Food New Zeal.* **2005,** *2005,* 1–12.
162. Vanderroost, M.; Ragaert, P.; Devlieghere, F.; De Meulenaer, B. Intelligent Food Packaging: The Next Generation. *Trends Food Sci. Technol.* **2014,** *39* (1), 47–62.
163. Vasile, C. Polymeric Nanocomposites and Nanocoatings for Food Packaging: A Review. *Materials* **2018,** *11* (10), 1834–1841.
164. Vermeiren, L.; Devlieghere, F.; van Beest, M.; de Kruijf, N.; Debevere, J. Developments in the Active Packaging of Foods. *Trends Food Sci. Technol.* **1999,** *10* (3), 77–86.
165. Vilela, C.; Kurek, M.; Hayouka, Z.; Röcker, B.; Yildirim, S.; Antunes, M. D. C.; Nilsen-Nygaard, J.; Pettersen, M. K.; Freire, C. S. R. A Concise Guide to Active Agents for Active Food Packaging. *Trends Food Sci. Technol.* **2018,** *80,* 212–222.
166. Wang, B.; Kerr, L. L. Dye Sensitized Solar Cells on Paper Substrates. *Solar Energy Mater. Solar Cells* **2011,** *95* (8), 2531–2535.
167. Wang, H.; Bian, L.; Zhou, P.; Tang, J.; Tang, W. Core–Sheath Structured Bacterial Cellulose/Polypyrrole Nanocomposites with Excellent Conductivity as Supercapacitors. *J. Mater. Chem. A* **2013,** *1* (3), 578–584.
168. Wei, Y. C.; Cheng, C. H.; Ho, Y. C.; Tsai, M. L.; Mi, F. L. Active Gellan Gum/Purple Sweet Potato Composite Films Capable of Monitoring pH Variations. *Food Hydrocoll.* **2017,** *69,* 491–502.
169. Weiss, J.; Takhistov, P.; McClements, D. J. Functional Materials in Food Nanotechnology. *J. Food Sci.* **2006,** *71* (9), R107–R116.
170. Wolfbeis, O. S.; List, H. AVL Medical Instruments AG. Method for Quality Control of Packaged Organic Substances and Packaging Material for Use with this Method. U.S. Patent 5,407,829, 1995.

171. Wu, C. L.; Zhang, M. Q.; Rong, M. Z.; Friedrich, K. Tensile Performance Improvement of Low Nanoparticles Filled-Polypropylene Composites. *Compos. Sci. Technol.* **2002,** *62* (10-11), 1327–1340.
172. Wyser, Y.; Adams, M.; Avella, M.; Carlander, D.; Garcia, L.; Pieper, G.; Rennen, M.; Schuermans, J.; Weiss, J. Outlook and Challenges of Nanotechnologies for Food Packaging. *Packag. Technol. Sci.* **2016,** *29* (12), 615–648.
173. Xiong, H.; Tang, S.; Tang, H.; Zou, P. The Structure and Properties of a Starch-Based Biodegradable Film. *Carbohydr. Polym.* **2008,** *71* (2), 263–268.
174. Yadav, H. M.; Kim, J. S.; Pawar, S. H. Developments in Photocatalytic Antibacterial Activity of Nano TiO_2: A Review. *Korean J. Chem. Eng.* **2016,** *33* (7), 1989–1998.
175. Yam, K. L.; Takhistov, P. T.; Miltz, J. Intelligent Packaging: Concepts and Applications. *J. Food Sci.* **2005,** *70* (1), R1–R10.
176. Yamamoto, O. Influence of Particle Size on the Antibacterial Activity of Zinc Oxide. *Int. J. Inorg. Mater.* **2001,** *3* (7), 643–649.
177. Yang, D.; Hu, Y.; Song, L.; Nie, S.; He, S.; Cai, Y. Catalyzing Carbonization Function of α-ZrP Based Intumescent Fire-Retardant Polypropylene Nanocomposites. *Polym. Degrad. Stab.* **2008,** *93*, 2014–2018.
178. Yang, Z.; Zhai, X.; Zou, X.; Shi, J.; Huang, X.; Li, Z.; Gong, Y.; Holmes, M.; Povey, M.; Xiao, J. Bilayer pH-Sensitive Colorimetric Films with Light-Blocking Ability and Electrochemical Writing Property: Application in Monitoring Crucian Spoilage in Smart Packaging. *Food Chem.* **2021,** *336*, 127634.
179. Yildirim, S.; Röcker, B.; Pettersen, M.K.; Nilsen-Nygaard, J.; Ayhan, Z.; Rutkaite, R.; Radusin, T.; Suminska, P.; Marcos, B.; Coma, V. Active Packaging Applications for Food. *Compr. Rev. Food Sci. Food Saf.* **2018,** *17*, 165–199.
180. Zeng, H.; Gao, C.; Wang, Y.; Watts, P. C.; Kong, H.; Cui, X.; Yan, D. In situ Polymerization Approach to Multiwalled Carbon Nanotubes-Reinforced Nylon 1010 Composites: Mechanical Properties and Crystallization Behavior. *Polymer* **2006,** *47* (1), 113–122.
181. Zeng, P.; Chen, X.; Qin, Y. R.; Zhang, Y. H.; Wang, X. P.; Wang, J. Y.; Ning, Z. X.; Ruan, Q. J.; Zhang, Y. S. Preparation and Characterization of a Novel Colorimetric Indicator Film Based on Gelatin/Polyvinyl Alcohol Incorporating Mulberry Anthocyanin Extracts for Monitoring Fish Freshness. *Food Res. Int.* **2019,** *126*, 108604.
182. Zhai, X.; Shi, J.; Zou, X.; Wang, S.; Jiang, C.; Zhang, J.; Huang, X.; Zhang, W.; Holmes, M. Novel Colorimetric Films Based on Starch/Polyvinyl Alcohol Incorporated with Roselle Anthocyanins for Fish Freshness Monitoring. *Food Hydrocoll.* **2017,** *69*, 308–317.
183. Zhang, X.; Sun, G.; Xiao, X.; Liu, Y.; Zheng, X. Application of Microbial TTIs as Smart Label for Food Quality: Response Mechanism, Application and Research Trends. *Trends Food Sci. Technol.* **2016,** *51*, 12–23.
184. Zheng, J. P.; Li, P.; Ma, Y. L.; Yao, K. D. Gelatin/Montmorillonite Hybrid Nanocomposite, I: Preparation and Properties. *J. Appl. Polym. Sci.* **2002,** *86* (5), 1189–1194.
185. Zhou, X.; Shin, E.; Wang, K. W.; Bakis, C. E. Interfacial Damping Characteristics of Carbon Nanotube-Based Composites. *Compos. Sci. Technol.* **2004,** *64* (15), 2425–2437.

CHAPTER 9

POTENTIAL OF INTELLIGENT PACKAGING FOR FERMENTED DAIRY PRODUCTS

S. SUPREETHA and NANDINI DUTTA

ABSTRACT

The advancement of technology in food product processing and packaging plays a primary role in food safety and preservation. With the impact of changes, active and intelligent packaging systems are gaining importance in modern food packaging trends. Packaging of fermented dairy products has crucial effect on microbial viability and product properties like acidity, pH, amount of oxygen present, storage temperature, and shelf life. Achieving proper monitoring and control over different factors affecting the safety and quality of fermented dairy products is difficult with the use of conventional packaging system. An introduction to intelligent packaging with different types of data processors, the properties of fermented dairy products, and the components of intelligent packaging for fermented dairy products with market status are briefly discussed in this chapter.

9.1 INTRODUCTION

Packaging of the food products is carried out to provide containment, protection, convenience, and to communicate with the consumers regarding the product. The advancement of technology has paved the way for food traceability systems implementation, which ensures legitimacy and safety of food products throughout the supply chain and has induced interest in development of accurate, noninvasive, cost-effective, consistent, rapid, and

Advances in Sustainable Food Packaging Technology. Megh R. Goyal, Preeti Birwal, Anand K. Pandey, & Ritesh B. Watharkar, (Eds.)

nondestructive methods, and instruments to evaluate the quality of the food products instantaneously.

Intelligent packaging system helps in communicating the information regarding the internal and external conditions prevailing with the product. Intelligent packaging refers to the ability of a packaging system that can sense and inform the environmental changes occurring inside the food product. The objective of intelligent packaging system is to enhance the quality, safety, and shelf life of fermented dairy products. It is carried out by the detection, measurement, and information transmission of the physical, chemical, and biological changes taking place in them. It also ensures more convenience to the consumers by providing information regarding the quality and safeguards the fermented dairy products against tamper and theft.

Intelligent packaging system is integrated with different instruments like sensors, amplifiers, transmitters, and indicators. It can monitor the internal conditions of the product when placed in direct contact, and external conditions can be measured when sensors are in indirect contact, that is, outside the package. It displays the external changes that affect the properties of the fermented dairy products and also provides information to the consumers regarding the changes taking place in the fermented dairy products. The tools and devices used in intelligent packaging undergo certain changes during sensing the physicochemical and microbial parameters.

The application of intelligent packaging has paved the way to obtain the solution for the enhancement of shelf life; reduction in complexity of the production process; enhancing of safety and quality of fermented dairy products; facilitating ease of storage, transportation, and distribution; and reduction in the use of preservatives in the product. In addition, it also provides convenience, helps to add various communicative features to the packaging, and reduces the occurrence of theft, pilfering, and malpractices, ensuring consumer safety. There are numerous intelligent packaging systems to determine the safety as well as the quality of fermented dairy products.

This chapter explores: types of intelligent packaging; properties of fermented dairy products; components of intelligent packaging used for fermented dairy products; applications, advantages, and disadvantages of intelligent packaging; market status and future prospects of dairy products using intelligent packaging.

9.2 DEFINITION OF INTELLIGENT PACKAGING

There are several proposed definitions of intelligent packaging. It can be defined based on the purpose of usage pattern, product packed in it, and the commercial conditions prevailing during storage. The definitions of intelligent packaging system are mentioned in Table 9.1.

TABLE 9.1 Different Definitions of Intelligent Packaging.

Author(s)	Definition
Brünnagel, D. (2014)	Intelligent packaging is a system that monitors the condition of the packed foods to give information about the quality of the packaged food during transport and storage [5]
Clarke (2001)	Packaging system which possesses logic capability of communicating the changes taking place inside the product [6]
Han, J. H. et al. (2005)	Systems attached as labels, incorporated or printed on the food packaging material which offer enhanced possibilities to monitor the product quality, trace critical points, and provide more detailed information throughout the supply chain [15]
Karel (2000)	Intelligent packaging is a package that contains sensor which notifies the consumer that the product is being impaired and it might begin to reverse the harmful changes that have occurred in the product [31]
Robertson, G. L. (2006)	Packaging which consists of an external or internal indicator to provide information about the aspects of the history of the package and/or the quality of the food [49]
Yam, K. L. et al. (2005)	A packaging system which is capable of carrying out the intelligent functions (like detecting, sensing, tracing, recording, communicating, and applying scientific logic) to facilitate the decision-making to extend shelf life, enhance safety, improve quality, provide information, and to warn about the possible problems [69]

The changes in the food product are affected by various parameters, affecting external as well as internal environment of the product. These parameters can be measured using the instruments and sensors mounted to the intelligent packaging to ensure the safety and quality of food by unique features like sensing, detecting, tracing, recording, and communicating the changes occurring in the food system during transportation and storage. Inside the food product, alterations in acidity, pH, moisture content, oxygen content, oxidation–reduction potential, etc. can predominantly affect the quality of food. In order to assess these changes, the packaging materials have quality indicators like freshness indicators, biosensors, and temperature sensors that are placed in direct contact with the product.[15,16]

Environmental conditions like temperature, relative humidity, leakage, gas content, etc. lead to physical, chemical, and microbial changes in dairy products. The sensors and other instruments used in these packaging systems can be different as well as numerous, since it is difficult to monitor and give complete information with the same sensor or instrument.[29]

9.3 CLASSIFICATION OF INTELLIGENT PACKAGING

The intelligent packaging systems are classified based on the technologies adopted for the manufacturing, sensors or processors used, and methods of data collection and interpretation. Based on the type of data processors, the intelligent packaging system can be classified as shown in Figure 9.1.

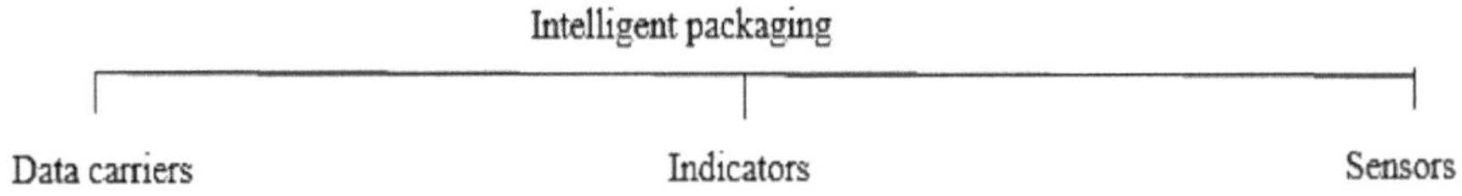

FIGURE 9.1 Classification of intelligent packaging.

9.3.1 DATA CARRIERS

The automatic identification devices that create the movement of information throughout the food supply chain are termed as data carriers. They improve the efficiency of data transfer throughout the food chain for monitoring food safety and quality. However, data carriers are integrated with tertiary packaging and do not directly provide the information of the changes taking place in the food product. Rather, they provide automated data regarding traceability, counterfeit, theft, and pilfering of the food product.[35] Radiofrequency identification tags (RFID) and barcodes are types of data carriers.

Barcodes are extensively used for tracing the products throughout the supply chain. Individually packed products like melons, apples, etc. can be identified for their exact location; similarly, it can be applied for individually packed yoghurt tubs, probiotic drinks, etc.[34] Product attributes depending on the conditions prevailing the storage and distribution system can be analyzed using barcodes and RFID tags. The food products assembled with global data synchronization in barcodes and RFID tags aid the manufacturers to access the data at stages if the product flows ~~id~~ between two or more manufacturers before it reaches the consumer. Hence, they can be used for data accession for products like sour cream, processed cheese or cheese spread if the product flow is between two or more production units.[34]

9.3.2 INDICATORS

Indicators aid in conveying information to the consumers with respect to food quality. They display information regarding the occurrence of a compound, concentration of the particular compound, and level of reaction among the compounds in the food product.[40] Indicators are placed directly in contact with the food product. The different types of indicators are time–temperature indicator (TTI), integrity indicator, freshness indicator, and gas indicator.[39] TTI is applied for monitoring various quality factors like acidity,[48] microbial load, and rheological properties of frozen yogurt, frozen vegetables, and fresh seafood. Freshness indicator is applied to determine the microbiological quality of the food product; hence, it is a very useful tool for assessment of viable microbial load in the fermented dairy products like dahi, yogurt, probiotic drinks, etc. The pathogens present in the product can be detected based on specific target metabolites associated with microbial-induced spoilage.[56]

9.3.3 SENSORS

Sensors are devices equipped with the processing and control electronics, a network with interconnections and software.[43] They are used to measure physical or chemical property by locating, detecting, and amplifying the energy or the signals sensed.[25] The innovative and highly efficient technology used in intelligent packaging systems for advanced applications is the sensors.[3,28] Sensors are placed either in direct contact or indirect contact with the food product. Biosensors, gas sensors, temperature sensors, and pressure sensors are the various types of sensors. Gas sensors for detection of CO_2 and oxygen are used in vegetables, fruits, and meat product packaging. Similarly, they can be used in kefir and koumiss packaging, where CO_2 has a major role in determination of its quality.[48] The microbial properties of the food product are determined by several enzymes-substrates-based biosensors.[37]

9.4 PROPERTIES OF FERMENTED DAIRY PRODUCTS

Fermentation is a method of preservation of food products that enhances its shelf life. It is carried out by inoculation of starter culture that brings about physical, chemical, and microbial changes. Fermented dairy products include dahi, yogurt, lassi, buttermilk, kefir, koumiss, shrikhand, etc. Fermentation of milk is carried out to improve palatability, flavor, bioavailability of nutrients, enhance shelf life, probiotic properties, and microbial safety.[9]

Fermentation of milk results in production of lactic acid from lactose, where the acidity of the product is increased and pH is reduced. The increase in acidity inhibits the proliferation of the pathogenic bacteria and other undesirable microorganisms. It also imparts an acidic flavor and sour taste to fermented dairy products.[8]

An increase in acidity and reduction in pH lead to a shift in the stability of protein that results in partial denaturation and coagulation of colloidal protein complex into insoluble curd matrix. Viscosity increases due to the denaturation of protein. Breakdown of proteins results in the release of the peptides and the amino acids that enhance the bioavailability of amino acids and improve protein digestibility.[14]

The starter culture will produce secondary metabolites, which include flavor compounds like diacetyl and acetaldehyde that improve the flavor of fermented dairy products. These flavor compounds will enhance the palatability and shelf life of the product. The metabolites also include antimicrobial compounds like organic acids, alcohol, hydrogen peroxide, carbon dioxide, diacetyl, and bacteriocins. The organic acids inhibit pathogenic bacteria, yeast, and mold. Hydrogen peroxide inhibits growth of microorganisms, especially in raw milk. Diacetyl controls the proliferation of gram-negative bacteria, yeasts, and molds.[9] Bacteriocins inhibit gram-positive bacteria, thus enhance the shelf life by reducing spoilage.[51]

9.5 COMPONENTS OF INTELLIGENT PACKAGING FOR FERMENTED DAIRY PRODUCTS

Intelligent packaging includes the use of certain components to detect and monitor the changes taking place during storage and distribution of the product. The sensors or instruments used for this purpose should be sensitive, precise, and inert when in direct contact with the product, and easily available. They should be flexible for usage in different types of product, and the changes should be easily accessed by proper display. Also, they should not affect the hygienic, nutritional, and sensory qualities of the product.[23] The various components of intelligent packaging system used for fermented dairy products are described in Figure 9.2.

9.5.1 BARCODES

Barcodes are the extensively used form of data carriers, which operate on symbologies that are commonly accepted languages for scanning and printing. The universal product code (UPC) is the most commonly applied barcode. It helps in inventory control, stock recording, and product tracking.[33] The data in the barcodes are represented by linear symbology with pattern of bars and spaces that is detected by an optical barcode scanner. The less storage capacity results in limited information storage with respect to manufacturer and product code on the barcodes. Hence, a number of new technologies, including global trade item number (GTIN) and reduced space symbology (RSS), are developed for better traceability.[39] Depending on the type of barcode, the storage capacity varies.[13] Two-dimensional barcode system, like quick response (QR) code, carries more information as compared to linear barcode.[12]

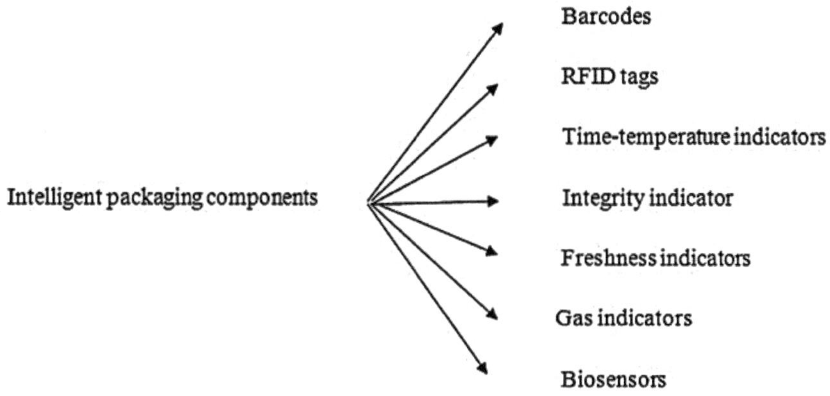

FIGURE 9.2 Components of intelligent packaging system.

The use of cytochrome c oxidase (COI) barcodes in fermented dairy products aids in detection of adulteration with animal fats and rendered fats using DNA genome tracing. The technique can also be used for determination of authenticity of ingredients used in the fermented milk products.[68] Barcodes infused with biosensors containing microbial genes are used for detection of certain strains of microorganisms with the principle of specific pathogenic antibody bound to the area formed by membrane in the barcode. If the specific microorganism is present, then the barcode turns into dark colored bar that renders the barcode unreadable while scanning.

The area in the barcodes is applied with specific antibodies to detect the antigens, like bacteria or spores or other microorganisms, present in the product that comes in contact with it. They are usually coated with pathogen-identifying agent, so that the presence of pathogen can be detected in the product. The detection of microorganisms using DNA barcoding method enables the phylogenetic hypothesis generation completely with the aid of universal primers. A fragment of ribosomal 16s rRNA is most commonly used as a barcode gene.[69]

9.5.2 RADIOFREQUENCY IDENTIFICATION TAGS (RFID)

RFID tags are an advanced type of data carriers. They have three major components, a tag that is fabricated by microchip and connected to an antenna, a sensor which emits and receives radio signals from a tag, and a hardware that connects sensors and signal network.[26,50] The ability to store a high number of different codes and transfer and communicate the information is the distinct feature of RFID tags that helps to simplify the identification and tracking of information.[44] RFID tags can be operated by active or passive technology with respect to battery usage during the operation.[45] They are mainly used for the product's batch identification, product tracing,[19] cold chain monitoring,[2] and shelf life prediction.[61]

The fermented dairy products quality can be determined and maintained during storage and distribution by application of RFID tags in the packaging material. It also helps the producer or manufacturer for efficient traceability and product recall.[38] RFID reader integrated with a semiactive tag of the gas sensor is used for the determination of gases along the storage and distribution chain to monitor the conditions during the entire supply chain.[11] Semi-passive RFID loggers are used in shelf life prediction of cheese spread.[20] Traceability of cheese during ripening, storage, and distribution is assessed commercially.[42,55]

9.5.3 TIME AND TEMPERATURE INDICATORS (TTI)

Temperature is a highly important factor affecting the properties like microbial or starter growth, acid production, protein denaturation, and flavor development in the fermented dairy products. Time–temperature indicators (TTI) are usually self-adhesive stickers applied to primary package of the product. They are considered as freshness indicators, since they indicate temperature of the product during various stages of storage and distribution. The change in temperature is usually measured in terms of chemical, electrochemical, mechanical, enzymatic, or microbial changes that can be detected by mechanical deformation or irreversible color change of the product as a response.[59] The response of the TTIs with changes in temperature is displayed by change in color intensity or diffusion of dye. They are also termed as integrators that monitor detrimental changes in temperature over the time during storage. There are different types of TTIs used commercially: critical temperature indicator that indicates only when the temperature of product exceeds threshold temperature, partial history indicator which changes the indicator color irreversibly after exceeding the threshold temperature, and complete history indicator that provides the time–temperature variation over a period of time throughout the distribution chain.[52,58]

The TTIs are used to examine the temperature changes taking place along the distribution chain of the fermented dairy products; as a result, microbial growth and resulting changes can be monitored. Temperature plays an important role during storage and distribution of fermented dairy products and its fluctuation results in impaired product properties. RFID tags along with global positioning system (GPS) and mobile communication, based on time–temperature theory, are used to monitor the temperature of the product as well as its environment during the entire supply chain.[70] The time–temperature history of the fermented dairy product determined using TTI will not provide the impact of heat treatment given to the product during processing.[48]

TTI is extensively used to study the optimization of cold chain management in fish and meat industry. Similarly, it can be used for fermented dairy products, since it is stored and transported at low temperature.[62] TTI is useful in monitoring temperature required for specific chemical reaction to take place. It is widely used to monitor the enzyme activity for achieving the energy of activation.[4] Hence, TTI can be used for culturing of short-set yogurt and long-set yogurt at different specific temperatures for optimum characteristics of the product. Formation of holes in cheese is characteristic to Swiss cheese, Gruyere cheese, etc., which is due to the production of CO_2 by certain bacteria. The starter culture requires specific temperature for production and distribution of CO_2 throughout the cheese block for obtaining optimum sized holes.[42] TTI plays an important role in cold chain management, which is extremely useful in storage and transportation of frozen yogurt and frozen popsicle.[4]

9.5.4 INTEGRITY INDICATOR

The shelf life and quality of the fermented food largely depend on the gases entrapped in the headspace and released due to metabolism during the process of fermentation. Integrity indicators are mainly used to detect the integrity of packages and leak seals.[18] Leaks in packaging result in invasion of extraneous matter like dust, dirt, microorganisms, etc., which leads to contamination of product and also causes loss of product. The volatile compounds like diacetyl and acetaldehyde produced during the fermentation of milk act as flavor compounds as well as pathogen inhibitors. Improper sealing or leaks in packaging will lead to loss of volatile compounds formed, which in turn reduces the acceptability of the product and also results in spoilage due to pathogen proliferation.[37]

The presence of oxygen plays an important role in fermentation process. A critically low oxygen content leads to growth of anaerobic microorganisms like *Clostridium botulinum*, *Bacillus subtilis*, etc., thus results in anaerobic fermentation of milk.[65] The sensory properties of kefir and koumiss depend on the optimum concentration of CO_2; due to leakage the CO_2 escapes, and this results in lower acceptability of the product.[24]

9.5.5 FRESHNESS INDICATORS

The superiority of the food is perceived with respect to freshness by the consumers. Freshness has a major role in monitoring the quality of food throughout storage and distribution. Loss of freshness of the fermented dairy product may be due to loss of moisture content, microbial spoilage, improper storage temperature, or chemical changes taking place in it.[53] The sensors used in freshness indicators will have different substrates depending on the concentration in the product.[56] During storage of fermented milk, certain changes take place in the product with respect to microbial and chemical factors where volatile compounds, like carbon dioxide, organic acids, ethanol, etc., are released into the headspace of the package.[57] The volatile compounds thus released are detected by compounds infused in the sensor and are indicated by color change due to pH alterations. The intensity of color change in indicator is measured to detect the extent of volatile compounds released from the product.[8]

Certain freshness indicators are also used to detect the microbial quality of the product. The sensors react with the toxic metabolites produced by the pathogenic microorganisms. The response for the reaction taking place can be a color change in chromogenic substances of the indicator due to substrate complex formed by pathogenic and spoilage microorganisms, nutrient loss in the product, enzymes, and other secondary metabolites released. Oxygen scavengers that are self-reactive and moisture-dependent aid in the detection of excessive microbial growth, oxygen content, production of off-flavor compounds, and rancidity in cheese, yogurt, lassi, etc.[28]

Moisture control aids like silicon gel, calcium oxide, calcium sulfide, and purge asbestos are used in detection and control of moisture loss from cheese and yogurt powder to maintain the texture. The ion-sensitive metal oxides are coupled with micro and nanostructured sensors, having biocompatible electrical and electrochemical properties that aid in the determination of pH and acidity in dahi, yogurt, lassi, yogurt shake, cheese, cheese spreads, etc.[32] The application of ion-sensitive metal oxides to acidity development in fermented dairy products can be controlled, which affects the protein denaturation, gelation, microbial growth, and gas production, especially during cheese ripening.

9.5.6 GAS INDICATORS

The chemical and microbial changes taking place during fermentation alter the initial gas composition in the product. Gas indicators are usually placed as labels on the packaging films that monitor the alteration of the gas composition and aid in determination of quality of the fermented dairy products. Oxygen indicators are the most commonly used type of gas indicators, especially in fermented dairy products, since oxygen is the major factor affecting the quality with respect to oxidative changes and rancidity, and also bring about color change, microbial growth, pathogen inhibition, and flavor changes. Alterations in gas compositions are measured by color changes in the detector, internally due to microbial metabolism and externally because of leakage, tampering, or pilfering[54]

Gas indicators are also used in detection of water vapor, carbon dioxide, ethanol, hydrogen sulfide, etc.[69] The probiotic microorganisms are usually anaerobic and microaerophilic, which proliferate in presence of very low oxygen. Hence, the oxygen level inside the package should be less to cease toxicity, reduce microbial inhibition, and improve shelf life of the probiotics.[24] Gas indicators are used to detect the amount of carbon dioxide and ethanol produced during fermentation of kefir and koumiss.[47] In certain cheeses like Swiss cheese or Emmental cheese, Gruyere cheese, Havarti-Tilsit cheese, etc., production of CO_2 forms small holes inside the cheese, which is its characteristic property. Optimum level of CO_2 should be produced to obtain small holes throughout the cheese. Gas indicators aid in determination of optimum level of CO_2 for good cheese quality.[42]

Concentration of CO_2 has an important role in sensory and storage properties of kefir; hence, incorporation of gas indicators is necessary for monitoring the quality of kefir.[23]

9.5.7 *BIOSENSORS*

The biochemical changes taking place in the fermented food are detected and recorded using biosensors. The information thus obtained is transmitted, which depends on the extent of metabolic changes. Biosensor helps to obtain rapid, accurate, and online results during processing and storage of fermented dairy products. It is basically a probe which converts biochemical components like enzymes, antibodies, cell organelles, or microorganisms into electronic signals. It is extensively used due to its accuracy, portability, reliability, specificity, sensitivity, and simplicity.[7,65] It consists of a bioreceptor that detects the analyte and a transducer that converts biochemical signals into an electrical response. The bioreceptor can be an antigen, enzyme, hormone, microorganism, or nucleic acid, and the transducer is an acoustic, electrochemical or optical component depending on the type of signals to be detected and measured. Biosensors are fused inside the packaging material, thus they come in direct contact with the product and are extensively used in pathogen detection.[1]

Biosensors are also used in estimation of composition of the product, toxins, pathogens, drug residues, pesticides, and antibiotics in the fermented dairy products. The antibodies present in biosensors, attached to the packaging material are aimed at detecting pathogens and toxins.[69] Biosensors are also used as freshness indicators, especially in lassi and yogurt shake type of products, for rapid identification of microbial contaminants or pathogens and sugar content in the beverages.[65] Amperometric biosensors are used for the determination of antioxidants in high fat fermented dairy products like sour cream or acid cream.[36] The presence of *Staphylococcus aureus* and *Escherichia coli* in fermented dairy products can be detected by the application of quantum dots, which are biotinylated complementary probes conjugated with streptavidin.[60]

Production of lactic acid and L-lactate during fermentation of dahi and yogurt can be monitored continuously and determined instantaneously using the biosensor equipped with flow injection analysis (FIA) system for online data acquisition and real-time monitoring.[60] The biosensor containing immobilized bacteria or yeast, gas permeable membrane, and an oxygen electrode is used in estimation of the alcohol content of kefir and koumiss during fermentation to minimize pathogen inhibition, increase yield, and to obtain optimum alcohol content.[37,47]

9.6 MERITS AND DEMERITS OF USING INTELLIGENT PACKAGING FOR FERMENTED DAIRY PRODUCTS

Intelligent packaging of fermented dairy products has both merits and demerits depending on the purpose of use, commercialization, cost of marketing, and type of product. Since the intelligent packaging is a novel concept in packaging, especially in fermented dairy products, the advantages and disadvantages play an important role in technical advancement for efficient commercialization.

The application of intelligent packaging in fermented dairy products enables manufacturers as well as consumers to determine the quality parameters such as pH and acidity of the products like dahi and yogurt, using metal oxides-based chemical sensors, potentiometric sensors,[32] and biosensors with FIA.[60] Potentiometric biosensors are also used for detecting presence of ethanol organophosphates, sucrose, penicillin, urea, tryptophan, and trichloroethylene. Whole-cell biosensor, having an immobilized *Pseudomonas alcaligenes* sensor, degrades the pathogens proliferating at low temperatures for fish storage. Similarly, it can be used for monitoring the quality of frozen yogurt, shrikhand, etc.[37]

The microbial quality of the fermented dairy products can be assessed with the help of biosensors compounded with pathogen detecting antigens or enzymes. The concentration of the gases produced during the fermentation process is analyzed by the gas indicators.[47] Biosensors based on bioluminescence are used for the detection of metal ions, phosphorus, heavy metals, naphthalene, genotoxicants, and chlorophenols in dairy industry.[60]

The rate of fermentation, acidity, pH, and amount of gases produced depend on the temperature maintained. Hence, the temperature indicators aid in the instantaneous assessment of the temperature of the internal and external environment of the product.[45] Intelligent packaging aids in monitoring and control of quality parameters of the food product, and thus results in the reduction of food spoilage and wastage to a great extent.[10,66]

The constant quality monitoring aims to reduce the cost, time, and product consumption involved in the analysis of packed products during storage and distribution.[64] By application of sensors and other instruments, the traceability of the product will become efficient. Intelligent packaging acts as a great packaging tool for monitoring the leakage and pilfering of the product during storage and distribution. Due to the precision in the determination of changes taking place in the product, the use of preservatives can be reduced or eliminated.[36]

Intelligent packaging system, with respect to TTI and RFID tags,[24] can provide access to the entire history of the supply chain of the product regarding details of raw materials used, conditions maintained during storage and distribution, manufacturer's details, and expiry date, to the consumers.[3] RFID tags have the capacity to store additional information and the distribution of this stored data will increase the fault tolerance in the entire supply chain system, which helps in inventory control and prevents the additional costs of damage and disposal.[21] Adoption and incorporation of interesting and attractive characters on the packaging material along with intelligent packaging components for providing information about the quality and safety aspects of the fermented dairy products increase the ease of communicating with consumers for taking appropriate decisions on the quality of the product.[38]

Intelligent packaging system has certain demerits for its application in the packaging of fermented dairy products. Studies are going on to determine the inertness of the intelligent packaging on the fermented dairy products packed in them. The intelligent packaging system has to be thoroughly validated for its efficiency of data collection, interpretation, and traceability.[15] The sensor or the packaging component has to be kept in direct contact with the fermented dairy product in the package during application of intelligent packaging. The physical or chemical agents present in these sensors might migrate into the fermented dairy product.[49] The migrated extraneous material can be intentional or unintentional and might lead to certain health effects in consumers. During storage, it is difficult to control the type and amount of the substance migrating into the product.[12]

The instruments and technologies used in intelligent packaging will add to the production cost of the product and limit its commercial usage.[24] Specific changes have to be made in the packaging material in order to suit the application of sensors and other instruments during its manufacture.[22] Similarly, the processing parameters, monitoring, and traceability process have to be modified to suit the intelligent packaging system. The quantity of pathogen or toxin required for the detection in the food product by biosensor is high as compared to the amount of toxin that can cause illness in human beings.[29] Intelligent packaging monitors only one parameter at a time. Hence, it is difficult to determine and control all the parameters simultaneously. If many parameters are to be measured, many sensors have to be installed in the packaging material.[26]

9.7 MARKET STATUS OF DAIRY PRODUCTS PACKED USING INTELLIGENT PACKAGING

The advancements in packaging technology are leading to the development of novel and efficient technologies. Intelligent packaging is currently in the blooming stage, and very few products have adopted the existing technology for commercial purposes.

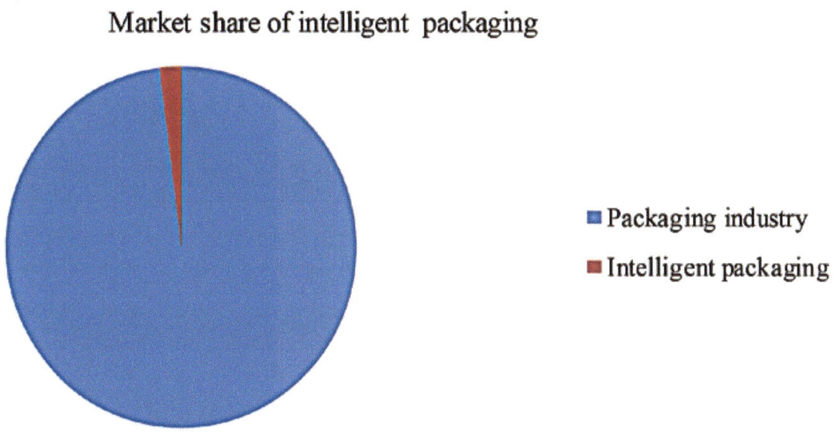

FIGURE 9.3 Market share of intelligent packaging.

The intelligent packaging system is not widespread due to the disadvantages caused by the additional cost of packaging, changes to be adopted while designing the packaging material, and ability of instruments to sense all the parameters. However, the advantages incurred with the use of intelligent packaging are high compared to the disadvantages occurring. To improve the extent of usage of intelligent packaging and its commercialization, research and improvements should be carried out to enhance the efficiency of detection of different parameters using the same components of intelligent packaging.[30]

Due to the disadvantages of increased cost, adaptation to suit the technology and processing, and knowledge of usage, the intelligent packaging system is not widely accepted for packing fermented dairy products.[22]

The intelligent packaging system has an insignificant market share globally for all the products (Fig. 9.3). It has a market share of less than 2% globally in the case of fermented dairy products, compared to the overall packaging industry.[38] The biosensors, oxygen scavengers, and moisture absorbers acquire the majority of the market share in the intelligent packaging industry.[67]

9.8 FUTURE PROSPECTS OF INTELLIGENT PACKAGING

To increase the interest in the intelligent packaging system among the consumers, various measures can be taken to enhance the technical advances in the production process. Further research in intelligent packaging systems with respect to changes occurring in the fermented dairy products, migration of components, and the suitability of instruments can lead to an improvement in manufacture of intelligent packaging.[17]

The sensors for the identification of foodborne pathogens, which are of great concern for the quality and safety of fermented dairy products, are being developed that are capable of detecting the presence of even very few microorganisms and a small amount of toxins. The safety of the fermented dairy product can be well maintained by the development of such sensors.

The synthetic polymers used on intelligent packaging material can be infused with antigens or enzyme substrates to detect the antibodies.[69] The fermentation process has to be monitored efficiently to reduce wastage due to high acid production and the proliferation of pathogens.

RFID tags can be used for the detection of multiple parameters at the same time. The information relating to temperature, relative humidity, acidity, pH, etc. can be determined using the same sensor.[50]

The integrated technology of data processors, scanners, and voice recognition has paved the way for development of devices that recognize voice and process it into electrical signals with certain characteristics and algorithms. The changes will be carried out according to the commands given by the consumer.[50] There are special seals being developed that are transparent until the package is opened. When the seal is opened, there will be a permanent color change which indicates tampering. It helps in informing the consumer about the possible contamination or damage to the product.

9.9 SUMMARY

Intelligent packaging has several advantages in terms of monitoring the quality and safety of fermented dairy products throughout the product supply chain. An efficient intelligent packaging system aids in reducing product loss due to spoilage. The supply chain can be monitored and maintained easily and efficiently via product traceability and recall by utilizing intelligent packaging.

KEYWORDS

- **fermented dairy products**
- **indicators**
- **intelligent packaging**
- **quality**
- **sensors**

REFERENCES

1. Alocilja, E. C.; Radke, S. M. Market Analysis of Biosensors for Food Safety. *Biosens. Bioelectron.* **2003,** *1* (5), 841–846.
2. Badia-Melis, R.; Ruiz-Garcia, L.; Garcia-Hierro, J.; Villalba, J. I. R. Refrigerated Fruit Storage Monitoring Combining Two Different Wireless Sensing Technologies: RFID and WSN. *Sensors* **2015,** *15*, 4781–4795.
3. Bagchi, A. Intelligent Sensing and Packaging of Foods for Enhancement of Shelf-life: Concepts and Applications. *Int. J. Sci. Eng. Res.* **2012,** *3* (10), 4–20.
4. Bobelyn, E.; Maarten, L.;Hertog, A. T. M.; Nicola, B. M. Applicability of an Enzymatic Time Temperature Integrator as a Quality Indicator for Mushrooms in the Distribution Chain. *Biol. Technol.* **2006,** *42*, 104–114.
5. Brünnagel, D. *Die Verpackungdenktmit* (The Packaging Thinks with You). *Verpackungsrundschau* (Packaging review). *VR Interpack* **2014,** *2014*, 14–16.
6. Clarke, R. Radio Frequency Identification: Smart or Intelligent Packaging. *J. Packag. Sci. Technol.* **2001,** *10* (5), 233–247.
7. D'Souza, S. F. Microbial Biosensors. *Biosens. Bioelectron.* **2001,** *16* (6), 337–353.

8. Dan, T.; Wang, D.; Jin, R. L.; Zhnag, H. P.; Zhou, T. T.; Sun, T. S. Characterisation of Volatile Compounds in Fermented Milk Using Solid-Phase Microextraction Methods Coupled with Gas Chromatography- Mass Spectrometry. *J. Dairy Sci.* **2017,** *100* (4), 2488–2500.
9. De Bok, F. A.; Janssen, P. W.; Bayjanov, J. R.; Sieuwerts, S.; Lommen, A.; Van HylckamaVlieg, J. E.; Molenaar, D. Volatile Compound Fingerprinting of Mixed-Culture Fermentations. *Appl. Environ. Microbiol.* **2011,** *77* (17), 6233–6239.
10. Elisa, P. R.; Jose, V. R.; Jos, L. V.; Ramon, M. M. Recent Advances on Intelligent Packaging as Tools to Reduce Food Waste. *J. Clean. Product.* **2018,** *172*, 3398–3402.
11. Eom, K. H.; Kim, M. C.; Lee, S.; Lee, C. W. The Vegetable Freshness Monitoring System Using RFID with Oxygen and Carbon Dioxide Sensor. *Int. J. Distrib. Sens. Netw.* **2012,** *8* (6), 472986.
12. Fang, Z.; Zhao, Y.; Warner, R. D.; Johnson, S. K. Active and Intelligent Packaging in Meat Industry. *Trends Food Sci. Technol.* **2017,** *61*, 60–71.
13. Ghaani, M.; Cozzolino, C. A.; Castelli, G.; Farris, S. An Overview of the Intelligent Packaging Technologies in the Food Sector. *Trends Food Sci. Technol.* **2016,** *51*, 1–11.
14. Ghosh, D.; Chattoraj, D. K.; Chattopadhyay, P. Studies on Changes in Microstructure and Proteolysis in Cow and Soy Milk Curd During Fermentation Using Lactic Cultures for Improving Protein Bioavailability. *J. Food Sci. Technol.* **2013,** *50*, 979–985.
15. Han, J. H.; Ho, C. H. L.; Rodrigues, E. T. Intelligent Packaging. Chapter 5; In *Innovations in Food Packaging*; Han, J. H., Ed.; Elsevier: Amsterdam, Netherlands, 2005; pp 138–155.
16. Heising, J. K.; Dekker, M.; Bartels, P. V.; Van Boekel, M. A. Monitoring the Quality of Perishable Foods: Opportunities for Intelligent Packaging. *Crit. Rev. Food Sci. Nutr.* **2014,** *54*, 645–654.
17. Huff, K. Active and Intelligent Packaging: Innovations for the Future. *Food Sci. Technol.* **2008,** *2008*, 1–13.
18. Hurme, E. Detecting Leaks in Modified Atmosphere Packaging. Chapter 13; In *Novel Food Packaging Techniques*; Ahvenainen, R., Ed.; Woodhead Publishing Ltd.: Cambridge, UK, 2003; pp 276–286.
19. Hwang, Y. M.; Moon, J.; Yoo, S. Developing a RFID-Based Food Traceability System in Korea Ginseng Industry: Focused on the Business Process Reengineering. *Int. J. Control Autom.* **2015,** *8*, 397–406.
20. Jedermann, R.; Schouten, R.; Sklorz, A.; Lang, W.; Van Kooten, O. In Linking Keeping Quality Models and Sensor Systems to an Autonomous Transport Supervision System, Cold Chain Management, 2nd Workshop; Bonn, 2006; pp 3–18.
21. Kaur, M.; Sandhu, M.; Mohan, N.; Sandhu, P. S. RFID Technology Principles, Advantages, Limitations & Its Applications. *Int. J. Comput. Electr. Eng.* **2011,** *3* (1), 1793–1810.
22. Keller, K. L. Ed.; In *Strategic Brand Management: Building, Measuring and Managing Brand Equity*, 2nd Ed.; Prentice-Hall: Englewood Cliffs, 2003; p 860.
23. Yam, K. L..; Takhistov, P. T.; Miltz, J. Intelligent Packaging: Concepts and Applications. *J. Food Sci.* **2005,** *1* (6), 1–10.
24. Korbekandi, H.; Mortazavian, A. M.; Iravani, S. Technology and Stability of Probiotic in Fermented Milks. In *Probiotic and Prebiotic Foods: Technology, Stability and Benefits to the Human Health*; Shah, N.; Cruz, A.G.; Faria, J. A. F., Eds.; Nova Science Publishers: New York, US, 2011; pp 131–169.

25. Kress-Rogers, E. Terms in Instrumentation and Sensors Technology. In *Instrumentation and Sensors for the Food Industry*; Kress Rodgers, E., Ed.; Woodhead Publishing Ltd.: Cambridge, UK, 1998; pp 673–691.
26. Kumar, P.; Reinitz, H. W.;Simunovic, J.; Sandeep, K. P.; Franzon, P. D. Overview of RFID Technology and its Applications in the Food Industry. *J. Food Sci.* **2009,** *74* (2),101–106.
27. Kuswandi, B.; Marysk, Ch.; Abdullah, A. J.; Heng, L. H. Real Time On-Package Freshness Indicator for Guavas Packaging. *Food Meas.* **2013,** *7*, 29–39.
28. Kuswandi, B.; Wicaksono, Y.; Jayus, Abdullah, A.; Heng, L.; Ahmad, M. Smart Packaging: Sensors for Monitoring of Food Quality and Safety. *Sens. Instrum. Food Qual. Saf.* **2011,** *5*, 137–146.
29. Lee, S. J.; Rahman, A. M. Intelligent Packaging for Food Products. *Innov. Food Packag.* **2014,** *2014*, 171–209.
30. Lord, J. B. The Food Industry in the United States. In *Developing New Food Products for a Changing Marketplace*; Brody, A. L., Lord, J. B., Eds.; CRC Press: Boca Raton, FL, 2008; pp 1–23.
31. Karel, M. Tasks of Food Technology in the 21st Century. *Food Technol.* **2000,** *2000*, 54–64.
32. Manjakkal, L.; Szwagierczak, D. Dahiya, R. Metal Oxides Based Electrochemical pH Sensors: Current Progress and Future Perspectives. *Prog. Mater. Sci.* **2020,** *109*, 100–635.
33. Manthou, V.; Vlachopoulou, M. Bar-Code Technology for Inventory and Marketing Management Systems: A Model for its Development and Implementation. *Int. J. Prod. Econ.* **2001,** *71* (1-3), 157–164.
34. McCathie, L. The Advantages and Disadvantages of Barcodes and Radio Frequency Identification in Supply Chain Management. *J. Eng. Inf. Sci.* **2004,** *2004*, 20–50.
35. McFarlane, D.; Sheffi, Y. The Impact of Automatic Identification on Supply Chain Operations. *Int. J. Logist. Manag.* **2003,** *14*, 1–17.
36. Mello, L. D.; Kubota, L. T. Biosensors as a Tool for the Antioxidant Status Evaluation. *Talanta* **2007,** *72*, 335–348.
37. Meshram, B. D.; Agrawal, A. K.; Adil, S.; Ranvir, S.; Sande, K. K. Biosensor and its Application in Food and Dairy Industry: A Review. *Int. J. Curr. Microbiol. Appl. Sci.* **2018,** *7*, 3305–3324.
38. Mohebbi, B. The Art of Packaging: An Investigation into the Role of Color in Packaging, Marketing, and Branding. *Int. J. Organ. Leadersh.* **2014,** *3*, 92–102.
39. Müller, P.; Schmid, M. Intelligent Packaging in the Food Sector: Brief Overview. *Foods* **2019,** *8* (1), 16–22.
40. O'Grady, M. N.; Kerry, J. P. Smart Packaging Technology. In *MeatPackaging Techniques*; Toldrà, F., Ed.; Woodhead Publishing Ltd.: Cambridge,UK, 2008; pp 276–286.
41. Papetti, P.; Costa, C.; Antonucci, F.; Figorilli, S.; Solaini, S.; Menesatti, P. A RFID Web-Based Infotracing System for the Artisanal Italian Cheese Quality Traceability. *Food Contr.* **2012,** *27* (1), 234–241.
42. Papetti, P.; Costa, C.; Antonucci, F.; Figorilli, S.; Solaini, S. A RFID Web-Based Info Tracing System for the Artisanal Italian Cheese Quality Traceability. *Food Contr*. **2012,** *27*, 234–241.
43. Patel, P. D.; Beveridge, C. Eds.; In *Rapid and On-Line Instrumentation for Food Quality Assurance*; Woodhead Publishing Ltd.: Cambridge, UK, 2003; pp 215–250.

44. Plessky, V. P. Review on SAW RFID tags. Frequency Control Symposium, 200 Joint with the 22nd European Frequency and Time Forum. *IEEE Int.* **2009,** *2009*, 14–23.
45. Preradovic, S.; Karmakar, N. C. Eds.; In *Multiresonator-Based Chipless RFID: Barcode of the Future.* Springer: New York, USA, 2012; p 380.
46. Quintero, A. V.; Molina-Lopez, F.; Smits, E. C. P.; Danesh, E.; Van den Brand, J.;Persaud, K.; Briand, D. Smart RFID Label with a Printed Multisensor Platform for Environmental Monitoring. *Flex. Print. Electron.* **2016,** *1* (2), 1–12.
47. Rajasekhar, T.; Upadhaya, J. B.; Shah, B. P. Role of Biosensors in Dairy and Food Industry. *Beverage Food World* **2005,** *2005*, 21–24.
48. Raymana, A.; Demirdövenb, A.; Baysala, T. Use of Indicators in Intelligent Food Packaging. *J. Food Sci.* **2016,** *70*, 1–10.
49. Robertson, G. L. Ed.; In *Active and Intelligent Packaging*, 2nd Ed.; CRC Press: Boca Raton, FL, 2006; p 280.
50. Sarac, A.; Absi, N.; Dauzère-Pérès, S. A Literature Review on the Impact of RFID Technologies on Supply Chain Management. *Int. J. Prod. Econ.* **2010**, *128*, 77–95.
51. Settanni, L.; Corsetti, A. Application of Bacteriocins in Vegetable Food Biopreservation. *Int. J. Food Microbiol.* **2008,** *121* (2), 123–138.
52. Singh, R. P. Scientific Principles of Shelf-Life Evaluation. In *Shelf-Life Evaluation of Food;* Man, D., Jones, A., Eds; Aspen Publishers: Gaithersburg, 2000; pp 3–22.
53. Siro, I. Active and Intelligent Packaging of Food. Chapter 4; In *Progress in Food Preservation*; Bhat, R.; Alias, A. K.; Paliyath, G., Eds.; John Wiley & Sons Inc.: New York, USA, 2012; pp 23–48.
54. Smiddy, M.; Fitzgerald, M.; Kerry, J. P.; Papkovsky, D. B.; O'Sullivan, C. K.; Guilbault, G. G. Use of Oxygen Sensors to Non-Destructively Measure the Oxygen Contentin Modified Atmosphere and Vacuum-Packed Beef: Impact of Oxygen Content on Lipid Oxidation. *Meat Sci.* **2002,** *61* (3), 285–290.
55. Smits, E.; Schram, J.; Nagelkerke, M.; Kusters, R.; Heck, G. V. Development of Printed RFID Sensor Tags for Smart Food Packaging, *Chem. Sens.* **2012,** *2012*, 403–406.
56. Smolander, M. Freshness Indicators for Direct Quality Evaluation of Packaged Foods. *Packag. Technol.* **2000,** *2000*, 7–11.
57. Smolander, M. The Use of Freshness Indicators in Packaging. In *Novel Food Packaging Techniques*; Ahvenainen, R. Eds.; Woodhead Publishing Ltd.: Cambridge, UK, 2003; pp 127–143.
58. Taoukis, P. S. Modelling the Use of Time Temperature Indicators in Distribution and Stock Rotation. In *Food Process Modelling*; Tijskens, L., Hertog, M., Nicolai, B., Eds.; Woodhead Publishing: Cambridge, U.K, 2001; p 512.
59. Taoukis, P. S.; Labuza, T. P. Eds.; In *Time Temperature Indicators (TTIs)*, 3rd Ed.; Woodhead Publishing Ltd.: Cambridge, UK, 2003; pp 103–126.
60. Thakur, M. S.; Ragavan, K. V. Biosensors in Food Processing. *J. Food Sci. Technol.* **2013,** *50* (4), 625–641.
61. Uysal, I.; Emond, J.; Bennett, G. In *Tag Testing Methodology for RFID Enabled Temperature Tracking and Shelf-life Estimation. RFID-Technologies and Applications (RFID-TA)*, IEEE International Conference, 2011; vol *2011*, pp 8–15.
62. Vaikousi, H.; Biliaderis, C. G.; Koutsoumanis, K. P. Development of a Microbial Time/Temperature Indicator Prototype for Monitoring the Microbiological Quality of Chilled Foods. *Appl. Environ. Microbiol.* **2008,** *74* (10), 3242–3250.

63. Vaikousi, H.; Costas, G.; Biliaderis.; Konstantinos, P.; Koutsoumanis. Development of a Microbial Time/Temperature Indicator Prototype for Monitoring the Microbiological Quality of Chilled Foods. *Cold Chain Manag.* **2008,** *2008*, 35–76.
64. Vanderroost, M.; Ragaert, P.; Devlieghere, F.; De Meulenaer, B. Intelligent Food Packaging: The Next Generation. *Trends Food Sci. Technol.* **2014,** *39* (3), 47–62.
65. Velasco-Garcia, M. N.; Mottram T. Biosensor Technology Addressing Agricultural Problems. *Biosyst. Eng.* **2003,** *84* (1), 1–12.
66. Verghese, K.; Lewis, H.; Lockrey, S.; Williams, H. Packaging's Role in Minimizing Food Loss and Waste Acrossthe Supply Chain. *Packag. Technol. Sci.* **2015,** *28*, 603–620.
67. Vermeiren, L.; Devlieghere, F.; De Meulenaer, B.; Debevere, J. Evaluating Safety, Effectiveness, Economic-Environmental Impact and Consumer Acceptance of Active and Intelligent Packaging. *Food Microbiol.* **1999,** *1999*, 7–18.
68. Wu, Y.; Yang, Y.; Liu, M.; Wang, B.; Wang, Y.; Wang, H. Applying COI Barcode to Identify Animal Origin of Food. *J. Food Sci.* **2019,** *84* (6), 1256–1265.
69. Yam, K. L.;Takhistov, P. T.; Miltz, J. Intelligent Packaging: Concepts and Applications. *J. Food Sci.* **2005,** *70*, 1–10.
70. Zhang, J.; Liu, L.; Mu, W.; Moga, L. M.; Zhang, X. Development of Temperature-Managed Traceability System for Frozen and Chilled Food during Storage and Transportation. *J. Food Agric. Environ.* **2009,** *7* (4), 28–31.

PART III

BIODEGRADATION AND BIO-BASED APPROACHES FOR MANAGEMENT OF PLASTIC WASTE

CHAPTER 10

BIODEGRADATION OF PLASTIC WASTE: MYTH OR A REALITY?

SHALJA VERMA and ANAND KUMAR PANDEY

ABSTRACT

With the increase in the variety of food products, the requirement for a variety of packaging and materials also increases. Plastic, metal, cardboard, paper packs, etc. are being increasingly used for different types of food products. But a most challenging aspect behind such frequent use is the degradation of the massive amount of packaging material waste generated after food consumption. All the packaging materials are discarded after consumption; some degrade over the due course of time in a biofriendly manner, like paper packaging, while others, mainly plastic waste, pose a great challenge to the environment as they are nondegradable and on accumulation at different sites lead to environmental pollution; thus indicates toward a highly adverse future. As plastic is essentially a high molecular weight polymer of hydrocarbons and petroleum derivatives and immensely covers the food packaging industry, there stands a great demand to find a solution for its degradation. Recent efforts of replacing conventional plastic with biodegradable plastics, like polycaprolactone, polyvinyl alcohol, polybutylene succinate, and polybutyrate adipate terephthalate, put forward highly promising results. These bioplastics are those which, when acted upon by microbial action, get degraded in a reasonable duration, finally producing water and CO_2. Some microbial species, which are capable of degrading specific plastic polymers, have also been identified and produce depolymerases leading to plastic depolymerization. Hence, biodegradation seems greatly promising to overcome the adverse upcoming conditions of plastic-based pollution.

Advances in Sustainable Food Packaging Technology. Megh R. Goyal, Preeti Birwal, Anand K. Pandey, & Ritesh B. Watharkar, (Eds.)

Millions of tonnes of plastic wastes are generated annually from food packaging materials. The practices which seem promising at the lab scale may or may not give desirable outcomes in the real world, thus tapping the most researched term "biodegradation" in controversy. Therefore, in the present chapter, the authors have clarified this controversial aspect of biodegradation by elaborating on significant research efforts by the scientific community to degrade and reduce the adverse effects of plastic waste and their potential to be applied in practical scenarios.

10.1 INTRODUCTION

Since the advent of globalization, the standard of living has improved in all aspects. The crosstalk between varied technologies developed in different parts of the world has led to significant global advancement. But along with these advances, here arose some major global challenges which can pose immense threats to society and the environment. The world is unable to counteract these challenges as they oppose the present technologies. Exponentially increasing global environmental pollution due to increasing plastic waste is one of such major issues that need great concern in today's scenario, as this will soon result in great environmental destruction. In the USA, one person uses approximately 365 plastic bags every year, whereas only 9% of total plastic gets recycled.[26,88]

The growing trends in the diversity of food products enhance the variety of processed foods that are healthy enough even when stored for a long duration. Packaging of food plays a major role in maintaining food quality and nutrition and avoiding contamination. Although many types of packaging are prevalent depending on the food product to be packed like metal cans, cardboard packs, paper packs, etc., but plastic packing has covered a major part of food packaging industry due to its strength, stability, flexibility, lightness, sterilization ease, impermeability, and resistance to variety of external factors which can adversely affect food items. Plastic food packaging accounts for 40% of the plastic market in Europe and is still exponentially increasing at the global level.

The most common polymers used for food packaging are polyethylene terephthalate, polypropylene, polyethylene, polystyrene, and polyvinyl chloride. But the resilient nature of these polymers packaging against degradation after being discarded is highly challenging.[42] These resistant plastic polymers when ingested by animals get accumulated in their digestive tract resulting in reduced stimuli of feeding, obstruction in the gastrointestinal tract, less secretion of digestive enzymes, and reduction in steroid hormone levels; thus lead to problems related to reproduction. When dragged into oceans, these plastic residuals produce toxic chemicals, such as bisphenol A, nonylphenol, aromatic polycyclic hydrocarbons, pesticides like dichloro diphenyl trichloroethane, polychlorinated biphenyls, and polybrominated diphenyl ethers, which can pose a great risk to aquatic and wildlife and can even cause perpetual damage to humans after being significantly biomagnified. These polymers also promote destructive and invasive fauna to explore new environments, in turn resulting in adverse effects to the environment.[32]

Degradation of these polymers is prolonged when subjected to environmental factors, which is one of the most valuable properties responsible for their widespread use.[109] The degradation mechanism of plastic present in the environment consists of photodegradation, thermooxidative degradation, degradation by hydrolysis, and lastly, biodegradation caused by environmental microbes. In general, initiation of this natural process of plastic degradation occurs by photodegradation, then shifting toward thermooxidative type in which UV rays from the sun deliver energy for activation and oxygen incorporation in polymers resulting in brittle plastic which progressively breaks down into smaller particles, which are as small to be metabolized via microbial action.[104,114]

This microbial action for plastic degradation is biodegradation, which either finally produces carbon dioxide or leads to polymer carbon incorporation into biomolecules. Though this process seems to be fairly easy and realistic, it takes more than 50 years to degrade these plastic polymers fully. In addition, the effect of photodegradation on plastic is insignificant due to reduced temperature and accessibility of oxygen in oceans, where 18 billion lb of plastic reaches each year, along with a null rate of hydrolysis.[64,114]

Biodegradation, the process of converting a substance into more minor new compounds via microbial or enzymatic action in the environment, is being investigated extensively to develop an eco-friendly process of plastic degradation. Still, the long duration associated with the process makes it insignificant.[55] The time required for sole biodegradation of a bottle of plastic (mainly polyethylene terephthalate) in the natural environment can range around 450 years, which is absolutely unbearable. Another approach of using natural or biopolymers for the preparation of biodegradable plastic also seems highly promising, but again, the long duration and precise conditions required for degradation can only be met in laboratory-defined conditions and not in conventional landfills, which reduce the effectiveness of this approach.[90] Hence, this presents a great dilemma that whether biodegradation is a myth or reality. Further, a recent approach of development of edible plastics from plant or seaweed carbohydrates and proteins, which are highly stretchable and flexible, presents an effective solution to minimize the usage of conventional plastic and puts forward a solely biodegradable form of plastic.[93]

This chapter presents details on the biodegradation of plastic and the development of biodegradable plastic to provide deep insights into the till date scientific expansions to deal with the problem of hazardous plastic waste via biological means. In conclusion, all these enumerations will clarify the quandary of biodegradation to a greater extent.

10.2 SYNTHETIC PLASTIC-BASED FOOD PACKAGING

The exploding consumer's demands for different food varieties are immensely transforming and increasing the food industry. Ready-to-eat processed food assortments, which can sustain their nutrition for long, are the major contributors to the improvement in the food sector. To serve the population with a healthy variety of food innovations, it is of utmost importance that the food products are packed in a sterilized and efficient packaging material that can withstand most of the environmental stresses and is flexible to be used for all types of products.[72] Plastic polymers fulfill most of these requirements and are being excessively used in food industries for packaging. Thus, the increasing trends of plastic usage in packaging and food varieties go hand in hand.

The malleability, ductility, unbreakable nature, lightness, resistance to moisture, impermeability, and ability to maintain sterile conditions inside packaging, along with aesthetic properties, make plastic a highly desirable material for packaging in food industries.[89] A wide variety of plastic polymers is being used in food packaging, mainly polyvinylidene chloride, polystyrene, polypropylene, polyethylene terephthalate, low-density and high-density polyethylene, polyamide, and polyethylene covinyl alcohol, depending on the shelf life and food quality, along with safety from toxic chemicals which leach out from packaging in contact with food. Sometimes the nature of food, like acidic food or more fat-containing food, also affects the selection of polymers.[25,47]

Based on chemical structure, these plastic polymers are classified as polyolefins, ethylene copolymers, substituted olefins, polyesters, polycarbonates, polyamides, and acrylonitriles.[47]

Polyolefins are polymers of ethylene or propylene having thermoplastic properties. This class includes low-density (915–939 kg/m^3) and high-density (greater than 940 kg/m^3) polyethylene and polypropylenes. Low-density polyethylene is most frequently used for rigid packaging of food as blow molded or film forms like squeezing flexible bottles, tubes, snap-on caps, dispensers, and spouts. High-density polyethylene is used only for the manufacturing of bottles, although at present, its use has gone down due to the takeover by polyethylene terephthalate bottles. Polypropylene has high-temperature resistance; thus is applied as a sealing layer in hot fluid-filled bottles, packaging compatible with microwave, and retort pouches.[16,77,83]

Ethylene copolymers contain ethylene conjugated with either alkene monomers like butene, propene, octene, or hexene or compounds containing polar groups like acrylic acid, ethyl acetate, vinyl acetate, vinyl alcohol, or methyl acrylate. Ethylene vinyl acetate, a polymer of randomly arranged vinyl acetate and ethylene monomers, finds its application mainly as stretchy films and cling wraps. However, it is also used to produce laminating materials in the coextrusion process.[2,45]

Ethylene vinyl alcohol acts as an excellent barrier to gases, fragrance, odor, and solvents, thus is actively replacing metal and glass containers used in food packaging. It retains the quality and flavor of food by preventing contact with oxygen and is highly effective for gas-filled packaging like modified atmosphere packaging (MAP). It resists organic vapors and oil; but as the penetrant polarity increases, this resistance reduces.[51,81] The associated drawback is that it absorbs moisture as ethylene vinyl alcohol has hydroxyl group, which makes it hydrophilic. This, in turn, affects its efficiency to prevent gas penetration; hence, the ethylene vinyl alcohol layer is usually covered by moisture-resisting polyolefins to prepare multilayer effective packaging. A copolymer of ethylene and acrylic acid is thermoplastic and flexible, and shows significant resistance to chemicals. It has high toughness, strength, adhesion, and hot tack force, thus is frequently used in blisters and adhesive laminations. It is used in the packaging of cheese, meat, snack, and in some skin packaging where they stick closely to the product surface.[36,51,53]

Substituted olefins consist of polystyrene, polyvinyl alcohol, polyvinyl chloride, polyvinylidene chloride, and poly tetrafluoroethylene. The catalyzed reaction of ethylene and benzene forms ethylbenzene, which on dehydrogenation converts to styrene; this styrene, when polymerized, forms polystyrene. Polystyrene was the first polymer to get commercialized because of its rigid clear plastic nature and being molded to form different shapes. Its recent applications involve coextrusion with ethylene vinyl alcohol or polyvinyl chloride–polyvinylidene chloride copolymer for the production of wide-mouthed thermoformed containers for food products that are shelf-stable and blow-molded multilayered bottles.[82] It has two more types, namely high-impact polystyrene and expandable polystyrene sheet, and is used for manufacturing tubs which are injection-molded and disposable packages, respectively.

Expandable polystyrene sheet is a form of foam and is used to prepare trays, disposable dinner sets, cartons of eggs, and packs or containers for takeaway meals. Polyvinyl alcohol, an atactic material, is formed by the alcoholysis of polyvinyl acetate but is not usually used for food packaging as it is water soluble and does not provide a barrier against it, though it prevents greases and oxygen. Polyvinyl chloride is another substituted olefin, made by dehydrochlorination of ethylene chloride to form vinyl chloride and then polymerizing it.[22] It is used in the production of containers that are blow molded, insert trays of biscuits or chocolate, and in the form of plasticized film, where it is employed for stretch wrapping of red meat tray. Polyvinylidene chloride is used for the manufacturing of barrier containers which are multilayered and shrink films. It has very less permeability for vapors, odors, alcohols, and grease. Polytetrafluoroethylene is a fluorinated polymer having properties like thermal stability, nonadhesiveness, inertness (chemically), resistance to heat, and small friction coefficient, which makes it ideal to be used as a nonstick separating layer in between the heat sealer jaws and thermoplastic layer in food packets.[68]

Polyesters are polymers having C–O–C bond with one carbon from carbonyl group. They are made by condensation polymerization, which results in the release of a water molecule. This class majorly consists of three types of polymers, namely polyethylene terephthalate, polyethylene naphthalate, and copolyesters. Polyethylene terephthalate is a copolymer of two compounds, namely ethylene glycol and terephthalic acid. It is used for various packaging in food industries like blister packs, films, food trays, bottles, and hot film containers. Films of this polymer have high resistance to chemicals, tensile strength, elasticity, least weight, and high temperature stability in the range of –60 to 220°C, which makes it widely acceptable for the products which are frozen and have to be boiled inside the bags before consumption, and for bags used in oven. It is also used to prepare trays used in prepared frozen foods, which can be directly used in the microwave.[49]

Polyethylene terephthalate in copolymerized form with polyvinylidene chloride is majorly used for wrapping single cheese slices. Metalized forms, when coextruded into films, are applied as ply inside packages of snacks.[50,86] Bottles of polyethylene terephthalate are widely used and manufactured by stretch blow mold technique to provide high tensile and barrier properties with low weight, which is highly economical. Polyethylene naphthalate is prepared from polycondensation of ethylene glycol and 2,6-dimethyl naphthalene dicarboxylate (which results from oxidation and esterification of 2,6-dimethyl naphthalene).[30] It has five-fold higher barrier properties than polyethylene terephthalate against oxygen, carbon dioxide, and moisture. Thus, it is used in the production of film flexible packaging and rigid containers of carbonated drinks and beer. Copolyester is synthesized by using two or more glycol or dibasic acid and is used majorly for manufacturing oven trays.[19,48,63]

Polycarbonates are carbonic acid polyesters having carbonate linkages prepared from the reaction of diphenyl carbonate or phosgene and bisphenol A. They are used to prepare coinjection and coextrusion moldings, where they contribute strength and transparency to the containers that encounter hot processing and filling. Multilayer bottles, having a central layer of ethylene vinyl alcohol and side layers of polycarbonates, are also used but are not economically feasible. Ovenable trays, boiling bags, and skin packagings are also manufactured from polycarbonates for prepared frozen meals, which require microwave treatment for boiling before consumption. Along with these products, reusable bottles are also prepared.[6,39]

Polyamides are thermoplastics made by condensation of carboxylic acid and amine, which give out amide linkage. This linkage contributes to high mechanical and barrier properties. A variety of polyamides has been developed, and some majorly used are nylon 6, nylon 6,6, nylon 11, and nylon 6,10. Films made of nylon 6 are more prevalent for packaging in the USA, whereas nylon 11 films are abundantly used in Europe. In general, the most commonly used polyamide films in food packaging are nylon 6, nylon 11, and nylon 6,6.[40,92] Polyamide is also employed in multilayer films used for the packing of meat and cheese to provide a vacuum environment. Biaxially-oriented films of this polymer are quite prevalent in the packaging of frozen foods, processed or fresh meat, and processed natural cheese.[28]

Acrylonitriles made by propylene ammoxidation contain a triple bond between carbon and nitrogen. These acrylonitriles polymerize to form mainly polyacrylonitrile and high nitrile resins. Polyacrylonitrile is a polymer of mainly acrylonitrile accompanied with some monomers like methyl methacrylate, methyl acrylate, and vinyl acetate. It provides excellent chemical and gas resistance; but at 220°C, it degrades. Because of this reason, its application in food packaging is insignificant, but high nitrile resins find some scope in the food industry. The copolymers of acrylonitrile–styrene and acrylonitrile–methyl acrylate constitute high nitrile resins and are used significantly for the packaging of carbonated drinks.[23,105,110]

The list of polymers is not limited to abovementioned. There are a variety of other polymers used for specific product packaging. Such a huge variety of plastic-based polymers and their highly resistant nature against chemicals, high temperature, moisture, and gases though prove them excellent in concerns to food packaging materials, but enforce a great burden on the environment, where these polymers get accumulated after getting discarded.[102] Scientific society is making numerous efforts to develop approaches to degrade these polymers, but the resisting powers of these polymers make the degradation process very slow, thus proving their efforts hopeless. Along with this, many polymers, when subjected to high temperatures or chemicals, break down to produce toxic chemicals, which in turn pose a great challenge.[20]

Biodegradation, a process of degradation by subjecting the matter to biological forces like microbes, shows a little hope in this regard, but its effectiveness is absolutely uncertain in the natural environment. The upcoming section will be introducing some of the aspects related to the biodegradation of plastic-based food packaging to explain related certainties and qualms.

10.3 BIODEGRADATION OF SYNTHETIC PLASTIC PACKAGING

Plastic production has increased significantly from 348 million t in 2017 to 359 million t in 2018, out of which only 10% was recycled. Packaging industry accounts for 90% of this production, with food packaging being the most significant contributor.[99] A variety of fossil fuel-based or biobased plastics are being used in food packaging, but their degradation after being discarded is a great challenge to the scientific society. Polyethylene, polypropylene, polystyrene, and polyvinyl chloride are some of the commonly used nonbiodegradable plastics. Even the plastic prepared from biological sources is nearly nonbiodegradable due to its complex chemical structure and very slow rate of degradation under environmental conditions, and requires specific optimized conditions.[100] Along with this, improper management of plastic waste and unrestrained cluttering are in turn posing ceaseless threats to the environment. Accumulation of such plastic polymers for a long term results in reduction of soil fertility, which additionally leads to ecological imbalance and health-related problems.

In oceans, vast accumulation of plastic polymers is ensuing very high mortality rates of fishes and aquatic birds. Such threats in the near future will soon convey highly fatal consequences to the human race too. The preliminary requirements before actuating toward degradation of plastic are education of society to increase awareness regarding proper management of waste to segregate different plastic types, controlled usage of plastics, and development of industrial-level facilities.[10] Although the researches till date are not so promising regarding biodegradation, the continuous efforts of scientists to develop biodegradation techniques around the globe are gradually providing anticipation to combat the major issue of plastic accumulation in the environment.

Biodegradation of plastic mainly specifies the use of microbes that can either just produce enzymes to disintegrate plastic polymers or can even consume these polymers for their nourishment. Several studies have reported the outstanding ability of microbes like fungi and bacteria in the degradation of plastic by producing exoenzymes under unfavorable growth conditions.[100,109] Most relevant enzymes in this case are lipases, cutinases, esterases, and proteases, and the widely studied microbial strains are *Pseudomonas aeruginosa*, *Pseudomonas fluorescens*, and *Penicillium simplicissimum*.[8,9,51,91,105] The process of biodegradation is not a single-step process but involves several steps, namely: biodeterioration, biofragmentation, bioassimilation, and biomineralization (Fig. 10.1).

Biodeterioration comprises microbial or nonmicrobial actions which superficially deteriorate the plastic and modify its chemical, physical, and mechanical properties. Often, abiotic factors lead to weakening of polymers to initiate biodegradation. Biofilm formation triggers biodeterioration by growing into and over the plastic surface. This biofilm development depends not only on the structure and composition of plastic but also on the surrounding environmental conditions. The hydrophobic nature of plastic polymers, like polyethylene and polystyrene, demands hydrophobic bacterial surface for the formation of highly stable biofilms.[29,94,103] *Rhodococcus ruber* C208 forms highly viable biofilm over polyethylene, which was stable even after 60 days without any carbon supplementation. Polymeric extracellular substances are secreted by the biofilms to strengthen biofilm cohesion and plastic adhesion.[9,94] This polymeric substance can enter the plastic pores encouraging microorganisms to grow into it, resulting in pore size increase and cracks, thereby weakening the plastic. Many biofilms of chemolithotrophic bacteria secrete acidic compounds like nitric acid by *Nitrobacter*, nitrous acid by *Nitrosomonas,* or sulfuric acid by *Thiobacillus* to promote chemical deterioration. Chemoorganic communities secrete organic acids like citric, oxaloacetic, oxalic, fumaric, glutaric, gluconic, and glyoxalic acids, which reduce pH in the pores on accumulation and lead to the progression of degradation by changing plastic microstructures.[11,104]

Biofragmentation involves the enzymatic or free radical-mediated cleavage of plastic polymers into small oligomers or monomers.[33] As very high molecular weight plastic polymers cannot go across the microbial cell wall, so to degrade these polymers, microorganisms need to secrete enzymes capable of initiating biochemical reactions to fragment the polymers at the boundary. To perform such reactions, a preliminary requirement is the presence of unbalanced charged species at the surface of polymers, but as these polymers are highly electrically stable, it poses a challenge for the microorganisms. To combat this issue of charge, bacteria capable of breaking plastic produce oxygenases, which work by adding oxygen to the long polymer carbon chains resulting into charge destabilization. Di or monooxygenase enzymes add two or one oxygen atom/s, respectively, thereby leading to the formation of peroxyl or alcohol groups and hence making the polymer less resistant to degradation. Further, transformation reactions on the destabilized state of the polymer occur by esterases and lipases or by endopeptidases.[34] Several microbial strains have been reported in many studies, which are capable of biofragmentation of polymers. *Pseudomonas* is the most well-known in this aspect as it can degrade polythene and polyvinyl chloride. *Streptomyces* is capable of degrading polyesters, whereas *Ochrobactrum* TD degrades polyvinyl chloride. Around 39 strains of proteobacteria and firmicutes are reported for the breakdown of plastic polymers. Some other reported bacterial species proficient in this regard are *Staphylococcus*, *Bacillus*, *Streptococcus*, *Micrococcus*, *Diplococcus*, *Pseudomonas,* and *Moraxella* sp.[11,33,34,66]

After the fragmentation of these polymers into low molecular weight oligomers or monomers, these fragments become proficient to cross the lipid cell wall of microorganisms. Thus, these fragments can be taken up by the microbe into their cytoplasm for utilization in their metabolism. This process of utilization of polymer fragments for metabolism by microbes is called bioassimilation.[33,66] Although, fragmentation of polymers does not ensure their assimilation by the microorganisms, as special transporters and carriers are required to be present in the microbial cell wall for the entry of fragments inside the cell. Those polymer fragments which find their way inside the cell mostly gets oxidized via catabolism pathways and result in the production of biomass, energy, and new cellular structures.

Several monomers are not completely degraded inside the cells and lead to the formation of secondary metabolites, which are often transported back

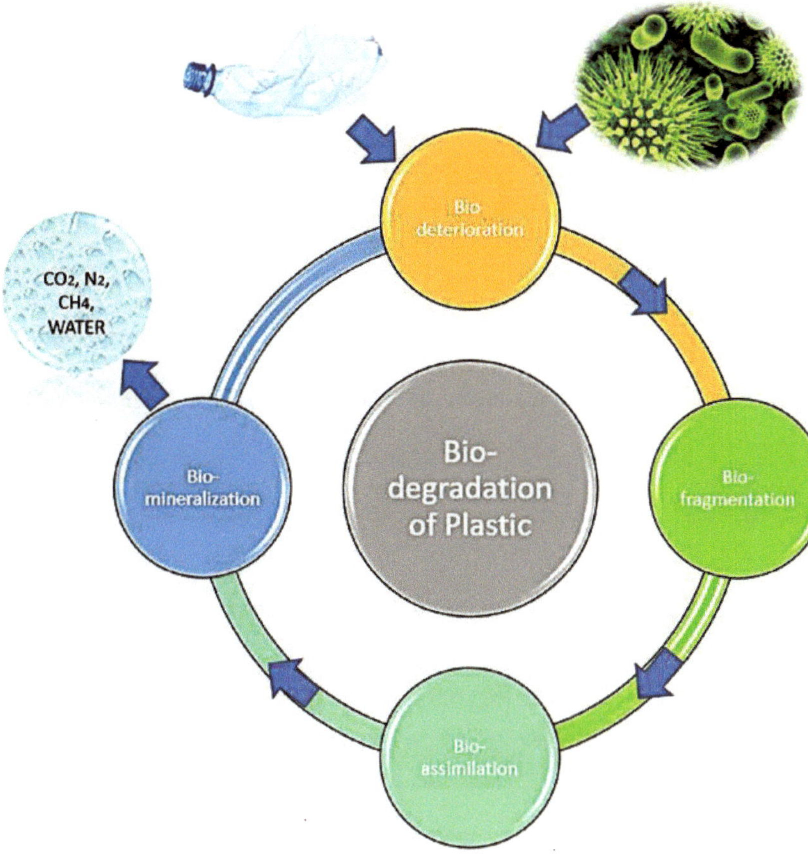

FIGURE 10.1 Process of biodegradation of plastic.

outside the cells as they cannot be utilized via metabolic transformations. But those metabolites which get completely degraded are excreted by the cells in fully oxidized forms like carbon dioxide, nitrogen, methane, or water. This complete oxidation degradation into completely oxidized moieties is called biomineralization.[33]

Considering these basic steps of biodegradation, many researchers around the world are engaged in finding out capable microorganisms to completely degrade the till date nonbiodegradable plastics. Numerous researches are battling with the great challenge of plastic pollution. The leading microbes discussed here in detail for the degradation of different synthetic plastic polymers have been summarized (Table 10.1).

TABLE 10.1 Prevalent Synthetic Plastic Polymers and Respective Microbes Investigated for Their Degradation.

Synthetic plastic polymer	Microbes investigated for biodegradation
Polyamides	*Arthrobacter* sp. strain KI72, rot fungus, *Pseudomonas sp.* ND11.
Polyethylene	*Rhodococcus, Bacillus, Streptomyces, Staphylococcus, Ralstonia, Pseudomonas, Stenotrophomonas, Penicillium, Cladosporium, Aspergillus, Brevibacillus borstelensis* strain B2,2, *Bacillus cereus* strain A5, *Aspergillus oryzae* strain A5,1, *Aspergillus fumigatus* strain B2,2, *Enterobacter* sp. D1, *Galleria mellonella, Pseudomonas* sp. E4, *Penicillium chrysogenum* NS10, and *Penicillium oxalicum* NS4.
Polyethylene terephthalate	*Actinobacteria* phylum, *Thermomonospora curvata, Thermobifida fusca, Ideonella sakaiensis* 201-F6, *Saccharomonospora viridis* AHK190, *Phaeodactylum tricornutum, Fusarium oxysporum, Fusarium solani, Pseudomonas palagia, Pseudomonas pseudoalcaligenes, Pseudomonas mendocina, Humilica insolens*, and *Fusarium cutinase.*
Polystyrene	*Gloeophyllum trabeum* DSM1398, *Gloeophyllum striatum* DSM9592, *Phanerochaete chrysosporium, Pleurotus ostreatus, Trametes versicolor, Gloeophyllum trabeum, Tenebrio molitor, Citrobacter sedlakii, Enterobacter sp., Brevundimonas diminuta, Alcaligenes* sp., *Gammaproteobacteria, Alphaproteobacteria*, and *Pseudomonas* sp.
Polyvinyl chloride and polypropylene	*Cochliobolus* sp., *Micrococcus* sp., *Bacillus cereus*, and *Sporosarcina globispora.*

Polyethylene terephthalate, which is mainly used for the production of foils and bottles, is formed by repeating the units made of ethylene glycol and terephthalic acid. Many researches have proved partial degradation of polyethylene terephthalate by hydrolases produced by some fungal and bacterial strains, but most of these enzymes have small turnover rate. Majority of the bacterial strains for degradation of this polymer belong to *Actinobacteria* phylum, which is categorized to contain gram-positive bacteria. The most highly effective genera known to degrade it are *Thermomonospora* and *Thermobifida.*

Serine hydrolases comprise most of the polyethylene terephthalate hydrolyzing enzymes, including PET hydrolase, MHETase, and tannase. These hydrolyzing enzymes are mainly lipases, cutinases, and carboxylesterases, which contain a fold of α/β hydrolase and a catalytically active triad of serine, histidine, and aspartate.[11,34] PETase enzyme from *Ideonella sakaiensis* 201-F6, which resembles cutinases structurally, also contains some disulfide bonds due to the presence of cysteine residues which promote stable binding to the polymer along with making the enzyme thermally stable.

Ideonella sakaiensis (phylum *Betaproteobacteria*; order *Burkholderiales*) has also been reported to use polyethylene terephthalate as a major carbon and energy source, as this strain contains an additional unique enzyme that is similar to tannase and is responsible for the degradation of mono-(2-hydroxyethyl) terephthalic acid (MHET), which is produced by the action of first enzyme PET hydrolase. After hydrolysis by PETase, then produced MHET is internalized and hydrolyzed by the enzyme MHETase. Monomers then produced are utilized for the metabolism of bacteria.

Researches have proved that simultaneous mutations W159H and S238F in PETase of *Ideonella sakaiensis* make the enzyme active site narrower, which increases the resemblance of the enzyme with cutinase of *Thermobifida fusca* and thus leads to improvement of the enzyme. The MHETase enzyme of *I. sakaiensis* contains a lid-like domain that conveys specificity for the substrate MHET with a K_{cat} of 11.1 s^{-1}. Although *I. sakaiensis* enzymes have been extensively studied for polyethylene terephthalate degradation, other strains have also revealed the ability to degrade this polymer.[61]

Furukawa et al. identified that activity of cutinase enzyme, efficient for hydrolysis of low-crystalline polyethylene terephthalate, from *Thermobifida fusca* was increased on the addition of positively-charged surfactant, which increases enzyme's affinity for the substrate via electrostatic interactions.[21] Numoto et al. analyzed the role of calcium binding in enhancing the functioning and stability of a polyesterase obtained from *Saccharomonospora viridis* AHK190 capable of breaking down polyethylene terephthalate.[58]

Moog et al. engineered *Phaeodactylum tricornutum* for producing exogenous PETase, which showed high degrading activity against polyethylene terephthalate and polyethylene terephthalate glycol. The turnover of the enzyme against polyethylene terephthalate glycol was 80 times higher than that for polyethylene terephthalate. Further, they found that PETase from such diatom source possesses activity against polyethylene terephthalate tattered from industrial source in salty water at mesophilic temperature also, which is beneficial to degrade the polymer not solely in lab conditions but also in the natural environment in which they are present.[52]

Dimarogona et al. performed expression and structural studies on cutinase FoCut5a enzyme found in *Fusarium oxysporum,* which had high similarity with cutinase enzyme of *Fusarium solani*. The maximum activity of FoCut5a for polyethylene terephthalate was attained at pH of 8 and temperature of 40°C.[14] Wei et al. identified and characterized two genes encoding polyester degrading hydrolases in *Thermomonospora curvata*. These two hydrolases, namely Tcur 0390 and Tcur 1278 shared a sequence identity of 61% with *Thermobifida fusca* hydrolase enzyme. Their study revealed that these enzymes had an optimal temperature of 55°C and 60°C for Tcur 0390 and Tcur 1278, respectively at pH of 8.5, thus proving high thermostability of Tcur 1278 compared to Tcur 0390, which can be advantageous for high-temperature zone regions with high polyethylene terephthalate load.[108]

Sulaiman et al., in their study on metagenome of leaf compost, isolated a novel enzyme of cutinase family having 59.7% identity with lipase of *Thermomonospora curvata* and 57.4% with cutinase of *Thermobifida fusca.* When this LC-cutinase was cloned with leader sequence pelB in *E. coli* cells for secretion in the periplasm, greater than 50% was excreted out as extracellular enzyme, which enhanced the applicability of the enzyme by reducing the cost of extraction. It was analyzed to have significantly high degradation activity of 12 mg per hour per mg of enzyme for polyethylene terephthalate at 50°C and pH 8.[95]

Haernvall et al. identified two polyester-degrading enzymes through their in silico search from isolates of the wastewater treatment plant. They identified lipase from *Pseudomonas palagia* and cutinase from *Pseudomonas pseudoalcaligenes*, which were capable of hydrolyzing various phallic acid polyesters having a molecular weight (number average) in the range of 1770–10,000 g/mol with crystallinity less than 1%. Further, they evaluated that 1,2-ethanediol was hydrolyzed more efficiently compared to 1,8-octanediol, as the former had high solubility and hydrophilicity along with short-chain diols.[24]

Rönkvist et al. performed a detailed comparative analysis between the cutinase enzymes produced from *Pseudomonas mendocina*, *Humilica insolens,* and *Fusarium solani* by employing biaxially-oriented and low-crystalline polyethylene terephthalate as a substrate. Out of the three, cutinase extracted from *Humilica insolens* was found to be thermostable with the highest activity in the temperature range of 70–80°C, while the other two had high activity at 50°C. In general, after keeping the pH stable for all the three cutinase enzymes of the considered strains, the affinity was revealed to be higher for low-crystalline (7%) substrate compared to biaxially-oriented form having 35% crystallinity. The reaction rate constant for *H. insolens* cutinase at 70°C was seven times higher than that of *P. mendocina* and *Fusarium cutinase* at 50°C and 40°C, respectively. Along with this, *H. insolens* cutinase showed 97% of weight reduction in low-crystalline polyethylene terephthalate film over a period of 96 h compared to only 5% in the other two at their respective temperatures. Hence, this study demonstrated the significant potential of *H. insolens* cutinase to be used for industrial degradation of the considered polymer.[79] Wallace et al. identified a new esterase (PpEst) from *Pseudomonas pseudoalcaligenes*, which was capable of hydrolysis of co-aliphatic-aromatic polyester namely poly (1-4-butylene adipate-co-terephthalate) in both milled and whole film form, but a feedback inhibition was detected due to the degradation product 4-(4-hydroxy butoxy carbonyl) benzoic acid.[106]

Although many microorganisms and their enzymes have been isolated for the degradation of the most widely used synthetic polymer, still many factors limit their application on a large scale. Inhibition of the enzymes by the products produced after incomplete degradation of polyethylene terephthalate (like BHET (bis-(2-hydroxy ethyl) terephthalate) and MHET) poses highly negative effect on the performance of the enzymes.[61] Though efforts have been made for designing reactors that can continuously eliminate the inhibiting products, but that will likely reduce the feasibility for whole cell use. Several researchers have used mixtures of enzymes for hydrolysis, which will pose synergistic effect to combat the inhibition, while some have tried inhibitor interacting amino acid modifications in the enzymes.[61]

Another widely used variety of polymers in food packaging is of polyethylene. These polymers account for 100 Mt of plastic produced every year at a global level.[76] Many bacterial genera are known for the degradation of polyethylene. Some gram-positive ones belong to genera *Rohodococcus, Bacillus, Streptomyces*, *Staphylococcus*, etc. and the gram-negative ones are of *Ralstonia*, *Pseudomonas,* and *Stenotrophomonas*. Few genera of fungal origin are also known for their ability to degrade this polymer, including *Penicillium*, *Cladosporium*, *Aspergillus,* etc. Researches have also connected polyethylene degrading ability to microbes found in complex microbiomes of the gut.[10]

Muhonja et al., in their study, evaluated the dumpsite of Nairobi for biochemical and molecular characterization of bacteria and fungi capable of degrading polyethylene of lower density. Their work resulted in the identification of 10 fungal and 20 isolates of bacteria through 18S and 16S rDNA sequences. Most active bacterial and fungal strains for polyethylene degradation found were *Brevibacillus borstelensis* strain B2,2 and *Bacillus cereus* strain A5, and *Aspergillus oryzae* strain A5,1, respectively. They screened the strains for the production of esterase and laccase enzymes where *Aspergillus oryzae* strain A5,1 displayed the highest esterase production, while *Aspergillus fumigatus* strain B2,2 showed the highest production of laccase enzyme. Along with this, four samples of bacteria were found to be positive when tested for the presence of alkane hydroxylase genes.[54]

Ren et al. isolated *Enterobacter* sp. D1 from the microbiome of *Galleria mellonella* gut and they cultivated the strain over the surface of polyethylene film for a span of 14 days. Analysis of the surface degradation via different techniques, including atomic force microscopy, scanning electron microscopy, and Fourier transforms infrared spectroscopy, was done. Depressions, roughness, and cracks were found on the film surface, and the presence of ether and carbonyl groups was also detected. Liquid chromatography tandem mass spectrometry, in turn, revealed an increase in acids, esters, and alcohol contents. All these results made it evident that *Enterobacter* sp. D1 is a highly significant polyethylene degrading strain and is highly promising for its use at an industrial scale.[75]

Yoon et al. reported a mesophilic bacterium capable of degrading polyethylene of low molecular weight. The isolated bacterium was characterized to be gram negative and rod shaped, and identified as *Pseudomonas* sp. E4 via 16S ribosomal DNA sequencing. The test of biodegradability showed 4.9–28.6% of carbon mineralization to CO_2 after a duration of 80 days at a temperature of 37°C. The trend of degradation proved that degradability was reduced with the increase in the weight of the polymer. In this study, the researchers reported degradation of an unoxidized form of low molecular weight polyethylene, and the leading enzyme responsible for the degradation was identified to be alkane hydroxylase.[113]

Ojha et al. performed a polyethylene degradation study including both low and high molecular weight polymers by fungal isolates. They isolated two fast-growing strains, namely *Penicillium chrysogenum* NS10 and *Penicillium oxalicum* NS4, which had polyethylene degrading ability and applied response surface methodology to optimize their growth medium. They subjected the unoxidized plastic sheet for a duration of 90 days and observed the degradative changes via scanning electron microscopy, atomic force microscopy, and Fourier transform infrared spectroscopy. Significant changes in morphology were observed, which made it evident that enzymatic degradation was effective. Along with this, these strains formed biofilm over the surface of polymers, which in turn enhanced the degradation of polyethylene. Gradually, reduction in polymer weight and the increase in dry mycelium weight over the duration of time proved these strains to be effective in natural conditions, as no pretreatment of polyethylene was needed for these strains to degrade the polymer.[60]

Polyamides are polymers with great monomer versatility, but the most common and popular ones are Kevlar and nylon. These polymers are used in multilayer food packaging to strengthen the packing and to provide toughness. Although no microorganism yet is known that can degrade polyamides fully, several bacteria have been reported which can act on cyclic or linear oligomers of nylon having small chains. Some oligomers can also fulfill the nitrogen and carbon demand of specific bacteria, which are adapted to the oligomers present in wastewater.[60] The first identified bacteria which can grow on polyamide oligomer mixtures was *Flavobacterium* sp. KI72, later named as *Achromobacter guttatus* KI72 and lately renamed to *Arthrobacter* sp. strain KI72. Isolates of *Arthrobacter* code for aminotransferases and hydrolases, which are responsible for the initiation of oligomer degradation and subsequently bacterial metabolism.[98]

Enzymes mainly responsible for hydrolysis initiation of the cyclic dimer, dimer, and endotype oligomer are coded as NylA, NylB, and NylC, respectively. NylC is typically an esterase though its tertiary structure revealed β-lactamase fold containing motifs. After oligomer hydrolysis, the resulting monomers are metabolized by 6-aminohexanoate aminotransferases (*nylD1*) and then adipate semialdehyde dehydrogenase (*nylE1*). Thus, 6-aminohexanoate aminotransferases metabolize 6-aminohexanoate to adipate semialdehyde, and then adipate semialdehyde dehydrogenase convert it to adipate. Manganese-dependent peroxidase extracted from rot fungus is the only enzyme reported till date for the degradation of nylon fiber of high molecular weight, mainly membrane of nylon-66. This enzyme is known to be affected by lactate and other α-hydroxy acids presence.[13,57,98]

Yamano et al., in their study on biodegradation of polyamide, identified a bacterial strain *Pseudomonas sp.* ND11, which was capable of degrading nylon-4 into its monomers of gamma-aminobutyric acid via NMR analysis. Their results proved that the strain produced some extracellular enzyme that hydrolyzed the amide bonding in between the monomers in polyamide-4. Although polyamides existed in the packaging industry for long, still studies related to degradation of high molecular weight polyamides are very few, and only some enzymes have been characterized for the same. Present studies are in preliminary state and are not eligible to be employed at a large scale to deal with global polyamide load. This limitation poses a great challenge to the scientific society and demands exhaustive research in this concern.[111]

Polystyrene plays a wide role in food packaging industry, and its production accounted for 14.7 million metric tons in 2016, thus significantly contributes to global petroleum-based plastic pollution. Regrettably, no enzymes are known for the degradation of polystyrene of high molecular weight.[70] Krueger et al. identified strains of brown rot fungi, which can degrade polystyrol via Fenton reaction driven by hydroquinone. In this study, they identified *Gloeophyllum trabeum* DSM1398 and *Gloeophyllum striatum* DSM9592 to be capable of substantially depolarizing at an incubation period of 20 days. Their results proved 50% molecular weight reduction in polystyrol, which seems highly promising to be employed at a large scale.[38]

In another study, white and brown rot strains of fungi were characterized for polystyrene depolymerization under conditions of coincubation with lignin. The identified fungi of white rot type were *Phanerochaete chrysosporium*, *Pleurotus ostreatus*, and *Trametes versicolor* and brown rot fungus was *Gloeophyllum trabeum.*

Although these reports were highly promising in the context of degradation of high molecular weight polystyrene, the enzymes utilized for polystyrene degradation have not yet been discovered, so the loss of polymer weight could be considered due to chemical additives degradation. In the same manner, many bacterial strains have been identified to form biofilms over the polystyrene polymers and hence degrade them, which was depicted by the reduction in polymer weight.[38]

Yang et al. have recently discovered that yellow mealworm larvae known as *Tenebrio molitor* display significant polystyrene degradation. These larvae eat the polystyrene foam and mineralize it to CO_2 through the microbial flora present in their gut within a duration of 15 h.[112] Shekhar et al. identified four bacterial cultures, namely *Citrobacter sedlakii*, *Enterobacter sp.*, *Brevundimonas diminuta*, and *Alcaligenes* sp., which were capable of growing in medium containing polystyrene of high impact and most often used as e-plastic. They performed flask experiments using supplementation of gelatin at a minimal concentration of 0.1%w/v and observed an increase in degradation to 12.4% in a duration of 30 days. They also revealed that effective degradation occurred due to depolymerase enzymes present in the strains, which became evident by the detection of intermediates in the supernatant of culture via HPLC analysis. This work showed the great potential of these strains to be used for degradation of polystyrene that is used in food packaging and can be analyzed further for the same.[84]

Syranidou et al. observed that member bacterial strains of *Gammaproteobacteria* and *Alphaproteobacteria* were significant in developed time-dependent biofilms. Their results supported reduction in polymer molecular weight along with functional group alterations showing promising degradation, which can be further researched.[97] Kim et al. isolated *Pseudomonas* sp. strain, which mostly survives in soil from superworms gut and has the capability to degrade polystyrene. They examined the changes in chemical structure by X-ray photoelectron spectroscopy (XPS), nuclear magnetic resonance (NMR), and Fourier transform infrared spectroscopy (FTIR), and found a significant reduction in hydrophobicity. Along with this, they found an eminent enzyme, serine hydrolase, which effectively took part in polystyrene degradation and commenced the search at the enzymatic level for the purpose.[35]

Polyvinyl chloride and polypropylene are the two most important polymers of the food packaging industry. Polypropylene and polyvinyl chloride procure second and third rank, respectively, among the most produced plastics. Although the cost of production of polyvinyl chloride is low with high production at the global level, barely any consistent microbial degradation research is present till date for both the abovementioned polymers making the situation highly challenging.[31] Sumathi et al. demonstrated the ability of fungal strain *Cochliobolus* sp. to degrade polyvinyl chloride of lower molecular weight by producing laccase enzyme at a temperature of 30°C. They observed significant results in their gas chromatography mass spectrometry (GC–MS), FTIR, and scanning electron microscopy (SEM) analysis.[96]

Patil and Bagde isolated potential strains having the capability to act and degrade polyvinyl chloride-rich plastic waste by applying culture enrichment technique. They reported *Micrococcus* sp. as an efficient one among the considered. They proved release of chloride after subjecting sole polyvinyl chloride to the strain, increase in density of microbial cell culture, and production of CO_2, but the percentage release of chloride was 0.36%, and mineralization was 8.87% over a 70 days period.[67]

Helen et al. studied biodegradation of highly resistant polypropylene polymer by isolates obtained from sediments of mangrove in the peninsular region of Malaysia. They found significant polymer weight loss of 12% and 11% from *Bacillus cereus* and *Sporosarcina globispora*, respectively over a period of 40 days. The rates of reduction achieved by respective strains were 0.003 and 0.002 g/day. Lack of significant studies in concerns to biodegradation of polyvinyl chloride and polypropylene prove their highly resilient nature, making them the biggest threat to the environment.[27]

Although many research studies have been conducted for biodegradation of synthetic plastic polymers, only a few of them seem to have the potential to be incorporated at a large scale under environmental conditions. The high production rate of synthetic polymers and the very slow rate of biodegradation are the most significant barriers, which are highly difficult to cope with. Along with this, many polymers require pretreatment, which makes the degradation process economically unfeasible. Many enzymatic degradation studies reported the issue of enzyme inhibition due to the production of intermediates and products, which in turn demand significant attention and make the process difficult to implement in the environmental niche. These aspects put the reality of biodegradation in a dilemma. A new approach to the production of bioplastic seems to be a bit promising, as this type of plastics is produced from biological renewable sources, and their biodegradability is highly probable, but still, the complex polymer structures pose some challenges. The upcoming section deals with the details in this concern to strengthen the foundation of biodegradation to some extent.[92]

10.4 BIOBASED PLASTIC AND BIODEGRADATION

Biobased plastics are either entirely or partly made by employing biomass as raw material. Transformation of biomass to bioplastic occurs by employing chemical, biological, and/or physical treatments on biomass. Some natural biopolymers present in plants like cellulose and starch can be extracted directly for the preparation of bioplastic. Fermentative procedures via microbial action on a suitable source of carbon can, in turn, lead to bioplastic production like polyhydroxy alkanoates. Even biomass from plants can be converted to unit cells for building polymers like polylactic acid. These bioplastics discharge low toxic emissions from the extraction of raw material and during production. These bioplastics can be either biodegradable or nonbiodegradable like synthetic plastics, but pose low environmental damage comparatively. At present, the production of bioplastic is around 1% of total annual plastics production, but with an increase in demand, it is expected to rise exponentially to 7.85 million t in 2019. In Europe, food packaging accounts for 80% of the bioplastic produced annually.[17,41]

Starch-based bioplastic is used for the preparation of disposable cutlery and tablewares, bottles, and coffee capsules used in machines. It is produced from potato, rice, and corn starch and is efficient in replacing polystyrene polymers. The major advantages of these starch polymers are their biodegradability, low cost, and compostable nature.[46]

Cellulose-based bioplastic is capable of replacing low-density and high-density polyethylene, polypropylene, and polystyrene. It is produced from wood, crop fiber, and herbaceous crops, and is compostable and biodegradable, depending on the material properties. Although this type of bioplastic can be used for packaging bread, meat, fruits, and dried products, but its low resistance to moisture and high production cost limit its application at a commercial scale.[40]

Another commercially most appropriate class of bioplastics is polyhydroxy alkanoates polymers, which are biopolyesters and have tunable crystallinity, stiffness, elasticity, and degradability. Their marine degradable nature has gained much importance, along with high biodegradability and compostable nature. They are composed of sugar crops, for example, sugarcane, fats, and oils from sunflower, soy, olive, and rapeseeds, and rarely contain residue from palm oil.[37,101] Polyhydroxy alkanoates have been employed for the production of bioplastic containers for food packaging, and these polymers show great potential to replace polyethylene and polypropylene (which are highly resilient to degradation).[74] The production cost of these polymers is 20–80% higher than the conventional synthetic plastics as their polymerization processes are still at the laboratory level.[78]

Polyhydroxy alkanoates are most commercially viable polymers and have attracted vast attention from scientific society. Their ability to get degraded in varied environmental conditions, like lake and seawater, soil, etc., seeks great importance, but the degradation rate varies with different environmental factors like microbial flora, presence of nutrients, pH, and temperature, and polymer properties consisting of additives, crystallinity, and surface area. Due to its high weight, its transportation inside the cell is difficult and requires extracellular enzymes for degradation. Several bacteria and fungi have been reported for effective polyhydroxy alkanoate degradation, including *Alcaligenes faecalis*, *Aspergillus fumigatus*, *Comamonas* sp., *Hyobacter delafieldii*, *Acidovorax faecalis*, *Variovorax paradoxus*, *Pseudomonas lemoignei*, *P. fluorescens*, *P. stutzeri,* etc.[4,43,59]

Polylactides prepared from natural sugars like sugarcane, corn and cassava, corn stover, wheat straw, bagasse, and chips of wood have comparable efficiency to low and high-density polyethylene, polyethylene terephthalate, and polystyrene and can replace them effectively. They can be used in the production of transparent, disposable, and rigid containers, jars, bags, and films, and represent 12% capacity of the global production of biobased plastics. The major drawback is that a separate infrastructure from synthetic fossil plastic is required for recycling. Biodegradability of biobased polymers is of great importance in concerns to environmental pollution due to excessive plastic usage. These plastics can be degraded in different natural conditions like marine, aquatic, compost, and soil environment.

Various studies have been conducted to isolate microbial species capable of degrading these polymers.[1,7]

Pranamuda et al. isolated polylactic acid degrading actinomycete *Amycolatopsis* HT32 from soil culture, and their results depicted 60% degradation within 14 days duration in liquid cultures.[71] Nair et al. isolated four strains capable of degrading polylactic acid polymers, including *Cladosporium sphaerospermum*, *Penicillium chrysogenum*, *Rhodotorula mucilaginosa,* and *Serratia marcescens*, and observed 44% w/w polylactic acid degradation efficiency of the consortium of mentioned strains over a duration of 30 days.[56]

Butbunchu and Aree reported members of actinobacteria phylum to be potent in polylactides degradation. Micromonosporaceae, Streptosporangiaceae, Streptomycetaceae, Thermomonosporaceae, and Pseudonocardiaceae families are contributors of polylactides degrading strains, out of which Pseudonocardiaceae family being the most potent one.[5] Bonifer et al. identified an isolate from the agricultural field having polylactic acid degrading ability. The novel strain identified was *Bacillus pumilus* B-12, which displayed significant degradation in around 48 h by releasing monomers of L-lactate. This study also illustrated potential regulation of degradation by supplementing the medium with different carbon sources proving high adaptability of strain at an enzymatic level based on environment, which can be employed to optimize the process of biodegradation.[3]

Biopolyethylene terephthalate can be a potent replacement for fossil-based polyethylene terephthalate. It has similar properties to the conventional polymer and is nonbiodegradable, but can be recycled via conventional techniques. It is produced from sugar crops (namely sugar beet and sugarcane), and is used for the production of water bottles, films, and containers. This biopolymer accounts for 35.4% capacity of production at the global level with respect to total bioplastic production. Several studies of biodegradation of polyethylene terephthalate have been recently reported as mentioned in the above section directing toward the possibility of conversion of nonbiodegradable polymer to biodegradable ones.[44]

Keeping concerns with the present studies and the latest findings of microbial or enzymatic degradation of synthetic as well as biobased plastics, biodegradation seems to be highly promising in the coming future to deal with the great threat of environmental plastic pollution. Though most of the biodegradation processes discussed so far are at laboratory scale, and very few show hope for industrial or large-scale application, still, continuous efforts to deal with plastic pollution through the weapon of biodegradation seem promising.

10.5 EXTENDING BIODEGRADATION: THE CONCEPT OF EDIBLE PLASTIC

Though much efforts have been made to develop microbial consortium and enzymatic processes for the degradation of plastic, yet several serious limitations like in situ degradation of accumulated plastic, complex degrading conditions, required pretreatments for plastic polymers, low rate of degradation, and inhibition related to enzymatic approaches encourage us to look for other alternatives.

Recently developed technology of edible plastic, which can be consumed as food after utilization in form of plastic, alleviates the need to search for microbial or enzymatic procedures. Such extended form of biodegradation, which uses the greatest life forms, "humans" as consumers or degrading species to deal with the problem of environmental plastic accumulation, provides an outstanding solution.[87]

Edible plastics can be a boom for the food packaging industry as the consumers will not only consume the food packed inside them but can also consume the packing, which too can be optimized to provide nutrition. Notpla, a startup, recently marketed sports drink capsules, called Ooho, which enclose liquid in a seaweed watertight film capsule. Consumers can quaff the drink and gulp the capsule. Even if this packing is discarded, this will biodegrade, leaving no traces within 4–6 weeks. Other sources which are being opted for the preparation of such edible packing are milk proteins and potato starch for producing tablewares, wrappers, straws, etc.[65]

Puscaselu et al. developed inulin-based edible biofilm incorporating biopolymers like sodium alginate, agar, and glycerol to be used in food packaging. As inulin was used, the preparation resulted in enhancement of physical and mechanical properties even over the duration of 3 months. Hence, their results proved that this kind of edible packaging is highly promising, being totally biodegradable, compostable, originated from a renewable source, economically feasible, and thus produce no waste.[73] Edible coatings of alginate have also attracted great scientific interest, as they enhance the shelf life and quality of vegetables, sea food, fruits, poultry, meat, and cheese by lowering water loss, regulating respiration, improving the appearance, and mechanical characteristics.[85]

Other types of edible packaging are made up of polysaccharide obtained from plants, animals, and algae. These packagings although have good resistance against O_2 and CO_2 and exceptional mechanical characteristics at low moisture conditions, but are not resilient against moisture because of their hydrophilic nature.[18] Polypetchara and Gohtani studied the effect of different sugars (allulose, maltose, and sucrose) on starch procured from varied sources (corn, waxy corn, rice, and waxy rice) to produce edible films by using sugar and starch as plasticizers. Sugar addition enhanced the crystallinity, shelf life, thickness, and homogeneity. Added sugar, in turn, improved breaking strain and reduced breaking stress during storage conditions, as well as just after preparation.[69]

Daza et al. applied artificial neural network and stepwise regression techniques to evaluate the effect of glycerol and starch concentrations and temperature of drying on ulluco starch used for edible biofilms. The properties like moisture permeability, solubility, transparency, tensile strength, and break elongation were found to be satisfactory. Their results depicted that glycerol content played a significant role in influencing the physical properties. Thus, this study concluded a highly promising scope of ulluco starch for the development of edible films.[12]

These edible films can be customized according to the need of packaging by changing the raw material used. Hence, edible plastic can be the best solution to reduce the plastic burden on earth in the upcoming future.

10.6 FUTURE PROSPECTS

Food packaging is of great importance for maintaining the nutrition, quality, and shelf life of food. It is a great barrier against the environmental factors, which can reduce the quality of food via oxidation or hydrogenation or by contamination. Plastic polymers have been widely used in the food industry, but such wide usage has also resulted in intolerable accumulation of these plastics in the environment and is leading to a deadly situation for the vast range of living species due to their toxic nature either by releasing toxic compounds or by entering the food chain and blocking the physiology of animals, birds, and fishes.

Biodegradation sounds to be a boon in this regard but vast major challenges mainly due to high resistant polymers are posing the scope of biodegradation in confusion. Though much research has been done, much more is required to find the biological solution to degrade the accumulated plastic. Isolation, characterization, and optimization of microbial enzymatic processes to develop high potency degradative procedure, which can be applied in situ, is the demand of the present scenario. Further development of biodegradable starch and cellulose-based biodegradable plastic and the most efficient polyhydroxy alkanoates can be a great solution to further deal with the growing plastic accumulation, due to their high efficiency to match the conventional petroleum-based plastic and replace them. The most sounding innovation is the development of edible plastic, which can be a great dose of relief to treat the problem of exponentially increasing toxic plastic accumulation. Thus, the future of biodegradation is highly promising and demands exhaustive research as the real world can be highly benefitted by further development of biodegradation and bioremediation approaches.

10.7 SUMMARY

Plastic has been a significant threat from production to degradation due to the release of harmful toxic emissions. Food packaging plays a major role in making the condition more adverse. Continuous use of conventional petroleum-based plastic can result in perilous outcomes. Extensive and exhaustive researches performed for biodegradation of these fossil-based polymers have shown a great scope for developing eco-friendly degradative processes. Many microbial strains have been identified, and their enzymatic pathways are being tracked to enhance and optimize the biodegradation processes for industrial or large-scale applications. Efforts are being made for identification and isolation of viable strains capable of in situ polymer degradation, which will reduce the cost of collecting, segregating, treating, and processing different plastic polymers separately. To date, findings show great scope, but due to the high resilient nature of some polymers or slow rate of degradation, incapability of in situ degradation, high industrial requirements, enzyme inhibition by intermediates or products of degradation, and associated high cost, the process of biodegradation is finding itself incapable and ineffective, but some solutions in this concern are also expected to come. Development of biodegradable plastics has a great potential to limit the ever-increasing plastic accumulation, and intensive research in this aspect to develop biodegradable forms of both biobased and fossil-based plastic and their customization to fulfill a variety of food packaging needs can significantly unburden the earth from resilient plastic pollution. The concept of edible biodegradable plastic, in turn, seems to be highly promising. Hence, although the process of biodegradation is at a preliminary stage of development, continuous efforts in this direction can surely be beneficial to work for the purpose. Thus, considering biodegradation a myth seems to be irrelevant as further findings in this regard can bring potential outcomes, which can be effectively applied to the real world.

KEYWORDS

- **biobased plastic**
- **biodegradation**
- **food packaging**
- **microbial activities**
- **plastic waste**
- **synthetic plastic**

REFERENCES

1. Ahmed, J.; Varshney, S. K. Polylactides: Chemistry, Properties and Green Packaging Technology: A Review. *Int. J. Food Prop.* 2011, *14* (1), 37–58.
2. Bhunia, K.; Sablani, S. S.; Tang, J.; Rasco, B. Migration of Chemical Compounds from Packaging Polymers During Microwave, Conventional Heat Treatment, and Storage. *Compr. Rev. Food Sci. Food Saf.* 2013, *12* (5), 523–545.
3. Bonifer, K. S.; Wen, X.; Hasim, S.; Phillips, E. K.; Dunlap, R. N.; Gann, E. R.; DeBruyn, J. M.; Reynolds, T. B. *Bacillus pumilus B12* Degrades Polylactic Acid and Degradation is Affected by Changing Nutrient Conditions. *Front. Microbiol.* 2019, *10* (2548), 1–13.
4. Boyandin, A. N.; Prudnikova, S. V.; Filipenko, M. L.; Khrapov, E. A.; Vasil'ev, A. D.; Volova, T. G. Biodegradation of Polyhydroxyalkanoates by Soil Microbial Communities of Different Structures and Detection of PHA Degrading Microorganisms. *Appl. Biochem. Microbiol.* 2012, *48*, 28–36.
5. Butbunchu, N.; Pathom-Aree, W. Actinobacteria as Promising Candidate for Polylactic Acid Type Bioplastic Degradation. *Front. Microbiol.* 2019, *10* (2834), 1–10.
6. Catala, R.; Muriel-Galet, V.; Cerisuelo, J. P.; Dominguez, I.; Lopez-Carballo, G.; Hernandez-Muñoz, P.; Gavara, R. Antimicrobial Active Packaging Systems Based on EVOH Copolymers. Chapter 22; In *Antimicrobial Food Packaging*; Barros Velázquez, J., Ed.; Elsevier Science: London, UK, 2016; pp 297–304.
7. Crevel, R. V.; Khristolyubova, V. In *Bio-based Food Packaging in Sustainable Development: Challenges and Opportunities to Utilize Biomass Residues from Agriculture and Forestry as a Feedstock for Bio-based Food Packaging*; Forestry Policy

and Resources Division, Food and Agricultural Organization (FAO) of the United Nations: Rome, 2016; p 87.

8. Dakal, T. C.; Cameotra, S. S. Microbially Induced Deterioration of Architectural Heritages: Routes and Mechanisms Involved. *Environ. Sci. Eur.* 2012, *24* (36), 1–13.
9. Danko, A. S.; Saski, C. A.; Tomkins, J. P.; Freedman, D. L. Involvement of Coenzyme M during Aerobic Biodegradation of Vinyl Chloride and Ethene by Pseudomonas putida Strain AJ and Ochrobactrum sp. Strain TD. *Appl. Environ. Microbiol.* 2006, *72* (5), 3756–3758.
10. Danso, D.; Chow, J.; Streit, W. R. Plastics: Environmental and Biotechnological Perspectives on Microbial Degradation. *Appl. Environ. Microbiol.* 2019, *85* (19), 1–14.
11. Danso, D.; Schmeisser, C.; Chow, J.; Zimmermann, W.; Wei, R.; Leggewie, C.; Li, X.; Hazen, T.; Streit, W. R. New Insights into the Function and Global Distribution of Polyethylene Terephthalate (PET) - Degrading Bacteria and Enzymes in Marine and Terrestrial Metagenomes. *Appll. Environ. Microbiol.* 2018, *84* (8), 1–13.
12. Daza, L. D.; Homez-Jara, A.; Solanilla, J. F.; Váquiro, H. A. Effects of Temperature, Starch Concentration, and Plasticizer Concentration on the Physical Properties of Ulluco (Ullucus tuberosus Caldas)-based Edible Films. *Int. J. Biol. Macromol.* 2018, *120* (B), 1834–1845.
13. Deguchi, T.; Kitaoka, Y.; Kakezawa, M.; Nishida, T. Purification and Characterization of a Nylon-degrading Enzyme. *Appl. Environ. Microbiol.* 1998, *64* (4), 1366–1371.
14. Dimarogona, M.; Nikolaivits, E.; Kanelli, M.; Christakopoulos, P.; Sandgren, M.; Topakas, E. Structural and Functional Studies of a *Fusarium oxysporum* Cutinase with Polyethylene Terephthalate Modification Potential. *Biochim. Biophys. Acta* 2015, *1850* (11), 2308–2317.
15. Dussud, C.; Ghiglione, J. F. In *Bacterial Degradation of Synthetic Plastics*, CIESM Workshop Monograph, 2014; vol 46, pp 49–54.
16. Emblem, A. Plastics Properties for Packaging Materials. Chapter 13; In *Packaging Technology: Fundamentals, Materials and Processes*; Emblem, A., Emblem, H., Eds.; Oxford Woodhead Publishing: Sawston, Cambridge, 2012; pp 287–309.
17. Farah, T. Bioplastics Continue to Blossom: Are They Really Better for the Environment? Replacing Plastic Looks Hard, but Alternatives to Fossil Fuel Materials Are on the Rise, 2020; p 9. https://arstechnica.com/science/2020/01/are-bioplastics-all-hype-or-the-future-of-textiles/ (accessed Feb 2, 2021).
18. Ferreira, A. R.; Alves, V. D.; Coelhoso, I. M. Polysaccharide-Based Membranes in Food Packaging Applications. *Membranes* 2016, *6* (2), 1–17.
19. Fresán, U.; Errendal, S.; Craig, W. J.; Sabaté, J. Does the Size Matter? A Comparative Analysis of the Environmental Impact of Several Packaged Foods. *Sci. Total Environ.* 2019, *2019* (687), 369–379.
20. Fu, Z.; Wang, J. Current Practices and Future Perspectives of Microplastic Pollution in Freshwater Ecosystems in China. *Sci. Total Environ.* 2019, *691* (2019), 697–712.
21. Furukawa, M.; Kawakami, N.; Tomizawa, A.; Miyamoto, K. Efficient Degradation of Poly(ethylene terephthalate) with Thermobifida fusca Cutinase Exhibiting Improved Catalytic Activity Generated using Mutagenesis and Additive-based Approaches. *Sci. Rep.* 2019, *9* (16038), 1–9.
22. Gausepohl, H.; Nießner, N. Polystyrene and Styrene Copolymers. In *Encyclopedia of Materials: Science and Technology*; Buschow, K. H. J., Cahn, R. W., Flemings, M. C.,

Ilschner, B., Kramer, E. J., Mahajan, S., Veyssière, P., Eds.; Elsevier: Pergamon, Turkey, 2001; pp 7735–7741.
23. Guillard, V.; Gaucel, S.; Fornaciari, C.; Angellier-Coussy, H.; Buche, P.; Gontard, N. The Next Generation of Sustainable Food Packaging to Preserve Our Environment in a Circular Economy Context. *Front. Nutr.* 2018, *5* (121), 1–13.
24. Haernvall, K.; Zitzenbacher, S.; Yamamoto, M.; Schick, M. B.; Ribitsch, D.; Guebitz, G. M. Hydrolysis of Ionic Phthalic Acid Based Polyesters by Wastewater Microorganisms and Their Enzymes. *Environ. Sci. Technol.* 2017, *51* (8) 4596–4605.
25. Hahladakis, J. N.; Velis, C. A.; Weber, R.; Iacovidou, E.; Purnell, P. An Overview of Chemical Additives Present in Plastics: Migration, Release, Fate and Environmental Impact During their Use, Disposal and Recycling. *J. Hazard. Mater.* 2018, *344* (2018), 179–199.
26. Han, J. W.; Garcia, L. R.; Qian, J. P.; Yang, X. T. Food Packaging: A Comprehensive Review and Future Trends. *Compr. Rev. Food Sci. Food Saf.* 2018, *17* (4), 860–877.
27. Helen, A. S.; Uche, E. C.; Hamid, F. S. Screening for Polypropylene Degradation Potential of Bacteria Isolated from Mangrove Ecosystems in Peninsular Malaysia International Journal of Bioscience. *Biochem. Bioinf.* 2017, *7* (4), 245–251.
28. Huang, T.; Qian, Y.; Wei, J.; Zhou, C. Polymeric Antimicrobial Food Packaging and Its Applications. *Polymers* 2019, *11* (560) 1–18.
29. Jacquin, J.; Cheng, J.; Odobel, C.; Pandin, C.; Conan, P.; Pujo-Pay, M.; Barbe, V.; Meistertzheim, A. L.; Ghiglione, J. F. Microbial Ecotoxicology of Marine Plastic Debris: A Review on Colonization and Biodegradation by the Plastisphere. *Front. Microbiol.* 2019, *10* (865), 1–16.
30. Jager, J. A.; Juijn J. A.; van den Heuvel, C. J. M.; Huijts R. A. J. Poly(ethylene-2,6-naphthalenedicarboxylate) Fiber for Industrial Applications. *J. Appl. Polym. Sci.* 2003, *57* (12), 1429–1440.
31. Johnson, T. PVC Plastics: Polyvinyl Chloride; Manufacture, Properties, Applications, and Environmental Concerns, 2020; p 5. https://www.thoughtco.com/what-is-pvc-plastics-820366 (acessed Jan 13, 2021).
32. Katsanevakis, S. Marine Debris, A Growing Problem: Sources, Distribution, Composition, and Impacts. Chapter 2; In *Marine Pollution: New Research*; Ofer, T. N., Ed.; Nova Science Publishers: New York, 2008; pp 53–100.
33. Kawai, F. Biodegradation of Polymers (Bioassimilation, Biomineralization, Biodisintegration, Compost), Overview. Chapter 27; In *Encyclopedia of Polymeric Nanomaterials*; Kobayashi, S., Müllen, K., Eds.; Springer: Berlin, Heidelberg, 2015; pp 1–6.
34. Kawai, F.; Kawabata, T.; Oda, M. Current Knowledge on Enzymatic PET Degradation and its Possible Application to Waste Stream Management and Other Fields. *Appl. Microbiol. Biotechnol.* 2019, *103* (2019), 4253–4268.
35. Kim, H. R.; Lee, H. M.; Jeon, E.; Yu, H. C.; Lee, S.; Li, J.; Kim, D. H. Biodegradation of Polystyrene by Pseudomonas sp. Isolated from the Gut of Superworms. *Environ. Sci. Technol.* 2020, *54* (11), 6987–6996.
36. Kim, Y. T.; Min, B.; Kim, K. W. General Characteristics of Packaging Materials for Food System. Chapter 2; In *Food Science and Technology, Innovations in Food Packaging*; Han, J. H., Ed.; Academic Press: New York, USA, 2014; pp 13–35.

37. Koller, M. Biodegradable and Biocompatible Polyhydroxy-Alkanoates (pha): Auspicious Microbial Macromolecules for Pharmaceutical and Therapeutic Applications. *Molecules* 2018, *23* (362), 1–20.
38. Krueger, M. C.; Hofmann, U.; Moeder, M.; Schlosser, D. Potential of Wood-Rotting Fungi to Attack Polystyrene Sulfonate and Its Depolymerisation by Gloeophyllum trabeum via Hydroquinone-Driven Fenton Chemistry. *PLoS ONE* 2015, *10* (7), 1–17./
39. Kyriacos, D. Polycarbonates. Chapter 17; In *Brydson's Plastics Materials*, 8th Ed.; Gilbert, M., Ed.; Elsevier: Butterworth, Heinemann, 2017; pp 457–485.
40. Lackner, M. Bioplastics - Biobased Plastics as Renewable and/or Biodegradable Alternatives to Petroplastics. In *Kirk-Othmer Encyclopedia of Chemical Technology*; Othmer, K., Ed.; John Wiley & Sons: New York, 2015; pp 1–40.
41. Lambert, S.; Wagner, M. Environmental Performance of Bio-based and Biodegradable Plastics: The Road Ahead. *Chem. Soc. Rev.* 2017, *46* (22), 6855–6871.
42. Lebreton, L.; Andrady, A. Future Scenarios of Global Plastic Waste Generation and Disposal. *Palgrave Commun.* 2019, *5* (6), 1–11.
43. Lim, J. H.; Rhie, H. G.; Kim, J. N. Identification and Analysis of Putative Polyhydroxyalkanoate Synthase (PhaC) in Pseudomonas fluorescens. *J. Microbiol. Biotechnol.* 2018, *28* (7), 1133–1140.
44. Luzi, F.; Torre, L.; Kenny, J. M.; Puglia, D. Bio- and Fossil-Based Polymeric Blends and Nanocomposites for Packaging: Structure-Property Relationship. *Materials* 2019, *12* (471), 1–49.
45. Maes, C.; Yuki, G. M.; Teniers, C.; Luyten, W.; Herremans, G.; Peeters, R.; Samyn, P.; Carleer, R.; Buntinx, M. Ethylene Vinyl Alcohol Copolymer (EVOH) as a Functional Barrier against Surrogate Components Migrating from Paperboard. *Innov. Strateg. Emerg. Technol. Food Saf.* 2019, *2019*, 1–7.
46. Marichelvam, M. K.; Jawaid, M.; Asim M. Corn and Rice Starch-Based Bio-Plastics as Alternative Packaging Materials. *Fibers* 2019, *7* (32), 1–14.
47. Marsh, K.; Bugusu, B. Food Packaging—Roles, Materials, and Environmental Issues. *J. Food Sci.* 2007, *72* (3), R39–R55.
48. Masmoudi, F.; Alix, S.; Buet, S.; Mehri, A.; Bessadok, A., Jaziri, M.; Ammar, E. Design and Characterization of a New Food Packaging Material by Recycling Blends Virgin and Recovered polyethylene terephthalate. *Polym. Eng. Sci.* 2020, *60* (2), 250–256.
49. McKeen, L. W. Polyesters. Chapter 4; In *Plastics Design Library, The Effect of Creep and Other Time Related Factors on Plastics and Elastomers*; McKeen, L. W., Ed.; William Andrew Publishing: Norwich, New York, 2009; pp 115–165.
50. McKeen, L. W. Polyvinyls and Acrylics. Chapter 10; In *Plastics Design Library, Film Properties of Plastics and Elastomers*; McKeen, L. W., Ed.; William Andrew Publishing: Norwich, New York, 2012; pp 219–254.
51. Mokwena, K. K.; Tang, J. Ethylene Vinyl Alcohol: A Review of Barrier Properties for Packaging Shelf Stable Foods. *Crit. Rev. Food Sci. Nutr.* 2012, *52* (7), 640–650.
52. Moog, D.; Schmitt, J.; Senger, J.; Zarzycki, J.; Rexer, K. H.; Linne, U.; Erb, T.; Maier, U. G. Using a Marine Microalga as a Chassis for Polyethylene Terephthalate (PET) Degradation. *Microb. Cell Factories* 2019, *18* (171), 1–15.
53. Morro, A.; Catalina, F.; Corrales, T.; Pablos, J. L.; Marin, I.; Abrusci, C. New Blends of Ethylene-Butyl Acrylate Copolymers with Thermoplastic Starch. Characterization and Bacterial Biodegradation. *Carbohydr. Polym.* 2016, *149*, 68–76.

54. Muhonja, C. N.; Magoma, G.; Imbuga, M.; Makonde, H. M. Molecular Characterization of Low-Density Polyethene (LDPE) Degrading Bacteria and Fungi from Dandora Dumpsite, Nairobi, Kenya. *Int. J. Microbiol.* 2018, *2018*, 1–10.
55. Müller, R. J.; Kleeberg I.; Deckwer, W. D. Biodegradation of Polyesters Containing Aromatic Constituents. *J. Biotechnol.* 2001, *86* (2), 87–95.
56. Nair, N. R.; Sekhar, V. C.; Nampoothiri, K. M. Augmentation of a Microbial Consortium for Enhanced Polylactide (PLA) Degradation. *Indian J. Microbiol.* 2016, *56* (1), 59–63.
57. Negoro, S.; Shibata, N.; Tanaka, Y.; Yasuhira, K.; Shibata, H.; Hashimoto, H.; Lee, Y. H.; Oshima, S.; Santa, R.; Oshima, S.; Mochiji, K.; Goto, Y.; Ikegami, T.; Nagai, K.; Kato, D.; Takeo, M.; Higuchi, Y. Three-Dimensional Structure of Nylon Hydrolase and Mechanism of Nylon-6 Hydrolysis. *J. Biol. Chem.* 2012, *287* (7), 5079–5090.
58. Numoto, N.; Kamiya, N.; Bekker, G. J.; Yamagami, Y.; Inaba, S.; Ishii, K.; Uchiyama, S.; Kawai, F.; Ito, N.; Oda, M. Structural Dynamics of the PET-Degrading Cutinase-like Enzyme from Saccharomonospora viridis AHK190 in Substrate-Bound States Elucidates the Ca^{2+}- Driven Catalytic Cycle. *Biochemistry* 2018, *57* (36), 5289–5300.
59. Ohura, T.; Kasuya, K. I.; Doi, Y. Cloning and Characterization of the Polyhydroxybutyrate Depolymerase Gene of *Pseudomonas stutzeri* and Analysis of the Function of Substrate-Binding Domains. *Appl. Environ. Microbiol.* 1999, *65* (1), 189–197.
60. Ojha, N.; Pradhan, N.; Singh, S.; Barla, A.; Shrivastava, A.; Khatua, P.; Rai V.; Bose S. Evaluation of HDPE and LDPE Degradation by Fungus, Implemented by Statistical Optimization. *Sci. Rep.* 2017, *7*, 1–13.
61. Palm, G. J.; Reisky, L.; Böttcher, D.; Müller, H.; Michels, E. A. P.; Walczak, M. C.; Berndt, L.; Weiss, M. S.; Bornscheuer, U. T.; Weber, G. Structure of the Plastic Degrading Ideonella sakaiensis MHETase Bound to a Substrate. *Nat. Commun.* 2019, *10*, 1–10.
62. Palmer, R. J. Polyamides, Plastics. In *Kirk-Othmer Encyclopedia of Chemical Technology*, 6th Ed.; Othmer, K., Ed.; John Wiley & Sons: New York, 2005; vol 19; pp 772–797.
63. Park, S. H.; Kim, S. H. Poly(ethylene terephthalate) Recycling for High Value Added Textiles. *Fash. Text.* 2014, *1* (1), 1–17.
64. Parker, L. How The Plastic Bottle Went from Miracle Container to Hated Garbage. National Geographic Society, 2019. https://www.nationalgeographic.com/environment/2019/08/plastic-bottles/ (accessed Jan 10, 2021).
65. Patel, P. Edible Packaging. *ACS Central Sci.* 2019, *5* (12), 1907–1910.
66. Pathak, V. M.; Navneet. Review on the Current Status of Polymer Degradation: A Microbial Approach. *Bioresour. Bioprocess.* 2017, *4* (15), 1–31.
67. Patil, R.; Bagde U. S. Isolation of Polyvinyl Chloride Degrading Bacterial Strains from Environmental Samples Using Enrichment Culture Technique. *Afr. J. Biotechnol.* 2012, *11* (31), 7947–7956.
68. Piergiovanni, L.; Limbo, S. Plastic Packaging Materials. Chapter 5; In *Food Packaging Materials. SpringerBriefs in Molecular Science*; Piergiovanni, L., Limbo, S., Eds.; Springer, Cham: Switzerland, AG, 2016; pp 33–49.
69. Ploypetchara, T.; Gohtani, S. Effect of Sugar on Starch Edible Film Properties: Plasticized Effect. *J. Food Sci. Technol.* 2018, *55* (9), 3757–3766.
70. Polystyrene (PS): Production, Market, Price and its Properties. *Plastics Insights*, 2020. https://www.plasticsinsight.com/resin-intelligence/resin-prices/polystyrene/ (accessed Jan 10, 2021).

71. Pranamuda, H.; Tokiwa, Y; Tanaka H. Polylactide Degradation by an Amycolatopsis sp. *Appl. Environ. Microbiol.* 1997, *63* (4), 1637–1640.
72. Priyadarshi, R.; Kumar, B.; Deeba, F.; Kulshreshtha, A.; Negi, Y. S. Food Packaging: Natural and Synthetic Biopolymers. In *Encyclopedia of Polymer Applications*; CRC Press: Boca Raton, Florida, 2018; vol 2.
73. Puscaselu, R.; Gutt, G.; Amariei, S. Rethinking the Future of Food Packaging: Biobased Edible Films for Powdered Food and Drinks. *Molecules* 2019, *24* (17), 1–14.
74. Reddy, C. S. K.; Ghai, R.; Rashmi; Kalia, V. C. Polyhydroxyalkanoates: An Overview. *Bioresour. Technol.* 2003, *87* (2), 137–146.
75. Ren, L.; Men, L.; Zhang, Z.; Guan, F.; Tian, J.; Wang, B.; Wang, J.; Zhang, Y.; Zhang, W. Biodegradation of Polyethylene by Enterobacter sp. D1 from the Guts of Wax Moth Galleria mellonella. *Int. J. Environ. Res. Public Health* 2019, *16* (1941), 1–11.
76. Ritchie, H.; Roser, M. Plastic Pollution. Published online at OurWorldInData.org.; 2020; p 42. https://ourworldindata.org/plastic-pollution (accessed Feb 20, 2021).
77. Robertson, G. L. Ed.; In *Food Packaging: Principles and Practice*, 3rd Ed.; CRC Press: Brisbane, Australia, 2013; p 733.
78. Rodrigues, P. R.; Nunes, J. M. N.; Lordelo, L. N.; Druzian, J. I. Assessment of Polyhydroxyalkanoate Synthesis in Submerged Cultivation of Cupriavidus necator and Burkholderia cepacia Strains Using Soybean as Substrate. *Braz. J. Chem. Eng.* 2019, *36* (1), 73–83.
79. Ronkvist, A. M.; Xie, W.; Lu, W.; Gross, R. A. Cutinase-Catalyzed Hydrolysis of Poly(ethylene terephthalate). *Macromolecules* 2009, *42* (14), 5128–5138.
80. Ruiz, J. A. R.; Trigo-López, M.; García, F. C.; García J. M. Functional Aromatic Polyamides. *Polymers* 2017, *9* (414), 1–44.
81. Sangatash, M. M.; Niazmand, R.; Jamab, M. S.; Modaressi, A. S. Development of Antioxidant Active Films Containing Sodium Ascorbate (SA) and Ethylene Vinyl Alcohol (EVOH) to Extend the Shelf-Life of Peanut. *J. Food Sci. Technol.* 2016, *53* (4), 1766–1783.
82. Satterthwaite, K. Plastics Based on Styrene. Chapter 12; In *Brydson's Plastics Materials*; Gilbert, M., Ed.; Elsevier: Butterworth, Heinemann, 2017; pp 311–328.
83. Sauter, D. W.; Taoufik, M.; Boisson, C. Polyolefins, a Success Story. *Polymers* 2017, *9* (185), 1–13.
84. Sekhar, V. C.; Nampoothiri, K. M.; Mohan, A. J.; Nair, N. R.; Bhaskar, T.; Pandey, A. Microbial Degradation of High Impact Polystyrene (HIPS), An E-plastic with Decabromodiphenyl Oxide and Antimony Trioxide. *J. Hazard. Mater.* 2016, *318*, 347–354.
85. Senturk Parreidt, T.; Müller, K.; Schmid, M. Alginate-Based Edible Films and Coatings for Food Packaging Applications. *Foods* 2018, *7* (170), 1–38.
86. Shimomoto, H. Poly(tetrafluoroethylene) and Other Fluorine-Containing Polymers. In *Encyclopedia of Polymeric Nanomaterials*; Kobayashi, S., Müllen K., Eds.; Springer: Berlin, Heidelberg, 2014.
87. Shit, S. C.; Shah, P. M. Edible Polymers: Challenges and Opportunities. *J. Polym.* 2014, *2014*, 1–13.
88. Single Use Plastic: A Road Map for Sustainability. UN Environment; Copyright United Nations Environment Programme, 2018. https://wedocs.unep.org/bitstream/handle/20.500.11822/25496/singleuseplastic_sustainability.pdf?Isallowed=y&sequence=1 (accessed Jan 10, 2021).

89. Siracusa, V. Food Packaging Permeability Behaviour: A Report. *Int. J. Polym. Sci.* 2012, *2012*, 1–11.
90. Siracusa, V. Microbial Degradation of Synthetic Biopolymers Waste. *Polymers* 2019, *11* (1066), 1–18.
91. Sivan, A.; Szanto, M.; Pavlov, V. Biofilm Development of the Polyethylene-degrading Bacterium Rhodococcus ruber. *Appl. Microbiol. Biotechnol.* 2006, *72* (2), 346–352.
92. Song, H. J.; Chang, Y.; Lyu, J. S.; Yon, M. Y.; Lee, H. S.; Park, S. J.; Choi, J. C.; Kim M. K.; Han, J. Migration Study of Caprolactam from Polyamide-6 Sheets into Food Simulants. *Food Sci. Biotechnol.* 2018, *27* (6), 1685–1689.
93. Song, J. H.; Murphy, R. J.; Narayan, R.; Davies, G. B. Biodegradable and Compostable Alternatives to Conventional Plastics. *Philos. Trans. R. Soc. B Biol. Sci.* 2009, *364*, 2127–2139.
94. Sriyapai, P.; Chansiri, K.; Sriyapai, T. Isolation and Characterization of Polyester-Based Plastics-Degrading Bacteria from Compost Soils. *Microbiology* 2018, *87*, 290–300.
95. Sulaiman, S.; Yamato, S.; Kanaya, E.; Kim, J. J.; Koga, Y.; Takano, K.; Kanaya, S. Isolation of a Novel Cutinase Homolog with Polyethylene Terephthalate-Degrading Activity from Leaf-Branch Compost by Using a Metagenomic Approach. *Appl. Environ. Microbiol.* 2012, *78* (5), 1556–1562.
96. Sumathi, T.; Viswanath, B.; Lakshmi, A. S.; SaiGopal, D. V. R. Production of Laccase by Cochliobolus sp. Isolated from Plastic Dumped Soils and Their Ability to Degrade Low Molecular Weight PVC. *Biochem. Res. Int.* 2016, *2016*, 1–10.
97. Syranidou, E.; Karkanorachaki, K.; Amorotti, F.; Franchini, M.; Repouskou, E.; Kaliva, M.; Vamvakaki, M.; Kolvenbach, B.; Fava, F.; Corvini, P. F.; Kalogerakis, N. Biodegradation of Weathered Polystyrene Films in Seawater Microcosms. *Sci. Rep.* 2017, *7*, 1–12.
98. Takehara, I.; Kato, D. I.; Takeo, M.; Negoro, S. Draft Genome Sequence of the Nylon Oligomer-Degrading Bacterium Arthrobacter sp. Strain KI72. *Genome Announc.* 2017, *5* (17), 1–2.
99. Thompson, R. C.; Moore, C. J.; vom Saal, F. S.; Swan, S. H. Plastics, The Environment and Human Health: Current Consensus and Future Trends. *Philos. Trans. Royal Soc. B Biol. Sci.* 2009, *364* (1526), 2153–2166.
100. Tokiwa, Y.; Calabia, B. P.; Ugwu, C. U.; Aiba, S. Biodegradability of Plastics. *Int. J. Mol. Sci.* 2009, *10* (9), 3722–3742.
101. Tsang, Y. F.; Kumar, V.; Samadar, P.; Yang, Y.; Lee, J.; Ok, Y. S.; Song, H.; Kim, K. H.; Kwon, E. E.; Jeon, Y. J. Production of Bioplastic Through Food Waste Valorization. *Environ. Int.* 2019, *127*, 625–644.
102. Tullo, A. H. The Growing Use of Plastic Food Packaging Benefits Consumers, But Critics Say Industry Isn't Doing Enough to Minimize the Negative Environmental Impact. *The Cost of Plastic Packaging, Chemical and Engineering News*, 2016, vol *94* (41), pp 32–37.
103. Urbanek, A. K.; Rymowicz, W.; Mirończuk, A. M. Degradation of Plastics and Plastic-Degrading Bacteria in Cold Marine Habitats. *Appl. Microbiol. Biotechnol.* 2018, *102* (18), 7669–7678.
104. Vert, M.; Doi, Y.; Hellwich, K. H.; Hess, M.; Hodge, P.; Kubisa, P.; Rinaudo, M.; Schué, F. Terminology for Biorelated Polymers and Applications (IUPAC Recommendations 2012). *Pure Appl. Chem.* 2012, *84* (2), 377–410.

105. Vidal J. 2019 the year of the Great Plastic Backlash. In *Plastic 2020*. https://commonseas.com/news/plastic-2020 (accessed Jan 15, 2021).
106. Wallace, P. W.; Haernvall, K.; Ribitsch, D.; Zitzenbacher, S.; Schittmayer, M.; Steinkellner, G.; Gruber, K.; Guebitz, G. M.; Birner-Gruenberger, R. PpEst is a Novel PBAT Degrading Polyesterase Identified by Proteomic Screening of Pseudomonas pseudoalcaligenes. *Appl. Microbiol. Biotechnol.* 2017, *101*, 2291–2303.
107. Webb, H. K.; Arnott J.; Crawford R. J.; Ivanova E. P. Plastic Degradation and Its Environmental Implications with Special Reference to Poly(ethylene terephthalate). *Polymers* 2013, *5*, 1–18.
108. Wei, R.; Oeser, T.; Then, J.; Kühn, N.; Barth, M.; Schmidt, J.; Zimmermann, W. Functional Characterization and Structural Modeling of Synthetic Polyester-Degrading Hydrolases from Thermomonospora curvata. *AMB Express*. 2014, *4* (44), 1–10.
109. Wei, R.; Zimmermann, W. Microbial Enzymes for the Recycling of Recalcitrant Petroleum-Based Plastics: How Far Are We? *Microb. Biotechnol.* 2017, *10* (6), 1308–1322.
110. Wu, M. M. Acrylonitrile and Acrylonitrile Polymers. In *Encyclopedia of Polymer Science and Technology*; John Wiley & Sons: New York, 2001; vol 1, pp 124–174.
111. Yamano, N.; Nakayama, A.; Kawasaki, N.; Yamamoto, N.; Aiba, S. Mechanism and Characterization of Polyamide 4 Degradation by Pseudomonas sp. *J. Polym. Environ.* 2008, *16*, 141–146.
112. Yang, S. S.; Brandon, A. M.; Flanagan, J. C. A.; Yang, J.; Ning, D.; Cai, S. Y.; Fan, H. Q.; Wang, Z. Y.; Ren, J.; Benbow, E.; Ren, N. Q.; Waymouth, R. M.; Zhou, J.; Criddle, C. S.; Wu, W. M. Biodegradation of Polystyrene Wastes in Yellow Mealworms (larvae of *Tenebrio molitor Linnaeus*): Factors Affecting Biodegradation Rates and the Ability of Polystyrene-Fed Larvae to Complete their Life Cycle. *Chemosphere* 2018, *191*, 979–989.
113. Yoon, M. G.; Jeon, H. J.; Kim, M. N. Biodegradation of Polyethylene by a Soil Bacterium and AlkB Cloned Recombinant Cell. *J. Bioremediat. Biodegrad.* 2012, *3* (145), 1–8.
114. Yousif, E.; Haddad, R. Photodegradation and Photostabilization of Polymers, Especially Polystyrene: Review. *Springerplus* 2013, *2* (398), 1–32.

CHAPTER 11

CHALLENGES IN BIODEGRADATION OF PLASTIC WASTE FROM FOOD PACKAGING

ANITA KUMARI, DEEPIKA PAWAR, SONIA SAINI,
ASHWANI KUMAR, SAVITA BUDHWAR, and
SUJATA PANDIT SHARMA

ABSTRACT

Massive use of synthetic and biodegradable plastics in food packaging leads to its disposal into landfills or ocean. Burning of plastic releases harmful gases like furans and dioxins, which are dangerous ozone-harming substances and cause exhaustion of the ozone layer. The commonly used plastics are polycarbonate, polypropylene, polyvinyl chloride, polyurethanes, nylons, polybutylene terephthalate, polyethylene, high-density polyethylene, and low-density polyethylene, etc. There are countless unfortunate effects of plastic on human wellbeing and the environment. Temperature, pH, UV radiation, and micro-organisms (e.g., *Pseudomonas* sp. and *Bacillus cereus*) help to initiate the degradation of plastics. This chapter deals with different kinds of plastics as well as highlights the ability of the micro-organism to degrade the plastics waste. Further, associated factors as greener solution for tackling the issue of continuously increasing plastic waste in the environment are also addressed.

11.1 INTRODUCTION

Plastics are undeniably prevalent materials in nature due to their associated expenses, capacity measure, and functional properties.[6] The aggregate sum of plastic manufactured on the planet from 1950 through 2019 is 8100 Mt,

Advances in Sustainable Food Packaging Technology. Megh R. Goyal, Preeti Birwal, Anand K. Pandey, & Ritesh B. Watharkar, (Eds.)

which had expanded from 2 Mt in 1950 to 450 Mt in 2019.[24] Elements like nitrogen, carbon, oxygen, bromine, hydrogen, and chlorine are used for manufacturing plastics, which are inert in nature.[83] Extensive industrial and domestic plastic applications lead to an increasing trend related to its consumption and thereby production.[2] Plastics are usually made up of man-made substances, which are extracted from the fossil fuel resources.

In the 19th and 20th centuries with the development of technology, the demand and popularity for plastics have increased significantly among consumers because of its customizable characteristics for market use like lightweight, durability, and unbreakable nature at a very low cost. It is mostly used in shelter, transportation, construction, food packaging, clothing etc. Plastic polymers are manmade polymers composed of long chains of molecular units called monomers. Many polymer materials like polyhydroxyalkanoate (PHA), polyhydroxybutyrate (PHB), polyurethane (PUR), polyvinyl chloride (PVC), polyethylene terephthalate (PET), polyethylene (PE), polypropylene (PP), polycaprolactone (PCL), polylactic acid (PLA), polystyrene (PS), and polybutylene succinate (PBS) are utilized for plastic production.[50,81]

Recently, bio-based and fossil-based plastics are used and considered under the category of nonbiodegradable plastics, which finally get accumulated in the environment, causing severe hazards to our ecosystem.[64] It causes adverse consequences resulting from diminishing soil richness that is polluted with plastics waste and harmful plastic compounds that enter into the food chain upon increased environmental accumulation.[36,49] Commonly used plastics and their applications in different sectors have been given in Table 11.1.

Decreased soil fertility is a result of accumulated nonbiodegradable plastic waste in the environment. Also, the plastic waste dumped into the ocean can lead to biomagnification of certain toxins in the food chains of aquatic fauna, thus causing severe harm to ocean biodiversity. Additives used in plastics also create health problems associated with individuals.[12] Therefore, it is a prime time to resolve the problem of nonbiodegradable plastics, as it drastically affects lifeforms as well as the environment. Bio-based biodegradable plastics seem promising for packaging, health, and agriculture sectors, but these plastics are still rarely utilized in the plastic industry. Biodegradation of plastic results in the breakdown of polymers into simple monomer units via biological processes (micro-organisms) and biodegradable materials upon degradation results in biomass, carbon dioxide (CO_2), and water.

Studies have stated that some micro-organisms (bacteria and fungi) produce exo-enzymes under different conditions and have a great ability to degrade plastics.[49,74] Proteases, lipases, and cutinases are important microbial enzymes that are generally used for biodegradation of polymers.[50,70] *Candida cylindracea*, *R. arrhizus*, *Rhizopus delemar*, and *Achromobacter sp.* produce lipases and esterases, which have been reported to work on polyethylene adipate and polycaprolactone complex polymers.[31,41] Complex polymers are degraded by various micro-organisms wherein plastic fragments are used as a direct nutritional source, or different microbial enzymes are used as an indirect mechanism. *P. aeruginosa*, *Penicillium simplicissimum*, and *Pseudomonas fluorescens* are the most broadly used fungal and bacterial strains.[52,58,65]

The Ministry of Agriculture under Government of India has taken some steps to stop the use of plastic completely. In India, the state government of Himachal Pradesh has completely prohibited the use of plastic in its regions. Other Indian states and higher institutions are also showing interest in minimizing the use of plastics as it is deteriorating the environment and causing health issues.

This chapter will give an overview of commonly used plastics, classification, biodegradation of plastic waste, factors affecting biodegradation, and health hazards.

TABLE 11.1 Commonly Used Types of Plastics and Their Applications.

Plastic types	Industrial applications
Acrylonitrile butadiene styrene (ABS)	Drainage pipes, printers, computers, monitors, etc.
Acrylonitrile butadiene styrene/ Polycarbonate (ABS/PC)	PC and ABS blends are used as interior and exterior parts of cars and mobile phones, respectively.
High impact polystyrene (HIPS)	Liners of a refrigerator, vending cups, and packaging of foods
High-density polyethylene (HDPE)	Milk jugs, detergent bottles
Low-density polyethylene (LDPE)	Curtains, furniture, floor tiles
Polyamides (PA)	Bristles of toothbrush, fibers
Polycarbonate (PC)	Windows, traffic lights, eyeglasses, etc.
Polyester (PES)	Textile industries
Polyethylene (PE)	Plastic bottles and bags
Polyethylene terephthalate (PET)	Plastic containers (food jars), bottles (carbonated drinks), plastic film (packaging of food items)
Polypropylene (PP)	Car (bumpers), caps (bottle), appliances, straws, yoghurt containers, pipes (plastic pipes used for irrigation purposes)
Polystyrene (PS)	Disposable cups, plates, CD, used as foam packaging, food containers, tableware plastics etc.
Polyurethanes (PU)	Used as thermal insulation, cushion, foams, printing rollers, and surface coatings
Polyvinyl chloride (PVC)	Guttering, pipes (plumbing), flooring, window frames
Polyvinylidene chloride (PVDC)	Packaging of foods

11.2 CLASSIFICATION OF PLASTICS (FIG. 11.1)

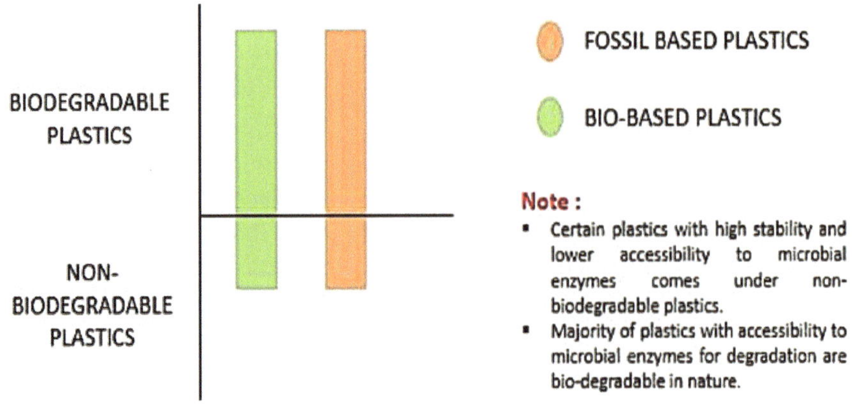

FIGURE 11.1 Distribution of plastics under different classes.

11.2.1 NONBIODEGRADABLE PLASTICS

Certain bio-based and fossil-based polymeric plastics are nonbiodegradable in nature due to their high stability and very slow rate of degradation. Commonly used fossil-based plastics are derivatives of hydrocarbon and petroleum. They are made up of several smaller units of monomers. Common examples of these types of plastics are PVC, PP, PS, PET, PUR, and PE. Plastics like PE are used as packaging material for mugs, carry bags, etc. Still, due to their poor degradability, the accumulation of these plastics is higher on the earth and has become a concern for the social problem of health and the environment. The rate of degradation of these plastics is very slow because of their high stability.[59] Hence, the waste management system for these polymers should be taken into consideration.[63] The properties of these polymers like inertness and resistance to microbial assault are deliberately reduced by adding prooxidants and starch to enhance disintegration.[59]

11.2.2 BIODEGRADABLE PLASTICS

Biodegradation of plastic transforms biochemicals to compounds which can be used by micro-organisms like fungi and bacteria.[65,83] The biodegradation and microbial assimilation rate determine the biodegradability of plastics which are categorized under bio and fossil-based polymers. Enzymatic and nonenzymatic hydrolysis are two modes involved in the biodegradation of plastics.[79] Factors like organism type, molecular weight, pretreatment nature, mobility, chemical components crystallinity, functional groups, and additives present in polymers affect the efficiency of biodegradation processes.[4] Exo-enzymes secreted by micro-organisms disintegrate polymers into small dimers and then into monomer units during the degradation process. These monomers are small enough to pass through a semipermeable membrane of bacterial cell which uses monomers as a carbon and energy source.[25,30]

11.2.2.1 FOSSIL-BASED BIODEGRADABLE PLASTICS

The major use of these plastics has been observed in pharmaceutical products, food items, the packaging industry, and cosmetics. The nonbiodegradable polymers have a slow degradation rate and create severe problem for their management.[28,76] However, for faster degradation of biodegradable polymers, live microbes and their associated enzymes are helpful up to a certain extent. Reddy et al.[59] described that research should be focused on developing novel enzyme-based degradation approaches, synthesis of gene copies that code biodegradation enzymes and micro-organism characterization is necessary for the degradation of fossil-based plastics.[59] Examples of fossil-based biodegradable polymers are as follows:

- **Biodegradable Polymer Blends:** The most commonly used biodegradable polymer blends are polyester/starch blends and PVA/starch blends. These blends are made up of different combinations and destruction of each of its constituents occurs during the biodegradation process.[57] The degradation process alters the organizational integrity of polymers, increasing the surface area for the enzymes to act upon. Due to the assimilation by various microbes, PVA/starch blends are supposed to be biodegradable.
- **Starch/ Polyester blends** are completely degradable and their chemical composition acts as an influencing factor during the process. A study conducted by Ratto et al.[57] reported that by enhancing the starch concentration, the degradation speediness of polyesters (PCL) in the blend was significantly amplified.[57] Anhydride polyesters added to straight-chain starch blends and polyesters provide superior characteristics.[44] *R. delemar* strains and *R. arrhizus* (microbes) secrete lipase enzyme, which helps hydrolyze polyester blends.[69]

- **PVA/ Starch Blends** have the efficiency of being soluble in water. The blending of starch with PVA increases its competitiveness. Several microbes through enzymatic hydrolysis degrade PVA/starch blends such as depolymerase secreted by *Alcaligenes feacalis* T1.[30,70] Due to good film-forming properties, they are extensively used for packaging and agricultural uses.[68,73]
- **Polycaprolactone** (PCL) is a biodegradable polyester that has a melting point of about 60°C.[6] Aerobic and anaerobic micro-organisms easily degrade the fossil-based biodegradable polymers. The other copolymers mixed with crystalline polyester have been used as packaging material in blood bags, catheters, biomedicals etc. Due to its biodegradability and flexibility, it is expensive.[80] Microbial esterases and lipases can degrade PCL.[32] In one study, *Aspergillus sp*. ST-01 has been reported to degrade this polymer into various products such as caproic, butyric, valeric, and succinic acids.[60]
- **Polyethylene succinate (PES)** is made through copolymerization of succinic anhydride and ethylene oxide.[27] Polyethylene succinate is used as a coating agent in paper packing. Strains obtained from the soil like *Pseudomonas sp*. AKS2 (Mesophilic bacterial strain) and *Bacillus sp*. TT96 (thermophilic strain) have been reported to degrade these polymers.[70,72]

11.2.2.2 BIO-BASED BIODEGRADABLE PLASTICS

These types of plastics are either directly consumed by micro-organisms or are acted upon by the enzymes and result in degradation. The most common examples are starch and cellulose. Starch is made up of amylose and amylopectin and is commonly used for the manufacturing of biodegradable plastics. Starch-based polymers are used because of their availability, inexpensiveness, and biodegradability, which depend on ecological conditions.[11,40,51] These polymers are classified into starch-filled and starch-based polymers.[30] Algae, fungi, and bacteria are responsible for the decomposition of these polymers.[33] Different micro-organisms (*P. lemoignei, Aspergillus fumigatus, Acidovorax faecalis*, and *Comamonas sp.*) isolated from the soil are helpful in the biodegradation of these polymers under aerobic and anaerobic conditions.[63] The two most commonly used bio-based degradable plastics are polyhydroxyalkanoates (PHA) and polylactic acid (PLA).[16] The attributes/characteristics of PHA and PLA are mentioned in (Table 11.2).

TABLE 11.2 Overview of Polyhydroxyalkanoates and Polylactic Acid.

Polymer material	Occurrence/ source in nature	Usages	Micro-organisms responsible for degradation	References
Polyhydroxy-alkanoates (PHA)	Sugars and lipids (bacterial fermentation)	Service materials (food companies), pharmaceutical Industries, paints, disposable medical tools, packaging materials etc.	*Bacillus, Burkholderia, Nocardiopsis,* and *Cupriavidus* (Mycobacterium and Micromycetes) Fungal genera	[3, 9, 20, 29]
Polylactic acid (PLA)	Corn starch, tapioca roots, sugarcane	Pharmaceutical Industries	*B. licheniformis* and *Amycolatopsis sp.* *Cryptococcus sp.* strain S-2. (Purified lipase from fungus)	[21, 46, 55, 65]

11.3 BIODEGRADATION OF PLASTIC WASTE

Ever since the enormous use of plastic, the controversy for its degradation was born simultaneously. The biodegradation of plastic is a peculiar solution to this problem. Biodegradation is the process where microbes, by several mechanisms, break down the organic material. Micro-organisms can carry out the deterioration and degradation of both natural and synthetic polymers.

Throughout plastic biodegradation, colonization by the microbes on the surface of plastic leads to a decrease in the molecular load of plastics. Afterward, the polymer is changed over to its monomers and finally the monomers are converted into simpler entities like water, carbon dioxide, and methane.[62,83]

Various research studies have been put forward showing the miraculous effect of the degradation of plastic using live microbes such as bacteria and fungi and enzymes secreted by them. These specific microbes were isolated from the landfills or pits where plastic was disposed of. However, the degradation of the long-chain polymers of plastic is not carried directly by micro-organisms. Several environmental factors are also involved. Stepwise degradation and deterioration of plastics occur via the involvement of several factors, which will be discussed later in this chapter. Steps for biodegradable of plastic via microbes are mentioned in Figure 11.2.

FIGURE 11.2 Flowchart depicting steps of biodegradation of plastics.

11.3.1 BIO-DETERIORATION

The onset of bio-deterioration (Fig. 11.3) starts when the plastic starts losing its physical properties like shape, color, brittleness, elastic strength, etc.[23] This is the initial stage of the bio-degradation of plastic. These changes happen due to weakening of plastic upon exposure to harsh environmental conditions or abiotic factors such as air, rain, snow, and animal tearing. Once the mechanical degradation has occurred, the photo-degradation begins, where UV rays from the sun become responsible for initiating the chemical reactions, destabilizing the polymers of plastic. After photo-degradation the thermo oxidative degradation starts where exposure to heat or high temperature of environment influences the stability of framework of polymers.[75] In the later stage, microbes or biotic factors (Table 11.3) cause the breakdown of plastic polymers into smaller ones by secretion of several enzymes. The synergistic effect of both the abiotic and biotic factors leads to the deterioration of plastics.[13] Some plastics are biologically unreactive and thus depend only on abiotic factors for their fragmentation.

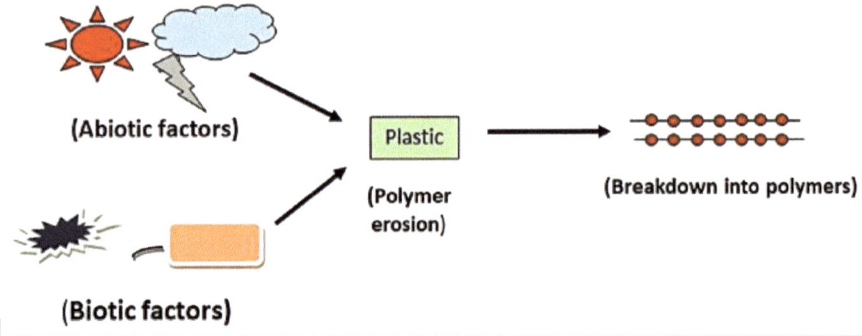

FIGURE 11.3 Process of bio-deterioration.

TABLE 11.3 Examples of Bacteria and Fungi Involved in Biodegradation of Different Kinds of Plastics.

Type of plastic	Micro-organism involved in biodegradation		Reference
	Bacteria	**Fungi**	
Degradable polyethylene/ Polyethylene films	*Nocardia steroids* *Bacillus sp.*	*Gliocladium viride* *Aspergillus awamori* *Mortierella subtilissima*	[34, 53]
HDPE	*Pseudomonas fluorescens* *Serratia marcescens* Bacillus sp.	*Aspergillus terreus* *Aspergillus tubingensis* VRKPT1 *Aspergillus flavus* VRKPT2 *Aspergillus niger*	[5, 7, 14, 15, 47]
LDPE	*Enterobacter sp.* *Pantoea sp.* *Pseudomonas sp.* *Staphylococcus epidermis* *Bacillus subtilis*	*Aspergillus oryzae* *Aspergillus clavatus* *Penicillium pinophilum* *Aspergillus niger*	[10, 22, 38, 66, 67, 71, 78]
Polyethylene carry bags	*Bacillus subtilis* *Staphylococcus* sp *Micrococcus* sp. *Listeria sp.Vibrio sp.*	*Aspergillus spp.* *Aspergillus terreus* *Aspergillus sydowii*	[39, 56, 61, 77]
Polyurethane	*Bacillus subtilis* *Arthrobacter* sp. *Pseudomonas sp.* *Micrococcus* sp. *Corynebacterium* sp.	*Aspergillus flavus* *Fusarium solani* *Candida ethanolica* *Aspergillus niger Cladosporium herbarum*	[18, 45, 48, 62, 82]
PVC powder (Poly Vinyl Chloride)	*Pseudomonas citronellolis* *Bacillus flexus*	*Aspergillus sp.* *Cladosporium sp.* *Aureobasidium pullulans*	[2, 74]

11.3.2 BIO-FRAGMENTATION

Once the plastic has been broken down into segments of polymers or oligomers, it becomes more accessible for enzymatic action by microbes. Enzymes break down these polymers into smaller fragments (Fig. 11.4). Enzymes do not degrade all polymers. Linear nonreactive segments are hard for enzymes to access and disrupt.[37] This step is a visual inspection for microbial attachment to plastic polymers. Once the microbes hydrolyze the plastics, gases are released, indicating the final stage of degradation. Rate-determining factors for this step are microbial load present as well as the amount of enzymes released by the microbes.

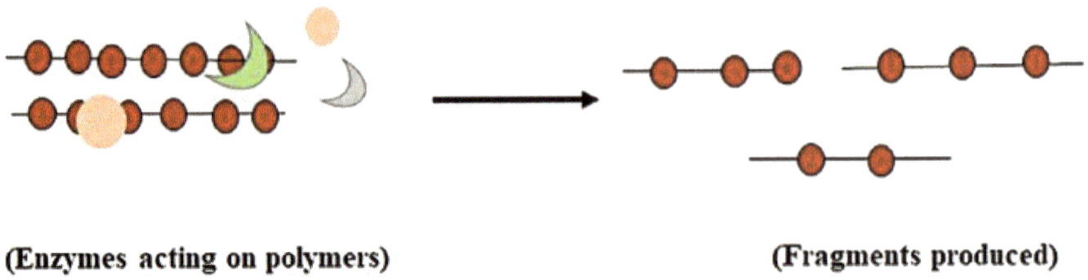

FIGURE 11.4 Enzymes acting on polymers and degrading them into smaller fragments.

11.3.3 ASSIMILATION AND MINERALIZATION

Assimilation (Fig. 11.5) is basically the act of "eating" and "digesting" of plastic parts by microbes for their growth and energy needs. Once the microbes have broken down the oligomers of plastics into monomers, these monomers can now be assimilated into carbon dioxide, water, and several other metabolic products.[37] The rate of conversion during this stage is directly proportional to the amount of gas evolved. To metabolize plastics at this stage, several microbes like fungi require oxygen while others like bacteria require anaerobic conditions.

11.4 FACTORS AFFECTING BIODEGRADATION

During the biodegradation process of both natural and synthetic plastic, several factors (Fig. 11.6) are involved, which leads to change in the rate, quality, and quantity of ongoing degradation. These factors can be abiotic factors or enzymes produced by micro-organisms. Not all types of plastics depend on abiotic factors for their degradation. Apart from the environmental conditions, the polymer characteristics are also a major factor on which the rate of biodegradation of plastic depends.

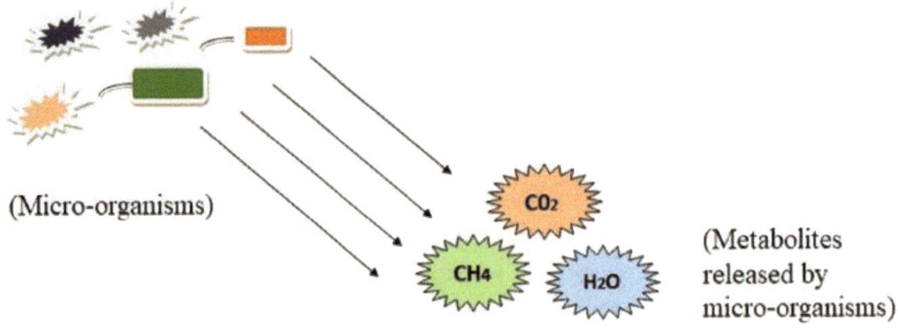

FIGURE 11.5 Process of assimilation and mineralization.

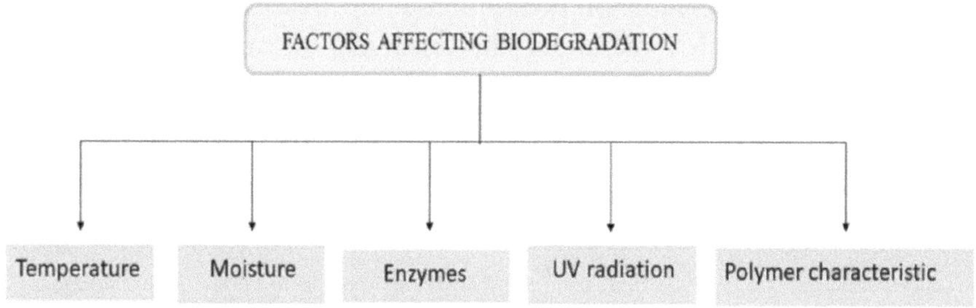

FIGURE 11.6 Factors affecting biodegradation

11.4.1 TEMPERATURE

Exposure to high temperatures or heat is important for the initiation of thermo-oxidative degradation of plastic. Thermo-oxidation, in turn, will influence the organized framework of polymers.[19] The high temperature will not directly affect biodegradation but will aid in the rapid degradation.

11.4.2 MOISTURE

Moisture can affect polymer biodegradation in various ways because of the fundamental necessity of water for development and increase of micro-organisms.[44] When high-moisture content is present in the environment, the hydrolysis reaction is initiated faster, producing more chain scission. In turn, there is a surge in the available sites of micro-organisms to assault the polymer chains, causing rapid degradation.

11.4.3 ENZYMES

Enzymes can be extracellular or intracellular. The microbes being used for the biodegradation produce these enzymes during the stage of bio-deterioration for the breakdown of large polymers of plastic into smaller fragments which microbes can easily attack for assimilation and mineralization.

11.4.4 UV RADIATION

UV radiations from the Sun have a wavelength of about 380 nm with high penetration power. The sunlight is essential for biodegradation as sunrays initiate the process of photo-degradation via UV radiation.[1] The method of photo-degradation can also be initiated by an artificial light source. The principle of photo-degradation is an initiation of chemical reactions to destabilize polymers via UV radiation. Photodegradation can disturb the plastic films by instigating random main chain scission via Norrish I or Norrish II mechanisms.[35]

11.4.5 POLYMER CHARACTERISTICS

Majority of the physical properties of plastics like elasticity, morphology, rigidity, and crystallinity are influenced by chemical structures and composition of its polymers.[4] These characteristics of polymers directly affect the rate of degradation of several plastics. The chemical structure affects the accessibility of degrading enzymes to carbonate bond in polycarbonates whose facile nature determines the better degradation possibility of their polymeric units. In case of polycarbonates like bisphenol, with inaccessible carbonate bond due to benzene ring or quaternary carbon presence in its structure, the degradation becomes challenging.[4] The effect of carbonate bond accessibility to enzymes for degradation process of plastic polymers has been illustrated in Figure 11.7.

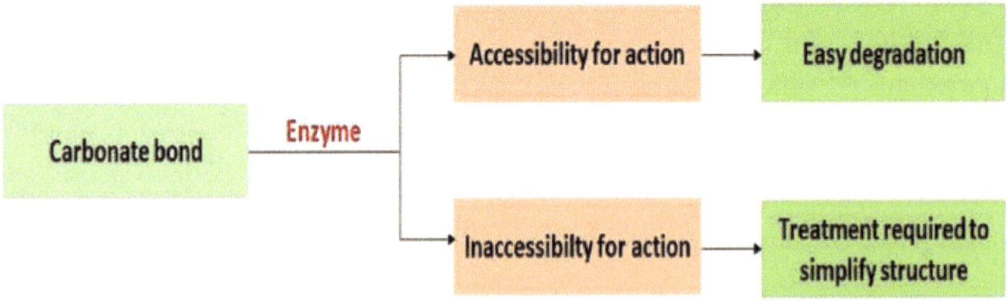

FIGURE 11.7 Effects of carbonate bond accessibility to enzymes for degradation process of plastic polymers.

11.5 HEALTH HAZARDS OF PLASTIC WASTE

Polymeric material of plastic is not considered toxic. Still, nonpolymeric compounds present in the plastic-like monomers, oligomers, polymerization solvents, low molecular weight fragments, catalyst remnants, and a wide range of additives (antioxidants, ultra violet (UV) stabilizers, or plasticizers) are hazardous to health as well as for the environment. Nonpolymeric compounds during the degradation process are released into water, air, or other contact media like food.[54] Hazardous monomer like BPA is commonly used in toys, food cans linings, beverage can linings, baby bottles, and other consumer products which interfere with human hormones.

In vivo studies showed results related to premature birth, intrauterine growth retardation, preeclampsia, and stillbirth.[8] A major effect on the sex hormones (progesterone) has been observed in females when exposed for a long time to BPA.[26] Many additives used in the production of plastics lead to cancer and endocrine interruption. Plasticizers like phthalates are used to make plastics soft and pliable. Studies showed that Di 2-ethylhexyl phthalate (DEHP), di-n-butyl phthalate (DBP), and butyl benzyl phthalate (BBP) cause reproductive and developmental toxicity.[43] A study of a phthalate-exposed population in Northern Mexico found an elevated breast cancer risk among women.[42]

Several metals, namely inorganic compounds cadmium, organic tin compounds barium, calcium, zinc carboxylates, and antimony compounds, are used as stabilizers and colorants. The lead compounds used as heat stabilizers are toxic for reproduction and may damage organs.[17] Skin sensitization, reproductive toxicity, skin corrosion, and sometimes acute toxicity have been observed due to some of the monomers like propylene oxide, phenol, and maleic anhydride.

11.6 SUMMARY

Stability, durable nature, optimal thermal, and mechanical properties of the plastic have made them useful in different sectors such as food, cosmetics, and electronics; however, in general, they are no more than long chains of polymeric units with several alterations respective to type of plastic. Majorly used plastics are PS, PP, PVC, PU, PET, nylons, polybutylene terephthalate, and PE (HDPE, LDPE, LLDPE, and MDPE). The growing demand for plastic over the past few decades has generated enormous plastic waste. The strategy of 3 R's (i.e., reduce, reuse, and recycle) is no longer enough to fight the issue. The biggest challenge is that the generation of tons of plastic waste and health hazards comes hand in hand. Nonpolymeric compounds present in the plastic-like monomers, oligomers, polymerization solvents, low molecular weight fragments, catalyst remnants, and a wide range of additives (antioxidants, Ultra Violet (UV) stabilizers, or plasticizers) are harmful to life forms. At this point, the biodegradation of plastics via microbes can serve as a savior. Several studies have given experimental proof for the same. The microbes, via four major steps, that is, biodeterioration, biofragmentation, assimilation, and mineralization, can degrade plastic waste over a duration of time. Several bacteria majorly used for biodegradation of plastic include *Corynebacterium, Pseudomonas sp., Bacillus sp.*, and *Arthrobacter sp.* The majorly used fungi are *Aspergillus sp, Penicillium sp, Cladosporium sp.*, and *Aureobasidium pullulans.* Much research is being done in this concern to implement the biodegradation of plastic worldwide on a large scale to eradicate the issue of growing plastic waste and health hazards caused by different plastic types.

KEYWORDS

- **bio-based plastics**
- **biodegradable plastics**
- **biodegradation**
- **fossil-based polymers**
- **health hazards**
- **microorganisms**
- **nonbiodegradable plastics**
- **plastics**

REFERENCES

1. Abd El-Rehim, H. A.; Hegazy, E. S.; Ali, A. M.; Rabie, A. M. Synergistic Effect of Combining UV-Sunlight–Soil Burial Treatment on the Biodegradation Rate of LDPE/Starch Blends. *J. Photochem. Photobiol. A: Chem.* **2004,** *163* (3), 547–556.
2. Ahmed, T.; Shahid, M.; Azeem, F.; Rasul, I.; Shah, A. A.; Noman, M.; Hameed, A.; Manzoor, N.; Manzoor, I.; Muhammad, S. Biodegradation of Plastics: Current Scenario and Future Prospects for Environmental Safety. *Environ. Sci. Pollut. Res.* **2018,** *25* (8), 7287–7298.
3. Anderson, J. M.; Shive, M. S. Biodegradation and Biocompatibility of PLA and PLGA Microspheres. *Adv. Drug Deliv. Rev.* **1997** *28* (1), 5–24.
4. Artham, T.; Doble, M. Biodegradation of Aliphatic and Aromatic Polycarbonates. *Macromol. Biosci.* **2008,** *8* (1), 14–24.
5. Baculi, R. Q.; Melegrito, J. B.; Sabaten, R. D.; Caranto, M. R. Biodegradation of High Density Polyethylene by Bacteria Isolated from Root Nodules of *Phaseolus vulgaris. Philippine Agric. Sci.* **2017,** *100*, S21–S31.
6. Bahl, S.; Dolma, J.; Singh, J. J.; Sehgal, S. Biodegradation of Plastics: A State-of-the-Art Review. *Mater. Today: Proc.* **2021,** *39*, 31–34.

7. Balasubramanian, V.; Natarajan, K.; Rajeshkannan, V.; Perumal, P. Enhancement of In Vitro High-Density Polyethylene (HDPE) Degradation by Physical, Chemical, and Biological Treatments. *Environ. Sci. Pollut. Res.* **2014,** *21*, 12549–12562.
8. Benachour, N.; Aris A. Toxic Effects of Low Doses of Bisphenol-A on Human Placental Cells. *Toxicology and Applied Pharmacology* **2009,** *241* (3), 322–328.
9. Boyandin, A. N.; Prudnikova, S. V.; Karpov, V. A.; Ivonin, V. N.; Đỗ, N. L.; Nguyễn, T. H.; Lê, T. M.; Filichev, N. L.; Levin, A. L.; Filipenko, M. L.; Volova, T. G. Microbial Degradation of Polyhydroxyalkanoates in Tropical Soils. *Int. Biodeterior. Biodegrad.* **2013,** *83*, 77–84.
10. Chatterjee, S.; Roy, B.; Roy, D.; Banerjee, R. Enzyme-Mediated Biodegradation of Heat Treated Commercial Polyethylene by *Staphylococcal* Species. *Polym. Degrad. Stab.* **2010,** *95* (2), 195–200.
11. Chattopadhyay, S. K.; Singh, S.; Pramanik, N.; Niyogi, U. K.; Khandal, R. K.; Uppaluri, R.; Ghoshal, A. K. Biodegradability Studies on Natural Fibers Reinforced Polypropylene Composites. *J. Appl. Polym. Sci.* **2011,** *121* (4), 2226–2232.
12. Comăniţă, E. D.; Hlihor, R. M.; Ghinea, C.; Gavrilescu, M. Occurrence of Plastic Waste in the Environment: Ecological and Health Risks. *Environ. Eng. Manag. J. (EEMJ)* **2016,** *15* (3), 675–685.
13. Corti, A.; Muniyasamy, S.; Vitali, M.; Imam, S. H.; Chiellini, E. Oxidation and Biodegradation of Polyethylene Films Containing Pro-Oxidant Additives: Synergistic Effects of Sunlight Exposure, Thermal Aging and Fungal Biodegradation. *Polym. Degrad. Stab.* **2010,** *95* (6), 1106–1114.
14. Devi, R. S.; Kannan, V. R.; Nivas, D.; Kannan, K.; Chandru, S.; Antony, A. R. Biodegradation of HDPE by *Aspergillus* spp. from Marine Ecosystem of Gulf of Mannar, India. *Marine Pollut. Bull.* **2015,** *96* (1–2), 32–40.
15. Devi, R. S.; Ramya, R.; Kannan, K.; Antony, A. R.; Kannan, V. R. Investigation of Biodegradation Potentials of High Density Polyethylene Degrading Marine Bacteria Isolated from the Coastal Regions of Tamil Nadu, India. *Marine Pollut. Bull.* **2019,** *138*, 549–560.
16. Elbanna, K.; Lütke-Eversloh, T.; Jendrossek, D.; Luftmann, H.; Steinbüchel, A. Studies on the Biodegradability of Polythioester Copolymers and Homopolymers by polyhydroxyalkanoate (PHA) Degrading Bacteria and PHA Depolymerases. *Arch. Microbiol.* **2004,** *182* (2), 212–225.
17. European Parliament and Council. Annex VI, Table 3.1 In: *Regulation (EC) No. 1272/2008 of the European Parliament and of the Council of 16 December 2008 on Classification, Labeling and Packaging of Substances and Mixtures, Amending and Repealing Directives 67/548/EEC and 1999/45/EC and Amending Regulation (EC) No. 1907/2006*, 80–82; Geneva: European Parliament, 2008; p 70.
18. Filip, Z. Polyurethane as the Sole Nutrient Source for *Aspergillus niger* and *Cladosporium herbarum*. *Eur. J. Appl. Microbiol. Biotechnol.* **1979,** *7* (3), 277–280.
19. Firestone, G.; Huang, H.; Bochinski, J. R.; Clarke, L. I. Photothermally-driven Thermo-Oxidative Degradation of Low-Density Polyethylene: Heterogeneous Heating Plus a Complex Reaction Leads to Homogeneous Chemistry. *Nanotechnology* **2019,** *30* (47), 475706.
20. Flieger, M.; Kantorova, M.; Prell, A.; Řezanka, T.; Votruba, J. Biodegradable Plastics from Renewable Sources. *Folia Microbiologica* **2003,** *48* (1), 27–44.

21. Fukushima, K.; Abbate, C.; Tabuani, D.; Gennari, M.; Camino, G. Biodegradation of Poly lactic acid and its Nanocomposites. *Polym. Degrad. Stab.* **2009,** *94* (10), 1646–1655.
22. Gajendiran, A.; Krishnamoorthy, S.; Abraham, J. Microbial Degradation of Low-Density Polyethylene (LDPE) by *Aspergillus clavatus* Strain JASK1 Isolated from Landfill Soil. *3 Biotech* **2016,** *6*, 52–60. DOI: 10.1007/s13205–016–0394-x
23. Gaur, R.; Darmwal, N.; Pandey, R. Favour the Growth of Microorganisms. However, Products Essentially. *Environ. Microbiol. Biotechnol.* **2004,** *2004*, 75–80.
24. Geyer, R.; Jambeck, J. R.; Law, K. L. Production, Use, and Fate of all Plastics Ever Made. *Sci. Adv.* **2017,** *3* (7), E-article 1700782.
25. Gu, J. D. Microbiological Deterioration and Degradation of Synthetic Polymeric Materials: Recent Research Advances. *Int. Biodeterior. Biodegrad.* **2003,** *52* (2), 69–91.
26. Hao, J.; Wang, J.; Zhao, W.; Ding, L.; Gao, E.; Yuan, W. Effect of Bisphenol a Exposure on Sex Hormone Level in Occupational Women. *J. Hygiene Res.* **2011,** *40* (3), 312–319.
27. Hoang, K. C.; Tseng, M.; Shu, W. J. Degradation of Polyethylene Succinate (PES) by a New Thermophilic Microbispora Strain. *Biodegradation* **2007,** *18* (3), 333–342.
28. Hoshino, A.; Tsuji, M.; Ito, M.; Momochi, M.; Mizutani, A.; Takakuwa, K.; Higo, S.; Sawada, H.; Uematsu, S. Study of the Aerobic Biodegradability of Plastic Materials under Controlled Compost. In: *Biodegradable Polymers and Plastics*; Springer: Boston, MA, 2003; pp 47–54.
29. Ikada, Y.; Tsuji, H. Biodegradable Polyesters for Medical and Ecological Applications. *Macromol. Rapid Commun.* **2000,** *21* (3), 117–132.
30. Jayasekara, R.; Harding, I.; Bowater, I.; Lonergan, G. Biodegradability of a Selected Range of Polymers and Polymer Blends and Standard Methods for Assessment of Biodegradation. *J. Polym. Environ.* **2005,** *13* (3), 231–251.
31. Jin, H. J.; Lee, B. Y.; Kim, M. N.; Yoon, J. S. Thermal and Mechanical Properties of Mandelic Acid-Copolymerized Poly(butylene succinate) and Poly(ethylene adipate). *J. Polym. Sci. B: Polym. Phys.* **2000,** *38* (11), 1504–1511.
32. Karakus, K.; Polycaprolactone (PCL) Based Polymer Composites Filled Wheat Straw Flour. *Kastamonu Univ. J. Forestry Faculty* **2016,** *16* (1), 264–268.
33. Kasirajan, S.; Ngouajio, M. Polyethylene and Biodegradable Mulches for Agricultural Applications: A Review. *Agron. Sustain. Dev.t*, **2012,** *32* (2), 501–529.
34. Kesti, S. S.; Shivasharana, C. T. The Role of Insects and Microorganisms in Plastic Biodegradation: A Comprehensive Review. *Int. J. Sci. Res. Biol. Sci.* **2018,** *5* (6), 75–79.
35. Kijchavengkul, T. *Design of Biodegradable Aliphatic Aromatic Polyester Films for Agricultural Applications Using Response Surface Methodology*. PhD Thesis at Michigan State University, 2010; p 219.
36. Kim, M. Y.; Kim, C.; Moon, J.; Heo, J.; Jung, S. P.; Kim, J. R. Polymer Film-based Screening and Isolation of Polylactic acid (PLA)-degrading Microorganisms. *J. Microbiol. Biotechnol.* **2017,** *27* (2), 342–349.
37. Kjeldsen, A.; Price, M.; Lilley, C.; Guzniczak, E.; Archer, I. *Review of Standards for Biodegradable Plastics*; Industrial Biotechnology Innovation Centre: London, 2019; pp 1–28.
38. Konduri, M. K.; Koteswarareddy, G.; Rohini Kumar, D. B.; Venkata Reddy, B.; Lakshmi Narasu, M. Effect of Pro-oxidants on Biodegradation of Polyethylene (LDPE) by Indigenous Fungal Isolate, *Aspergillus oryzae*. *J. Appl. Polym. Sci.* **2011,** *120* (6), 3536–3545.

39. Kumar, S.; Hatha, A. A.; Christi, K. S. Diversity and Effectiveness of Tropical Mangrove Soil Microflora on the Degradation of Polythene Carry Bags. *Revista de Biología Trop.* **2007,** *55* (3–4), 777–786.
40. Kyrikou, I.; Briassoulis, D. Biodegradation of Agricultural Plastic Films: A Critical Review. *J. Polym. Environ.* **2007,** *15* (2), 125–150.
41. Lam, C. X.; Hutmacher, D. W.; Schantz, J. T.; Woodruff, M. A.; Teoh, S. H. Evaluation of Polycaprolactone Scaffold Degradation for 6 months In Vitro and In Vivo. *J. Biomed. Mater. Res. A* **2009,** *90* (3), 906–919.
42. López-Carrillo, L.; Hernández-Ramírez, R. U.; Calafat, A. M.; Torres-Sánchez, L.; Galván-Portillo, M.; Needham, L. L.; Ruiz-Ramos, R.; Cebrián, M. E. Exposure to Phthalates and Breast Cancer Risk in Northern Mexico. *Environ. Health Perspect.* **2010,** *118* (4), 539–544.
43. Lyche, J. L.; Gutleb, A. C.; Bergman, Å.; Eriksen, G. S.; Murk, A. J.; Ropstad, E.; Saunders, M.; Skaare, J. U. Reproductive and Developmental Toxicity of Phthalates. *J. Toxicol. Environ. Health B* **2009,** *12* (4), 225–249.
44. Mani, R.; Bhattacharya, M. Properties of Injection Molded Blends of Starch and Modified Biodegradable Polyesters. *Eur. Polym. J.* **2001,** *37* (3), 515–526.
45. Manohar, S.; Kim, C.; Karegoudar, T. Enhanced Degradation of Naphthalene by Immobilization of *Pseudomonas* sp. strain NGK1 in Polyurethane Foam. *Appl. Microbiol. Biotechnol.* **2001,** *55* (3), 311–316.
46. Masaki, K.; Kamini, N. R.; Ikeda, H.; Iefuji, H. Cutinase-like Enzyme from the Yeast *Cryptococcus* sp. strain S-2 Hydrolyzes Polylactic acid and other Biodegradable Plastics. *Appl. Environ. Microbiol.* **2005,** *71* (11), 7548–7550.
47. Mathur, G.; Mathur, A.; Prasad, R.; Colonization and Degradation of Thermally Oxidized High-Density Polyethylene by *Aspergillus niger* (ITCC No. 6052) Isolated from Plastic Waste Dumpsite. *Bioremed. J.*, **2011,** *15* (2), 69–76.
48. Mathur, G.; Prasad, R. Degradation of Polyurethane by *Aspergillus flavus* (ITCC 6051) Isolated from Soil. *Appl. Biochem. Biotechnol.* **2012,** *167* (6), 1595–602.
49. Mohanty, A. K.; Misra, M. A.; Hinrichsen, G. I. Biofibers, Biodegradable Polymers and Biocomposites: An Overview. *Macromol. Mater. Eng.* **2000,** *276* (1), 1–24.
50. Muhamad, W. N.; Othman, R.; Shaharuddin, R. I.; Irani, M. S. Microorganism as Plastic Biodegradation Agent Towards Sustainable Environment. *Adv. Environ. Biol.* **2015,** *9*, 8–14.
51. Nanda, S.; Sahu, S.; Abraham, J. Studies on the Biodegradation of Natural and Synthetic Polyethylene by *Pseudomonas* spp. *J. Appl. Sci. Environ. Manag.* **2010,** *14* (2), 57–60.
52. Norman, R. S.; Frontera-Suau, R.; Morris, P. J. Variability in *Pseudomonas aeruginosa* Lipopolysaccharide Expression during Crude Oil Degradation. *Appl. Environ.Microbiol.* **2002,** *68* (10), 5096–5103.
53. Nowak, B.; Pająk, J.; Drozd-Bratkowicz, M.; Rymarz, G. Microorganisms Participating in the Biodegradation of Modified Polyethylene Films in Different Soils under Laboratory Conditions. *Int. Biodeterior. Biodegrad.* **2011,** *65* (6), 757–767.
54. OECD, (Organization for Economic Coopera-tion and Development) Emission Scenario Document on Plastic Additives. Series on Emission Scenario Documents, No. 3. Paris: Paris Environmental Directorate, OECD Environmental Health and Safety Publications, 532538, 2004; p 81.
55. Philip, S.; Keshavarz, T.; Roy, I. Polyhydroxyalkanoates: Biodegradable Polymers with a Range of Applications. *J. Chem. Technol. Biotechnol.* **2007,** *82* (3), 233–247.

56. Raaman, N.; Rajitha, N.; Jayshree, A.; Jegadeesh, R. Biodegradation of Plastic by *Aspergillus* Spp. Isolated from Polythene Polluted Sites Around Chennai. *J. Acad. Ind. Res.* **2012,** *1* (6), 313–316.
57. Ratto, J. A.; Stenhouse, P. J.; Auerbach, M.; Mitchell, J.; Farrell, R. Processing, Performance and Biodegradability of a Thermoplastic Aliphatic Polyester/Starch System. *Polymer* **1999,** *40* (24), 6777–6788.
58. Raziya-fathima, M.; Praseetha, P. K.; Rimal-Isaac, R. S. Microbial Degradation of Plastic Waste: A Review. *J. Pharma. Chem. Biol. Sci.* **2016,** *4* (2), 231–242.
59. Reddy, R. M. Impact of Soil Composting Using Municipal Solid Waste on Biodegradation of Plastics. *Indian J. Biotechnol.* **2008,** *7* (2), 235–239.
60. Sanchez, J. G.; Tsuchii, A.; Tokiwa, Y. Degradation of Polycaprolactone at 50° C by a Thermotolerant Aspergillus sp. *Biotechnol. Lett.* **2000,** *22* (10), 849–853.
61. Sangale, M. K.; Shahnawaz, M.; Ade, A. B. Potential of Fungi Isolated from the Dumping Sites Mangrove Rhizosphere Soil to Degrade Polythene. *Sci. Rep.*, **2019,** *9* (1), 1–11.
62. Shah, A. A.; Hasan, F.; Akhter, J. I.; Hameed, A.; Ahmed, S. Degradation of Polyurethane by Novel Bacterial Consortium Isolated from Soil. *Ann. Microbiol.* **2008,** *58* (3), 381–386.
63. Shah, A. A.; Hasan, F.; Hameed, A.; Ahmed, S. Biological Degradation of Plastics: A Comprehensive Review. *Biotechnol. Adv.* **2008,** *26* (3), 246–265.
64. Sharma, M.; Dhingra, H. K. Poly-β-hydroxybutyrate: A Biodegradable Polyester, Biosynthesis and Biodegradation. *Br. Microbiol. Res. J.* **2016,** 14 (3), 1–11.
65. Shimao, M. Biodegradation of Plastics. *Curr. Opin. Biotechnol.* **2001,** *12* (3), 242–247.
66. Skariyachan, S.; Manjunatha, V.; Sultana, S.; Jois, C.; Bai, V.; Vasist, K. S. Novel Bacterial Consortia Isolated from Plastic Garbage Processing Areas Demonstrated Enhanced Degradation for Low Density Polyethylene. *Environ. Sci. Pollut. Res.* **2016,** *23* (18), 18307–18319.
67. Tadimeti, A. The Effects of Different Aquatic Environments on the Rate of HDPE and LDPE Degradation by Bacillus Subtilis. *Columbia Junior Sci. J. Spring* **2020,** 19, 1–2.
68. Tang, S.; Zou, P.; Xiong H.; Tang, H. Effect of Nano-SiO2 on the Performance of Starch/ Polyvinyl Alcohol Blend Films. *Carbohydr. Polym.* **2008,** *72* (3), 521–526.
69. Tokiwa, Y.; Calabia, B. P. Biodegradability and Biodegradation of Polyesters. *J. Polym. Environ.* **2007,** *15* (4), 259–267.
70. Tokiwa, Y.; Calabia, B. P.; Ugwu, C. U.; Aiba, S. Biodegradability of Plastics. *Int. J. Mol. Sci.* **2009,** *10* (9), 3722–3742.
71. Tribedi P, Sil AK. Low-Density Polyethylene Degradation by Pseudomonas sp. AKS2 Biofilm. *Environ. Sci. Pollut. Res.* **2013,** *20* (6), 4146–4153.
72. Tribedi, P.; Sil, A. K. Cell Surface Hydrophobicity: A Key Component in the Degradation of Polyethylene Succinate by *Pseudomonas* sp. AKS 2. *J. Appl. Microbiol.* **2014,** *116* (2), 295–303.
73. Tudorachi, N.; Cascaval, C. N.; Rusu, M.; Pruteanu, M. Testing of Polyvinyl Alcohol and Starch Mixtures as Biodegradable Polymeric Materials. *Polym. Test.* **2000,** *19* (7), 785–799.
74. Upreti, M. C.; Srivastava, R. B. A potential Aspergillus Species for Biodegradation of Polymeric Materials. *Curr. Sci.* **2003,** *84* (11), 1399–1402.
75. Vasile, C. Degradation and Decomposition. *Plastics Eng. NY* **2000,** *59*, 413–476.

76. Vert, M.; Santos, I. D.; Ponsart, S.; Alauzet, N.; Morgat, J. L.; Coudane, J.; Garreau, H. Degradable Polymers in a Living Environment: Where do You End Up? *Polym. Int.* **2002,** *51* (10), 840–844.
77. Vimala, P. P.; Mathew, L. Biodegradation of Polyethylene Using *Bacillus Subtilis. Procedia Technol.* **2016,** *24*, 232–239.
78. Volke-Sepúlveda, T.; Saucedo-Castañeda, G.; Gutiérrez-Rojas, M.; Manzur, A.; Favela-Torres, E. Thermally Treated Low Density Polyethylene Biodegradation by *Penicillium Pinophilum* and *Aspergillus niger. J. Appl. Polym. Sci.* **2002,** *83* (2), 305–314.
79. Wackett, L. P.; Hershberger, C. D. Microbial Transformation of Organic Compounds. In: *Biocatalysis and Biodegradation*; ASM Press: Washington, DC, 2001; pp 216–228.
80. Wu, C. S. A Comparison of the Structure, Thermal Properties, and Biodegradability of Polycaprolactone/Chitosan and Acrylic Acid Grafted Polycaprolactone/Chitosan. *Polymer* **2005,** *46* (1), 147–155.
81. Yoshida, S.; Hiraga, K.; Takehana, T.; Taniguchi, I.; Yamaji, H.; Maeda, Y.; Toyohara, K.; Miyamoto, K.; Kimura, Y.; Oda, K. A Bacterium that Degrades and Assimilates Poly (Ethylene terephthalate). *Science* **2016,** *351* (6278), 1196–1199.
82. Zafar, U.; Houlden, A.; Robson, G. D. Fungal Communities Associated with the Biodegradation of Polyester Polyurethane Buried Under Compost at Different Temperatures. *Appl. Environ. Microbiol.* **2013,** *79* (23), 7313–7324.
83. Zheng, Y.; Yanful, E. K.; Bassi, A. S. A Review of Plastic Waste Biodegradation. *Crit. Rev. Biotechnol.* **2005,** *25* (4), 243–250.

CHAPTER 12

MODERN APPROACHES IN FOOD PACKAGING WASTE MANAGEMENT

NITYA SHARMA, AASTHA BHARDWAJ, SUKIRTI JOSHI, J. K. SAHU, and SANA FATMA

ABSTRACT

Due to the current state of eco-pollution produced by plastic packaging wastes that are usually found contaminating soils and fresh water, it is critical to develop a long-term solution, particularly for the food industry. As previously said, packaging wastes from foods makes about two-thirds of total volume of the packaging wastes, which is driven by a variety of factors such as rising economies, new products, public perception changes, consumerism, increases in income, population growth, and so on. Therefore, methods of treating and processing of food packaging wastes must be relooked to understand the environmental, economic, and social ramifications involved with food packaging waste management. Source reduction, recycling, composting, burning, and landfilling are all part of the traditional integrated waste management method. Although just by altering the process design of these conventional food packaging waste treatment methods, yet various modern green waste management approaches are now being developed. Therefore, this chapter presents system of waste generation in the sector of food and then provides an insight on the conventional as well as modern food packaging waste management approaches, finally giving a detailed comparison thereof. Additionally, the goal of this chapter would be to widen understanding of packaging waste treatment techniques in the food sector.

Advances in Sustainable Food Packaging Technology. Megh R. Goyal, Preeti Birwal, Anand K. Pandey, & Ritesh B. Watharkar, (Eds.)

12.1 INTRODUCTION

Rapid industrialization, population explosion, and developing lifestyles have followed in a rise in the demand for processed and packaged goods. Despite the fact that different food components require varied levels of processing, the semiprocessed and ready-to-eat packaged food sector is worth over four thousand crores rupees and is evolving at a rate of over 20% per year. Food packing, like any other type of packaging, is an external method of keeping food fresh during storage, transportation, and distribution. Containment, protection, convenience, and communication are the four basic objectives of food packaging. Many materials have been discovered by man for use as food packaging materials to achieve these goals, including wood and paper, glass, metals, polymers, composites, and so on. These packaging materials' inherent qualities make them either very suitable or unsuitable for a specific food product or package level. The food industry products are packed either for retailing or wholesaling. Transport packaging required for sales is considered as waste of industry generally at the food production site itself. Depending on the size and location of the setup, the transport packages end up in retail, where they are managed as commercial or industrial waste. The sales packaging is discarded in households, where it is considered and treated as household waste.[1] As a result, it is an essential component of both product manufacture and distribution.

For packaging materials, the food sector has relied primarily on petroleum-based polymers. Plastic packaging materials are popular for a variety of reasons, including their outstanding functionality, ease of processing, light weight, and low cost. Despite their versatility, most plastics in food packaging have the drawback of being poor barriers to gases and vapors such as oxygen, carbon dioxide, and organic vapors. Furthermore, because they are not biodegradable, they are considered a threat to the environment once they reach municipal solid waste (MSW) and/or landfills.

The food packaging sector is the largest user of plastics (~40%). In the year 2011, India generated approximately 15,300 tons/day of plastic waste according to Central Pollution Control Board [The Hindu, 2011], compared to an alarming 660,787 tons/day in the year 2017–2018. As stated by the Department of Science and Technology (Government of India), about 60% of the collected and segregated plastic waste is recycled back into usable plastic packaging, while the remaining percentage is left unutilized. The amount of plastics discarded each year generates a significant amount of waste, posing a serious environmental danger.

Metals like tin and aluminum are used as packaging materials in canning and perforated trays. As a result, the concept of creating materials from biodegradable components that can be composted or recycled has gained traction.

In Indian cities, garbage generation is surpassing the increase of the urban population. To prevent future problems, we must think of taking pressing steps to control waste generation, augment recycling recovery and reuse, and to warrant improved collection and sustainable disposal. In recent times, the processed food industry, with great support from its packaging aide, has been able to deliver the public with options for quicker, less frequent shopping, more choices and lower prices apart from the outstanding better quality. Modern supermarkets cannot function without the sophistication of packaging. Approximately, 75% of packaging is used for convenience food and drinks and home-delivered products. This trend, however, has led to a 'throw-away' society where much of the packaging has been reduced to a single trip and then discarded. In these respects, the packaging, of course, has to be effective in complying with legal, health, as well as safety requirements, while due emphasis should also be given to it being kept to a minimum.

Other interesting and major changes have been observed in the attitudes of the governments globally to reduce packaging consumption, control its disposal, and encourage its recovery, reuse and recycling through legislation. Owing to its high profile, used packaging is regarded as one of the significant constituents of waste elements. Besides, people are also becoming aware of the high levels of post-consumer food packaging waste.

In this chapter, the authors have emphasized on the conventional and modern packaging waste treatment techniques that are employed in the food sector. EPA's solid waste management guidelines[11] focus on using an ordered, integrated approach of management which involves source reduction, composting, landfilling, recycling, and combustion. Henceforth, this chapter discusses various strategies, their features, and drawbacks.

12.2 FOOD PACKAGING WASTE GENERATION

In the food sector, the shelf-life of foods is largely dependent on the package design and construction of the food product. Using the right materials and technology, the freshness and quality of the product can be easily retained during storage and transportation.

FIGURE 12.1 Packaging waste generation in food industry.

The conventional packaging materials are used for foods including metals (tinplate, aluminum, laminates and foils, and tin-free steel), paper and its boards, glass, and plastics. Except these, a wide range of both flexible and rigid plastics are also utilized for food packaging. Several other materials are now increasingly being exploited to improve the functional as well as aesthetic properties of the food packages, leading to advancements that may have environmental impacts.

Therefore, to create sustainable packaging the packaging material and design are decided based on the following factors—marketing considerations (including consumer and distribution needs), characteristics of food products, waste and environmental management issues, and the cost. Most of the times, these factors are related to each other: for example, the food types and packaging material properties that determine food–package interactions nature during the storage. While on the other hand, these factors oppose each other: for example, consumer needs are met by single-serving packaging, and bulk packaging works well in protection against environment.[24] The post-consumer packaging waste stream is a diverse waste source and requires sorting and segregation, to recover valuable materials. The packaging waste in the food industry is generated at almost every step of operation, starting from the packaging industry, food industry, retail outlets, and finally consumer households (Fig. 12.1). Thus, proper management of packaging wastes requires adequate planning, financing, collection, segregation, and transportation.

12.3 PLASTIC WASTE TREATMENT TECHNIQUES

12.3.1 CONVENTIONAL FOOD PACKAGING WASTE TREATMENT TECHNIQUES

12.3.1.1 SOURCE REDUCTION, COLLECTION, AND SORTING OF WASTES

Source reduction implies preventing, minimizing, or avoiding generation of surplus waste, as a result, saving the expenses incurred during manufacture, operation, and maintenance of waste treatment and disposal alternatives. It saves energy in terms of collection and segregation of waste materials, production, as well as transportation of the materials to consumers. Gradual drifting toward circular economy approach presents an efficient way for food waste management. A circular economy intends to keep a continuous circulation of packaging materials in a high value state of use, through supply chains, for the longest time possible. It emphasizes on sustainable development and achieving economic, social, and environmental well-being through cleaner technology.

Reducing, reusing, and recycling packaging materials are vital for diverting packaging materials from waste streams. Reusable glass beverage bottles and shipping crates have been the mainstays of direct food-contact packaging reuse. The utilization of reusable standardized shipping crates and modified environment flexible packaging for distribution in tertiary packaging has a lot of possibilities. To properly handle packaging via recycling, reuse, landfilling, composting, degrading, or incineration, a greater focus on collecting and sorting is also required. Sorting is done to eliminate/separate hazardous components from the packaging to generate products suitable for downstream recycling operations. Usually, all forms of food packaging (tinplate, aluminum, liquid cartons, paper cartons, and mixed plastics) and household wastes are subjected to conditioning, sieving, air separation, magnetic sorting, and sensor and eddy current sorting.[19]

Plastics are most difficult to visually segregate due to presence of multilayers and coatings; thus, novel techniques have been implemented lately for efficient sorting of plastic wastes such as Near Infrared (NIR) sorting, Laser-induced breakdown spectroscopy, Triboelectric separation, Fourier transform infrared spectroscopy (FTIR), Froth floatation, and Ultrasound.[27,29,34] Another new identification and sorting system based on fluorescent tracers and a detecting device is tracer-based sorting.[12] Separation and sorting of paper and plastic waste by the consumers reduces the overall cost of recycling as compared to when done at a material recovery facility due to the mixed nature of the waste and lower quality and quantity of sorted fractions.[30]

Consumer separation and sorting behavior is governed by a range of elements, including sociodemographics, age, gender, level of education, and knowledge about which materials are recyclable, moral attitudes, and distance between home and recycling station, as well as government intervention, collection services, and costs.[10] Materials that are viewed as recyclable, such as paper and metal, urge customers to sort and separate their recyclable packaging, whereas materials like plastic might have a detrimental impact on consumer attitudes.[20]

12.3.1.2 COMPOSTING

Composting, in general, is a natural process of decomposition and recycling of organic matter to humus-rich soil known as compost. The complex role as well as the varied standards and types of compostable packaging all work together to promote better sustainable food packaging. Passive composting (piling), Aerated static piles, Bins, In-vessel systems, and Vermi-composting are the few approaches of efficient composting. PLA, polyhydroxyalkanoates, starch, polybutylene succinate, and polybutylene adipate terephthalate are some of the compostable polymers commonly used for food packaging. Paper and paperboards, derived primarily from cellulosic fiber, are also compostable.

During the last decade, an immense interest has been evolved in deriving compostable and biodegradable packaging materials from natural agro-based residues and food wastes such as sugarcane bagasse, wheat straw, rice straw, corn cobs, barley straws, coconut coirs, carrot and potato peels, etc. Compostable packaging's environmental impact and material qualities differ depending on where it comes from. However, food packaging materials that cannot be composted include plastics, glass, and metals; plastic wrappings and bags, condiment packages, foil, silverware, drinking straws, bottles, polystyrene, and other synthetic petroleum-derived materials.

12.3.1.3 INCINERATION (COMBUSTION)

It involves burning of the packaging waste with or without energy recovery in specially designed furnaces or cemented kilns. With incineration, waste volume can be reduced by 80–90%.[5] Conventionally, municipal solid waste, hazardous and toxic wastes, and medical wastes are subjected to incineration. Incineration of plastic waste releases carbon dioxide, a substantial contributor to global climate change. The incineration process, on the other hand, can be integrated as part of a "Waste to Energy" (WtE) system. WtE is a type of energy recovery in which energy from plastics is saved and used to generate energy.

The energy embodied in flexible packages, such as laminated pouches, can be recovered by a clean burning WtE incineration setup to produce electricity and heat, where the flexible packaging being incinerated contains an aluminum foil layer for increased barrier; only a minute portion of this aluminum is converted into energy. The leftover aluminum melts and can be collected using eddy current technology from the incinerator's "bottom ashes" near the incinerator's exit. Incineration of packaging materials, like wood, paper, and plastics, has a significant fuel value. Presently, it is a widely accepted technology used particularly for the disposal of plastic; however, designed incinerators poorly release highly toxic compounds (furans and chlorinated dioxins), and therefore are inadequate according to various industrial setups and environmental organizations. Unless efficiently controlled, toxic combustion by-products and pollutants that may be emitted into the air, land, and water can cause detrimental effects on human health and the environment.

12.3.1.4 LANDFILLING

Landfill is the commonly used method of disposing package waste where the waste is deposited onto land for a longer duration of time. It has been considered detrimental for the environment because of leaching of toxic materials to the soil. Methane has accumulated up in ageing landfills and caused major explosions in the past. This accumulating gas may be efficiently tapped and converted into energy. Modern landfills are now constructed, engineered, and maintained to reduce the likelihood of these issues occurring. Items dumped in landfills should be monitored, and dangerous materials should be disposed of through controlled incineration for proper landfill management.

Household packaging waste can be compacted more than 10% but in later stages in the absence of oxygen, methane and carbon-dioxide may be produced. Since these gases can pose major health risks, the site must be managed to allow them to vent to the atmosphere, or a system of wells can be drilled to extract the gas for use as an energy source.

To reduce global greenhouse effect, provisions will have to be made to reduce the land filling of biodegradable municipal waste. In coming years, all waste will need to be treated before land filling, and landfills will become waste-specific regarding nonhazardous waste, hazardous waste, and inert waste. Landfill of waste plastics is the least favored option.[35] Furthermore, landfilling is also demonstrated as the worst method for waste disposal of composite packaging material wastes.[41,42] Composite packages (such as Aseptic Tetra Pak) comprise varying proportions of paper, polyethene, and aluminum. The polyethylene and aluminum remnants produced by cellulose recovery from the composite packaging are difficult to separate and reuse. As a result of the lack of effective separation technology, these residues are handled in sanitary landfills, which has a bigger environmental impact.

12.3.1.5 RECYCLING

Food packaging wastes are generally subjected to mechanical or chemical recycling. Mechanical recycling of plastic wastes involves collection, sorting, washing, and grinding of the wastes. Figure 12.2 shows the steps for the mechanical recycling process.

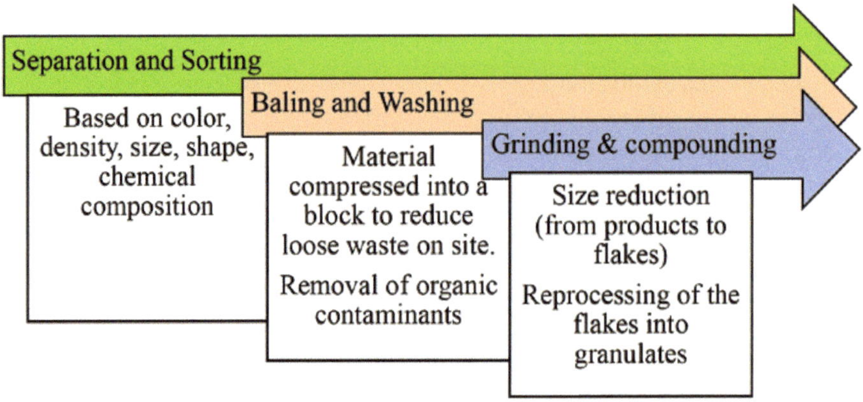

FIGURE 12.2 Mechanical recycling process.

The process of mechanical recycling is complicated owing to poisonous and noxious dioxin emissions, the usage of coatings and paints, insoluble pollutants, and so on.[28] All these limitations have promoted an interest in chemical recycling of plastic wastes, which generate beneficial value-added components as a by-product and in much less harsher conditions. Chemical recycling of plastic wastes aims at transforming plastic packaging into valuable chemical products (e.g., petrochemical feedstocks) and is prescribed as a large-scale boulevard to reduce the exhaustion of fossil resources and emission of greenhouse gases. Chemical recycling refers to the breakdown of a plastic polymer into monomers, which are then rebuilt into new polymers for use in fresh plastic manufacture.[36] Chemical recycling, on the other hand, has sparked heated discussion about its environmental consequences.[25] Table 12.1 shows the various chemical recycling technologies employed for plastic packaging and household wastes, along with their advantages and limitations.

TABLE 12.1 Advantages and Disadvantages of Chemical Recycling Technologies for Plastic Packaging and Household Wastes.

Packaging waste	Chemical recycling method	Advantages	Limitations
Mixed plastic wastes	Conventional gasification	Complete breakdown to hydrogen and methane; Syngas produced (used for production of new plastic products)	Cost and energy intensive;
Most plastic wastes	Catalytic cracking	Compared to conventional cracking, it has a lower operating temperature and a higher oil output.	Catalyst can be deactivated by chloride and nitrogen components in wastes; pretreatment required.
Multilayered plastic packaging	Thermal cracking (conventional pyrolysis)	Pyrolytic oil with high calorific value obtained.	High energy requirement; pretreatment required
Plastic wastes	Hydrocracking	High-quality product	Cost of hydrogen; high cost of operation
Plastic wastes	Plasma gasification	Higher purity of product gas; reduced level of tars	High cost of operation
Plastic wastes	Pyrolysis with in-line reforming	Product gas free of tars; low process temperature	Catalyst deactivation

Chemical recycling can produce a high-quality product by breaking down polymers into single monomers, while mechanical recycling is comparatively less tolerant of complex plastic waste streams than chemical recycling.[36]

Plastic packaging has relatively lower recycling rates and a large amount of packaging is left unrecycled in the current processes, specifically, packaging with multi-material layers. Recycling of multilayer plastic package wastes is utmost challenging owing to their inherent compositional heterogeneity among all forms of packaging wastes. To avoid landfilling, various other strategies have been sought after such as recovery of cardboard for use in paper products, recovery of polyethylene-aluminum, recycling (together or separately), and pyrolysis of the entire container, eliminating the need for separation. Table 12.2 shows the recent researches involving mechanical and chemical processing methods undertaken at laboratory scale as alternative approaches to manage post-consumer multi-layer food packaging wastes:

TABLE 12.2 Alternative Approaches to Manage Multilayer Plastic Wastes.

Packaging waste	Treatment/ approach	Results	Challenges/ limitations	References
Recycled multilayer Tetra Pak	Hydropulping, filtration, delamination using formic acid	Retrieved Microcrystalline cellulose as by-product; separation of constituent layers	De-inking the polyethylene, recycling of organic acid used.	[9]
Post-consumer beverage cartons	Hydropulping, selective dissolution precipitation process.	High recovery of paper, aluminum, PE fractions	Outer plastic fraction still contained ink, suggesting further purification.	[13]
Colored multilayer flexible packaging films	Delamination with carboxylic acid (formic acid)	Constituent polymer layer separation	Recovery of delamination medium	[37]
Tetra Pak	Thermal and catalytic pyrolysis	Production of aromatic hydrocarbons and recovery of aluminum	-	[33]

Another interesting intervention, which involves recycled Tetra Pak aseptic multilayered food packaging wastes, has been used to extract nanocrystalline cellulose to find potential application in bio-nanocomposite development.[10]

12.3.2 MODERN APPROACHES FOR FOOD PACKAGING WASTE TREATMENT

12.3.2.1 PLASMA PYROLYSIS TECHNOLOGY

This may be recognized as a clean, non-incineration process, and is a well-proven commercially viable, novel technology, currently utilized in many industrial plants worldwide. Plasma pyrolysis provides various advantages over thermal cracking/incineration. For example, the generated gas has a low tar content and a high heating value, making it useful for generating energy in turbines or producing hydrogen.[36]

In Plasma Pyrolysis, high temperature is produced using a graphite plasma torch in an oxygen-deficient setting to annihilate plastic wastes such as thin carry bags, thermoset plastic wraps, and metalized and multilayer food pouches competently and in an eco-friendly manner.[6] High temperature (close to >1000°C) is achieved by the plasma torch where gases such as methane, carbon monoxide, and hydrogen are produced. These gases are further ignited to a temperature between 800 and 1000°C that give rise to nonharmful reaction products (primarily CO_2 and water vapors) in the presence of air. Plasma pyrolysis technology has proven lower emission rates and quantities of toxic compound, for example, dioxins and furans, below the levels prescribed under regulatory requirements and also in comparison to those for hazardous waste incinerators. However, in the long-term operation of this technology, particle agglomeration can substantially undermine the stability of fluidized bed reactors used to pyrolyze plastic waste, resulting in inadequate fluidization or de-fluidization.

Arena et al.[4] proposed a basic model for the sintering of polyolefin particles during fluidized pyrolysis. Pyrolysis process is one of the most commonly used technologies for utilization of plastic waste materials for energy recovery, in the form of solid, liquid, and gaseous fuels. Table 12.3 provides a summary of the approaches undertaken for the conversion of various packaging wastes involving pyrolysis and thermal treatment to produce high-value components.

This technology also tends to make way for the reduction in fossil fuel consumption, and reducing or restoring landfills. However, the high cost

TABLE 12.3 Pyrolysis for Conversion of Packaging Wastes to High Value By-Products.

Food packaging wastes	**Approach**	**Products obtained**	**Application**	**References**
HDPE and LDPE	Hydrothermal processing	Clean waxes and ultra-low-sulfur fuels	As potential wax product and transportation fuel respectively	[16]
Metallized food packaging plastics waste	Pyrolysis, mechanical and chemical treatments	Biooil, Biogas, and aluminum mixed with char (Al/char)	Renewable energy applications; decrease carbon footprint	[43,44]
Mono and multi-layer flexible plastic packaging (PET) waste	Pyrolysis and catalytic cracking of volatile hydrocarbons	Oil and multi-walled carbon nanotubes	As a replacement material for conventional electrode material	[39]
Multi-layer cardboard packaging wastes and plastic bottles (PET).	Pyrolysis	–	Significant decrease in solid residue	[38]
Polyethylene (HDPE, LDPE), polypropylene and PET wastes	Pyrolysis-gasification (with metal modified zeolite catalysts)	Syngas	Renewable energy applications	[2]

of investment in a relatively newer technology, and any potential impacts of the process on local communities still need ardent investigation. Recent years have witnessed the application of microwave heating in pyrolysis that provides numerous advantages such as rapid heating rate and uniform heating regimes, lower operating temperatures, and energy savings, in comparison to conventional pyrolysis process.[17]

12.3.2.2 HYDROTHERMAL CARBONIZATION (HTC)

Present plastic waste management which includes mechanical sorting and reprocessing seems unsustainable; thus, an integration of thermo-chemical processes to produce petrochemical products, such as oil and gasoline from the plastic wastes, can be considered worthy to improve the overall waste management system. Hydrothermal carbonization is a novel, thermo-chemical process that involves dehydration, polymerization, and carbonization to convert food packaging wastes such as plastics, into hydro-char and other valuable compounds, at relatively low temperatures (180–250°C) and saturated pressures (2–10 MPa).[21,32] The process allows material recovery, mitigates greenhouse gases emission, and undoubtedly minimizes plastic packaging waste intended for landfilling alone.[26] Paper-based food packaging subjected to HTC pretreatment produced carbon densified hydro-char, thus improving energy recovery potential from wet biodegradable wastes.[14] Various researchers have aimed at investigating the application of HTC to generate valuable, energy-rich by-products from food wastes,[15] textile wastes,[22] and municipal solid wastes[23]; however, the applicability of HTC for food packaging wastes should be explored further.

12.4 BIODEGRADABLE PACKAGING: A WAY FORWARD FOR FOOD PACKAGING WASTE MANAGEMENT

Food packaging materials such as paper and paperboards, metals, steel, tin, glass, plastics etc. are the most common forms of bulk generated litter by the community, for obvious reasons. Significant achievements in recycling or reusing of materials like glass, metal, and paper through landfill have been sought after; however, reduction in plastic packaging lags way behind, due to difficulties involving their collection, cleaning, sorting, and identification. High-barrier, multi-layer, biodegradable food packaging could be a helpful solution for present multi-layered packaging that is nonrecyclable and nondegradable from a waste management standpoint.[8]

Development of biodegradable packaging has paved a way forward for efficient waste management in terms of landfilling, incineration, recycling/reusing, and composting. For instance, compostable liner bags made from corn starch can be easily separated from heaps of wastes to be transported to commercial composting facilities where they bio-degrade rapidly and safely.[18] Another technique, known as oxo-biodegradation, which involves progressive oxidation and microbial biodegradation, has been reported by some manufacturers that claim compost ability of polyolefin-based materials due to addition of pro-degradant additives.[18] Figure 12.3 summarizes the process of development of such plastics and their degradation.

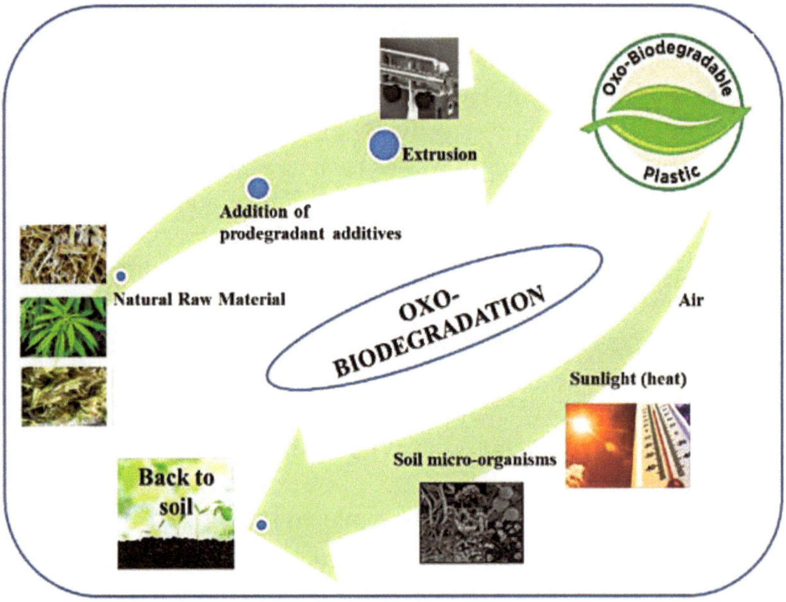

FIGURE 12.3 Development and degradation of oxo-biodegradable plastics.

Another class of biodegradable plastics is the "Bioplastics" that are obtained from biodegradable and renewable materials like, sugars, corn, hemp, potato, sugarcane etc. (such as PLA) or as a blend (such as PHBV/PLA), and can be reused, mechanically or organically recycled, and also employed for energy recovery. Mechanical recycling includes waste recovery through separation, crushing, washing, aerating or drying, re-granulating, and combining to form recyclates, which could give new products while chemical and organic recycling aims at reusing bio-based and plastic waste as carbon and biogenic substances source for synthesizing new materials.[40]

Chemical recycling of blends of bioplastic has been also researched due to the limitations associated with mechanical method of recycling, especially the sorting, degradation by thermomechanical means of the polymers, and sensitiveness to impurities in material.[40] Organic recycling of bio-based packaging waste involves mechanical pretreatment before the process. Another method for organic recycling is to compost bio-based products with bio-waste. However, the research related to recycling of bio-blends and bio-composites are still at a pilot stage and require further scrutiny for validation.

Ample studies have reported the bio-based waste biodegradability in a composting environment. For example, micro-organisms can degrade the PLA or corn-based bioplastic in compost, quite efficiently due to the presence of biodegradable corn husk.[31] Figure 12.4 shows the flowchart for the biodegradation of PLA-based corn plastic through composting.

FIGURE 12.4 Compostable and biodegradable PLA-based corn plastic (bioplastic).[3]

Source: Reprinted courtesy of Zeus Packaging Group. https://www.zeuspackaging.com/uk/blog/Compostable-Sustainable-and-Biodegradable-Packaging-Materials/

Biodegradable packaging is considered unsuitable for landfill due to their tendency to liberate methane under anaerobic conditions; however, they are rendered most suitable for domestic or municipal composting.[7] In fact, bio-based compostable packaging products reinforce industrial level composting or organic recycling as options for the management of waste.

12.5 SUMMARY

Food packaging waste management is a debatable issue pertaining to the environment and requires an urgent solution. Source reduction, collection, efficient sorting, material reuse, and mechanical/chemical recycling are commonly used strategies to manage packaging wastes. Newer interventions such as conversion of plastic wastes to valuable products through pyrolysis and other thermo-chemical treatments have shown to provide economic and environmental benefits. The last decade has also shown immense interest in deriving compostable and biodegradable packaging materials from natural agro-based residues and food wastes. To mitigate the burden of plastic packages, an integrated approach to food packaging waste management with the employment of alternative modern waste recovery technologies is quintessential.

KEYWORDS

- **Conventional**
- **food packaging waste**
- **modern techniques**
- **sustainable waste treatment**
- **waste treatment**

REFERENCES

1. Aarnio, T.; Hämäläinen, A. Challenges in Packaging Waste Management in the Fast Food Industry. *Resour. Conserv. Recyc.* **2008,** *52* (4), 612–621.
2. Al-asadi, M.; Miskolczi, N.; Eller, Z. Pyrolysis-Gasification of Wastes Plastics for Syngas Production using Metal Modified Zeolite Catalysts Under Different Ratio of Nitrogen/Oxygen. *J. Clean. Prod.* **2020,** *271*, 1–10.
3. Anonymous, 2019. http//www.zeuspackaging.com/uk/blog/Compostable-Sustainable-and-Biodegradable-Packaging-Materials/ (accessed on July 31, 2021).
4. Arena, U.; Mastellone, M. L. The Phenomenology of Bed Defluidization During the Pyrolysis of a Food-Packaging Plastic Waste. *Powder Technol.* **2001,** *120* (1–2), 127–133.
5. Arvanitoyannis, I. S.; Bosnea, L. A. Recycling of Polymeric Materials Used for Food Packaging: Current Status and Perspectives. *Food Rev. Int.* **2001,** *17 (3*), 291–346.
6. Central Pollution Control Board (CPCB). *Study on Plastic Waste Disposal through Plasma Pyrolysis Technology*; Unpublished Report by Central Pollution Control Board, Government of India, New Delhi, 2016; pp 1–51.
7. Davis, G.; Song, J. H. Biodegradable Packaging Based on Raw Materials from Crops and their Impact on Waste Management. *Ind. Crops Prod.* **2006,** *23* (2), 147–161.

8. Dilkes-Hoffman, L. S.; Lane, J. L.; Grant, T.; Pratt, S.; Lant, P. A.; Laycock, B. Environmental Impact of Biodegradable Food Packaging When Considering Food Waste. *J. Clean. Prod.* **2018,** *180*, 325–334.
9. Diop, C. I. K.; Lavoie, J. M. Isolation of Nanocrystalline Cellulose: A Technological Route for Valorizing Recycled Tetra Pack Aseptic Multilayered Food Packaging Wastes. *Waste Biomass Valorization* **2017,** *8* (1), 41–56.
10. Diop, C. I. K.; Lavoie, J. M.; Huneault, M. A. Separation and Reuse of Multilayer Food Packaging in Cellulose Reinforced Polyethylene Composites. *Waste Biomass Valorization* **2017,** *8* (1), 85–93.
11. EPA. *The Solid Waste Dilemma: An Agenda for Action. EPA530-SW- 89–019*; Environmental Protection Agency (EPA): Washington, DC, 1989; p 29.
12. Gasde, J.; Woidasky, J.; Moesslein, J.; Lang-Koetz, C. Plastics Recycling with Tracer-Based-Sorting: Challenges of a Potential Radical Technology. *Sustainability* **2021,** *13* (1), 1–16.
13. Georgiopoulou, I.; Pappa, G. D.; Vouyiouka, S. N.; Magoulas, K. Recycling of Post-Consumer Multilayer Tetra Pak® Packaging with the Selective Dissolution - Precipitation Process. *Resour. Conserv. Recycl.* **2021,** *165*, 1–9.
14. Gupta, D.; Mahajani, S. M.; Garg, A. Effect of Hydrothermal Carbonization as Pre-Treatment on Energy Recovery from Food and Paper Wastes. *Bioresour. Technol.* **2019,** *285*, 1–9.
15. Idowu, I.; Li, L.; Flora, J. R.; Pellechia, P. J.; Darko, S. A.; Ro, K. S.; Berge, N. D. Hydrothermal Carbonization of Food Waste for Nutrient Recovery and Reuse. *Waste Manag.* **2017,** *69*, 480–491.
16. Jin, K.; Vozka, P.; Kilaz, G.; Chen, W. T.; Wang, N. H. L. Conversion of Polyethylene Waste into Clean Fuels and Waxes via Hydrothermal Processing (HTP). *Fuel* **2020,** *273*, 1–11.
17. Jing, X.; Wen, H.; Gong, X.; Xu, Z.; Kajetanowicz, A. Recycling Waste Plastics Packaging to Value-Added Products by Two-Step Microwave Cracking with Different Heating Strategies. *Fuel Process. Technol.* **2020,** *201*, 1–17.
18. Kale, G.; Kijchavengkul, T.; Auras, R.; Rubino, M.; Selke, S. E.; Singh, S. P. Compostability of Bioplastic Packaging Materials: An Overview. *Macromol. Biosci.* **2007,** *7* (3), 255–277.
19. Krahling, H.; Sartorius, I. Plastics after Use: Sustainable Management of Material and Energy Resources. Chapter 10.31; In: *Polymer Science: A Comprehensive Reference*; Matyjaszewski, K., Möller M., Eds., Vol. 10; Elsevier: San Diego, 2012; pp 581–595.
20. Langley, J.; Turner, N.; Yoxall, A. Attributes of Packaging and Influences on Waste. *Pack. Technol. Sci.* **2011,** *24*, 161–175.
21. Li, L., Diederick, R.; Flora, J. R.; Berge, N. D. Hydrothermal Carbonization of Food Waste and Associated Packaging Materials for Energy Source Generation. *Waste Manag.* **2013,** *33* (11), 2478–2492.
22. Lin, Y.; Ma, X.; Peng, X.; Yu, Z. A Mechanism Study on Hydrothermal Carbonization of Waste Textile. *Energy Fuels* **2016,** *30* (9), 7746–7754.
23. Lucian, M.; Volpe, M.; Gao, L.; Piro, G.; Goldfarb, J. L.; Fiori, L. Impact of Hydrothermal Carbonization Conditions on the Formation of Hydrochars and Secondary Chars from The Organic Fraction of Municipal Solid Waste. *Fuel* **2018,** *233*, 257–268.
24. Marsh, K.; Bugusu, B. Food Packaging: Roles, Materials, and Environmental Issues. *J. Food Sci.* **2007,** *72* (3), R39–R55.

25. Meys, R.; Frick, F.; Westhues, S.; Sternberg, A.; Klankermayer, J.; Bardow, A. Towards a Circular Economy for Plastic Packaging Wastes: The Environmental Potential of Chemical Recycling. *Resour. Conserv. Recycl.* **2020,** *162*, 3–10.
26. Mastellone, M. L. Technical Description and Performance Evaluation of Different Packaging Plastic Waste Management's Systems in a Circular Economy Perspective. *Sci. Total Environ.* **2020,** *718*, 1–15.
27. Ncube, L. K.; Ude, A. U.; Ogunmuyiwa, E. N.; Zulkifli, R.; Beas, I. N. An Overview of Plastic Waste Generation and Management in Food Packaging Industries. *Recycling* **2021,** *6* (1), 1–12.
28. Ragaert, K.; Delva, L.; Van Geem, K. Mechanical and Chemical Recycling of Solid Plastic Waste. *Waste Manag.* **2017,** *69*, 24–58.
29. Rahimi, A.; García, J. M. Chemical Recycling of Waste Plastics for New Materials Production. *Nat. Rev. Chem.* **2017,** *1* (6), 1–11.
30. Rousta, K.; Dahlén, L. Source Separation of Household Waste: Technology and Social Aspects. Chapter 4; In: *Resource Recovery to Approach Zero Municipal Waste*; Taherzadeh, M. J., Richards T., Eds., Vol. 1; CRC Press: Boca Raton, **2015**; pp 61–76.
31. Sarasa, J.; Gracia, J. M.; Javierre, C. Study of The Biodisintegration of a Bioplastic Material Waste. *Bioresource Technology*, **2009**. *100* (15), 3764–3768.
32. Shen, Y. A Review on Hydrothermal Carbonization of Biomass and Plastic Wastes to Energy Products. *Biomass Bioenergy*, **2020,** *134*, 1–18.
33. Siddiqui, M. Z.; Han, T. U.; Park, Y. K.; Kim, Y. M.; Kim, S. Catalytic Pyrolysis of Tetra Pak over Acidic Catalysts. *Catalysts* **2020,** *10* (6), 602.
34. Singh, N.; Hui, D.; Singh, R.; Ahuja, I. P. S.; Feo, L.; Fraternali, F. Recycling of Plastic Solid Waste: A State of Art Review and Future Applications. *Comp.B: Eng.* **2017,** *115*, 409–422.
35. Song, J. H.; Murphy, R. J.; Narayan, R.; Davies, G. B. H. Biodegradable and Compostable Alternatives to Conventional Plastics. *Phil. Trans. R. Soc. B: Biol. Sci.* **2009,** *364* (1526), 2127–2139.
36. Solis, M.; Silveira, S. Technologies for Chemical Recycling of Household Plastics–A Technical Review and TRL Assessment. *Waste Manag.* **2020,** *105*, 128–138.
37. Ügdüler, S.; De Somer, T.; Van Geem, K. M.; Roosen, M.; Kulawig, A.; Leineweber, R.; De Meester, S. Towards a Better Understanding of Delamination of Multilayer Flexible Packaging Films by Carboxylic Acids. *ChemSusChem.* **2021,** 14, 1–17.
38. Urbaniak, D.; Bala-Litwiniak, A.; Wyleciał, T.; Wawrzyniak, P. Benefits of the Utilization of Waste Packaging Materials in the Pyrolysis Process. In: *Renewable Energy Sources: Engineering, Technology, Innovation*; Wróbel, M., Jewiarz, M., Szlęk, A., Eds.; Springer: Cham, 2020; pp 787–795.
39. Veksha, A.; Yin, K.; Moo, J. G. S.; Oh, W. D.; Ahamed, A.; Chen, W. Q.; Weerachanchai, P.; Giannis, A.; Lisak, G. Processing of Flexible Plastic Packaging Waste into Pyrolysis Oil and Multi-Walled Carbon Nanotubes for Electrocatalytic Oxygen Reduction. *J. Hazard. Mater.* **2020,** *387*, 1–32.
40. Wojnowska-Baryła, I.; Kulikowska, D.; Bernat, K. Effect of Bio-Based Products on Waste Management. *Sustainability* **2020,** *12* (5), 1–12.
41. Xie, M.; Qiao, Q.; Sun, Q.; Zhang, L. Life Cycle Assessment of Composite Packaging Waste Management—A Chinese Case Study on Aseptic Packaging. *Int. J. Life Cycle Assess.* **2013,** *18* (3), 626–635.

42. Xie, M.; Bai, W.; Bai, L.; Sun, X.; Lu, Q.; Yan, D.; Qiao, Q. Life Cycle Assessment of The Recycling of Al-PE (A Laminated Foil Made from Polyethylene and Aluminum Foil) Composite Packaging Waste. *J. Clean. Prod.* **2016,** *112*, 4430–4434.
43. Yousef, S.; Eimontas, J.; Striūgas, N.; Zakarauskas, K.; Praspaliauskas, M.; Abdelnaby, M. A. Pyrolysis Kinetic Behavior and TG-FTIR-GC–MS Analysis of Metallized Food Packaging Plastics. *Fuel* **2020,** *282,* 1–12.
44. Yousef, S.; Eimontas, J.; Zakarauskas, K.; Striūgas, N. Microcrystalline Paraffin Wax, Biogas, Carbon Particles and Aluminum Recovery from Metallized Food Packaging Plastics Using Pyrolysis, Mechanical and Chemical Treatments. *J. Clean. Prod.* **2021,** *290*, 1–10.

CHAPTER 13

BIOBASED FOOD PACKAGING MATERIALS: CLASSIFICATION, EDIBLE FILMS, APPLICATIONS, AND SUSTAINABLE APPROACH

CHAITRADEEPA G. MESTRI, PREETI BIRWAL, PAVAN M. GUNDU, and P. G. GAUTHAMI

ABSTRACT

The migration of species from primary packaging to confined product can affect the packed product's quality, shelf-life, organoleptic, or medicinal performance. Sorption can result in stress-cracking, squeezing, or container panelling. When it comes to food-grade polymers, edible films are viable alternative packaging materials since these allow for the use of otherwise discarded film-forming polysaccharides and polypeptides to create edible packaging with biodegradability, nutritional, and sensory properties. This chapter presents a literature review on the classification of biobased films, interaction between product and package, nanotechnology and edible films, applications.

13.1 INTRODUCTION

The development of food packaging materials and its approach has evolved to maintain the quality of the packed food, transfer knowledge, and communicate to the consumers regarding the food product packed. The primary function of the packaging material includes communication, shield, and

Advances in Sustainable Food Packaging Technology. Megh R. Goyal, Preeti Birwal, Anand K. Pandey, & Ritesh B. Watharkar, (Eds.)

resistant. The major concern is that there are varied amounts of microorganism which toxifies the food and hence is vulnerable to various diseases.[5]

There are various issues with respect to food packaging materials and these are needed to be addressed and overcome by continuous research. Most efficiently and repeatedly used food packaging material is plastic packaging material; its interaction with the food is not acceptable for consumption. The disposal methods of various packaging materials including plastics directly into the environment will cause pollution and nonbiodegradability of the packaging material. There is a need to overcome this problem and develop biodegradable and sustainable food packaging materials. The various issues with the packaging materials are: Material and food interaction, environment impact, and economic implications.

When we find it difficult to use packaging material sustainably, we will switch to the use of renewable packaging material usage, which material is eco-friendly too in nature.[3] The characteristics of these packaging materials are degradability, renewability (gelatin, PLA, cellulose, and starch) and they are identified as innovative packaging material.[4]

Starch is one of the most abundantly used polysaccharide material for the development of biobased polymer after cellulose; various sources of starch have been depicted in Figure 13.1. These starches are available in the form of granules[4,29] and consist mainly of amylose and amylopectin.

The starch is having three structural arrangements, which are linear, branched, and α-glucose forms. And the crystallinity feature of the starch material is because of the presence of linear and branched arrangement of amylopectin and amylose.[5]

Cellulose is another abundantly used polysaccharide which is inherent and peculiar polymer on the earth. The developed films have the ability to be a hygroscopic in nature and are dependent on relative humidity. The hygroscopic nature of the cellulose-based films affects the quality and temperature of the substituents.[5] Other polysaccharides (Fig. 13.2), such as hemicellulose, serve as an alternative material for petrochemical and can be used as current interest topic. Hemicellulose serves as the active alternative for petrochemical and is the topic of interest nowadays. It has various advantages from crop residues or from low value wastewater.[17] Pectin is a polysaccharide and is acidic in nature occurring in various natural by-products. It is mostly present in the intercellular regions of the plant cell walls. The major source of pectin is a lemon or lemon peel.[38]

For marketing and preservation of the food material packaging plays a significant and crucial activity. To overcome various packaging issues due to nonbiodegradability, a lot of research has undergone to develop biodegradable food packaging materials from renewable resources.

This chapter will convey concepts regarding interactions of packaging materials and the product, various testing methods, various sources for the development of biobased packaging films, sustainable approach toward the environment, and any future scope.[5]

13.2 INTERACTION BETWEEN PACKAGING MATERIAL AND PRODUCT

The migration of chemicals and inks from the primary packaging materials to the product packed could lead to many consequences. Alterations that take place on migration are changes in the quality, organoleptic, and the therapeutic efficacy of product packed. Absorption might cause a paneling, squeeze, or cracking of the container or packaging material.[6,42] The concept of mass transfer between the product and packaging material is a base for several research studies; the rate of packaging deterioration in terms of quality leads to decline in the image of the brand. To overcome these problems, the company has to look after the quality of the packaging material and also cost of manufacturing and transport cost. Further, stresses occur on the packaging and food product during handling, transportation undergoes deformation of the product which is irreversible. To overcome this problem knowing the container and content is most important, and this can be carried out by coupling physical-chemical, thermo mechanical studies on interaction of product and packaging material.[8]

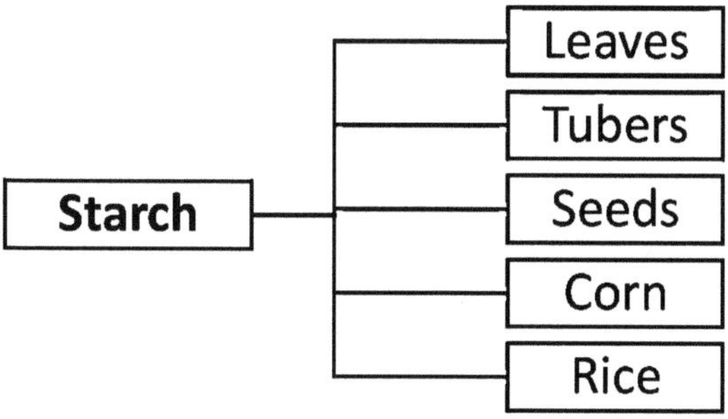

FIGURE 13.1 Sources of starch.

Source: Adapted from Ref. [4]

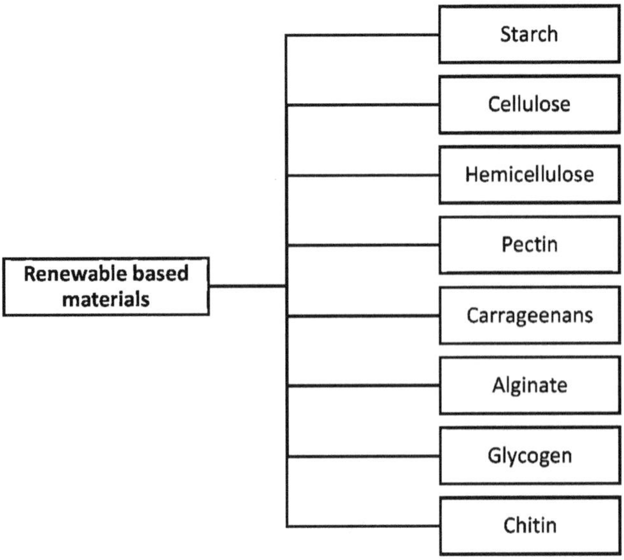

FIGURE 13.2 Polysaccharides.
Source: Adapted from Ref. [14]

When the product comes in contact with the packaging material especially thermoplastic container, various interactions might take place from product to packaging material or packaging material to product. The migration might take place either during processing, manufacturing, or during storage.

- Migration of constitutive substances of the packaging material to the product[32]
- Gas permeation: Generally, oxygen (O_2) diffuses into the food, or carbon dioxide (CO_2) will diffuse out of the packaging[19,47]
- The sorption (or absorption) of the product components of packaging (for example, flavors)[6,20]

The interaction between packaging material and product might have several consequences, visual or not under normal conditions of use: by modification of primary packaging material, chemical and organoleptic modifications and many more. Figure 13.3 represents the interaction phenomenon of container and product. Constituents of packaging materials (additives and other low molecular weight substances) can diffuse package content, such as food or medication, during the migration process. Food's proclivity for interacting with its packaging is a crucial aspect that can influence quality, appearance, and shelf-life. Food residue adhesion to packaging may reduce product acceptance, increase oxidation and off tastes, increase waste, and diminish overall product quality.

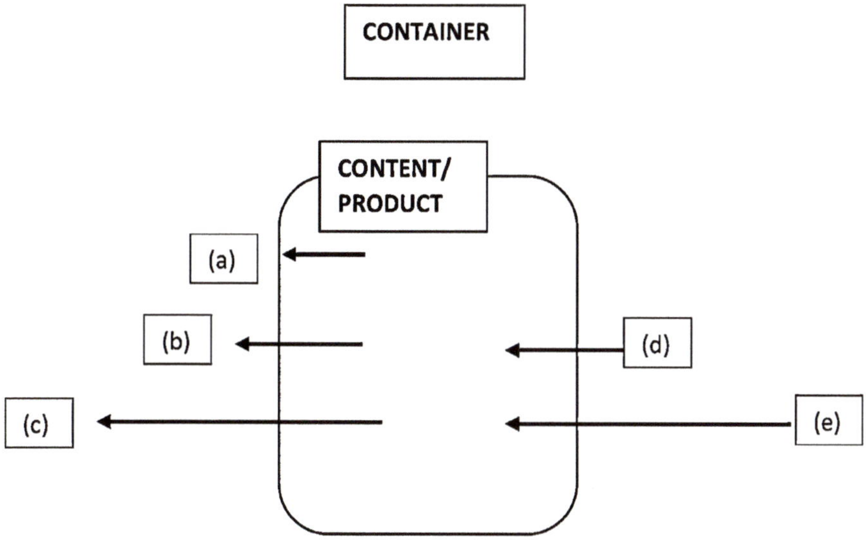

FIGURE 13.3 Schematic diagram for container/content, interaction phenomena: (a) adsorption, (b) absorption, (c) permeation, (d) migration, and (e) permeability.

Source: Adapted from Ref. [14]

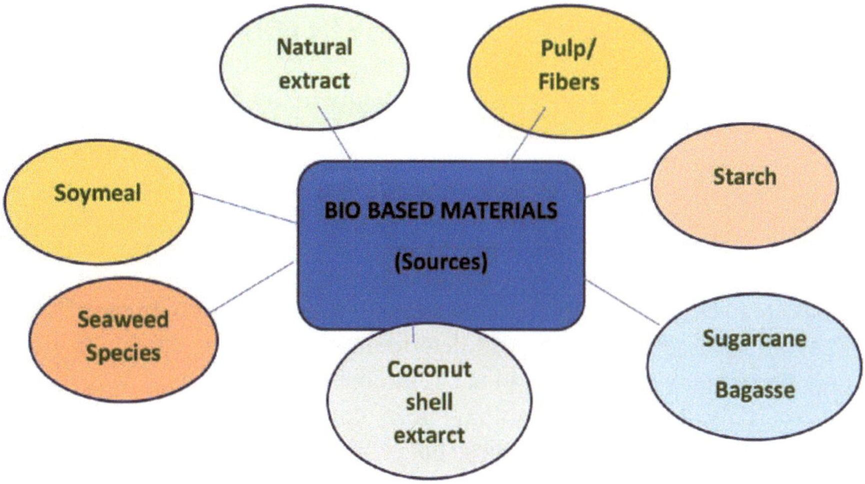

FIGURE 13.4 Natural sources of biobased materials.

Source: Adapted from Ref. [8]

13.3 CLASSIFICATION OF BIOBASED MATERIALS

Bioplastics or biobased packaging materials are not a single class of polymers; there are various classes of source materials which differ from one another. The bioplastics also denotes packaging materials that are biodegradable, biobased ones.[37] The production and utilization of biobased packaging materials have increased in recent times. Their application can also be seen in surgical implant devices, drug delivery system, food containers, packaging materials, and agriculture. Natural sources of biobased packaging material have been depicted in Figure 13.4.

The most widely used bioplastic classification distinguishes between biobased and/or bio-degradable materials. Plastics can be classified into four types based on their biodegradability and biobased content:

- Group 1: Nonbiodegradable bioplastics that are biobased or partially biobased, such as bio-PE, bio-PET, bio-PA, and bio-PTT.
- Group 2: Biobased and biodegradable bioplastics, such as PLA, PHA, PBS, and starch blends.
- Group 3: Nonbiodegradable bioplastics derived from fossil fuels, such as standard PE, PP, and PET.
- Group 4: Fossil-based and biodegradable bioplastics, such as PBAT and PCL.

13.4 SCOPE OF NANOTECHNOLOGY IN EDIBLE FILMS

Food packaging has long figured prominently in the food supply chain, not only by unifying commodities and expanding the customer base through marketing appeal, but also by shielding packed products from external spoiling agents, resulting in improved food safety, quality, and shelf-life. Oxygen, moisture, light, heat, and bacteria are all abundant in the environment. When it comes to food, there is universal consensus that these components are toxic, necessitating the use of packing materials as efficient barriers to avoid food contamination and deterioration. Packaging materials are the only barrier between food and the environment.[12,34]

Not only has nanotechnology been combined with edible films to enhance their physical properties, but it has also been used to give extra protection against food contamination and spoiling. These nanocomposites are in keeping with the growing trend of active packaging, which includes packaging that has functions other than physical protection, such as antibacterial and antioxidant effects, to name a few.[36]

Nanomaterials have been shown to improve the physical–mechanical properties of polymers intended to be used as film-forming matrices for food packaging by acting as fillers. Other characteristics (for example, biological activity against foodborne pathogens and/or spoilage microorganisms) could be created and employed as major features in biopolymer-based food preservation materials.[51]

When it comes to food-grade biopolymers, edible films are a viable alternative packaging material since it allows for the use of otherwise discarded film-forming polysaccharides and polypeptides to create edible packaging with biodegradability, nutritional, and sensory properties. When appropriate nanoscaled additions are employed, these consumable materials may have enhanced active roles (antimicrobial and antioxidant, in particular).[12,34,35]

13.5 PROPERTIES OF BIOPOLYMERS FOR FOOD PACKAGING

The attributes that are required for the enhanced functionality and properties of biobased materials must be effective, clean, and recyclable in nature. Bio-based materials support green production by successfully enclosing and protecting items, which are important packaging characteristics. Moreover, the packaging system is designed to use materials and energy efficiently throughout the product life cycle, resulting in minimal waste. When it comes to food bio-based materials, properties such as gas and water vapor permeability, sealing capability, thermoforming properties, transparency, antifogging capacity, printability, availability, UV resistance, water resistance, and acid, alkali, and grease resistance must all be considered. One of the most essential aspects of bio-based products is their cost.[37]

Advances in bio-based sustainable materials with increased functionality and features such as moisture permeability, porosity, and adjustable gas permeability for storing and packing wet and dry foods and fresh vegetables bio-based polymers, plastics, biodegradable plastics, and composites are becoming increasingly popular as viable alternatives to petrochemical-based materials. Jute, hemp, flax, banana, wheat straw, and other natural fibers are important sources for biodegradable composites with commercial value as food packaging materials.[41,46]

Environmentally friendly and biodegradable biocomposites manufactured from plant-based fiber sources and biopolymers/biomass-derived polymers have mechanical strength and flexibility comparable to petroleum-based polymers. Bio-based packaging materials have improved mechanical resistance, thermal insulation, and physico-chemical properties, all of which are important for barrier and permeability features. Protein-based materials have also been examined viable food packaging materials because they have high barrier qualities and are impermeable to oxygen (in the absence of moisture) and aromatic chemicals.

The physical properties of a biopolymer material mainly depend on molecular weight, density, and molar volume of the packaging material. The higher the density of the packaging material, the higher the transportation cost and this in turn affects the mechanical properties of the packaging material too.[15,44] The packaging material should possess barrier properties against gases (O_2 and CO_2) and water vapor transmissions depending upon the humidity of the surrounding atmosphere.[11,22] The water vapor transmission rates vary with different materials used to develop biofuels. Carbon dioxide permeation coefficient and oxygen permeation coefficient are used for quantifying the carbon dioxide and oxygen barrier capabilities respectively. This indicates that the oxygen and carbon dioxide permeation rate per unit area and time ($kg/(mm^2s^1Pa^1)$).[10,50]

When polymers are exposed to atmospheric relative humidity, the water molecules are absorbed and transported through packaging material due to the hydrophilic nature.[14,15] The amount of water transported is quantified by water vapor permeability coefficient; this rate indicates the actual amount of water vapor that has permeated through per unit area and time ($kg/(mm^2s^1Pa^1)$) required into the material. The disturbance in the fiber or polymer is due to the migration of water, which in turn reduces the strength of the composite.[15,16]

Thermal properties include melting temperature and glass transition temperature. The amorphous region or part of the polymer determines the property of glass transition temperature, whereas the melting point property is determined by the crystalline region of the polymer. The intermolecular forces, molecular weight, and plasticizer depict the glass transition temperature of the polymer. If the food is to be stored at lower temperature, the polymer used for packaging should possess a higher transmission temperature and also it is applicable for those foods to be stored at higher temperature should possess higher melting point.[25,26]

Mechanical properties include tensile strength, percentage elongation break, toughness, Young's modulus, and viscoelasticity. During transportation, storage, and handling process, the food product might undergo several stresses. So, the packaging material should possess good mechanical properties like stiffness and modulus, and is having a positive response as compared to other conventional polymers. These mechanical properties depend on the molecular mass of the polymer, materials used, and concentration of film-forming solution.[2,10,15,17]

For a sustainable environment the biodegradable property is the most useful one which involves biological activity for the complete degradation process. During the composting process, the organic matter is converted into a soil-like material, that is, humus and carbon dioxide upon the activity of microorganisms.[16,17] Here, the degradation process has been carried out in three steps and they are biodeterioration, bio fermentation, and assimilation. The time taken for degradation is less as compared to conventional packaging materials. This process is affected by several factors such as chemical chain, chemical structure, crystallinity, pH, temperature, oxygen, and moisture content.[16,43]

13.6 ENVIRONMENTALLY FRIENDLY BIOPOLYMERS

The use of petrochemical-based polymers like polypropylene (PP), polyethylene (PE), and polystyrene (PS) has been in various sectors such as decoration, packaging, automobile parts production, and so on. The plastic material, when disposed off in the environment, will take longer time to degrade under normal environmental conditions as they are resistant to chemical, enzymatic, and microbial reactions.[48]

The demand for sustainable packaging material is increased to maintain the ecological balance, and to solve the problems related to food safety and environment. To combat the usage of plastic and develop a sustainable packaging material, packaging material made out of biopolymer is of utmost important concern.[13,14,24]

Protein, carbohydrate, and other long-chain compounds originating from plants and animals are known as biopolymers. Biopolymers ensure biodegradability and biocompatibility while also allowing for environmentally acceptable applications.[21,31,33] In food packaging, biopolymers such as starch, protein, cellulose, chitosan, PLA, and their derivatives are employed. These polymers are utilized as an edible film or coating over food to reduce moisture loss, prevent oxidation, and decrease lipid migration, among other things.[9,39]

Microbial, enzymatic, and chemical processes can degrade starch, protein, and PLA-based packaging materials. Biodegradable and biocompatible materials make up the majority of bio-based packaging materials.[1,27] As a result, biodegradable food packaging helps to reduce plastic waste and the global plastic crisis.[31]

The rate of biodegradation is determined by the environment and the qualities of the packing material.[11,20,32] By reducing nonbiodegradable petroleum-based polymers and waste disposal costs, biodegradable and renewable polymers are used to replace nonbiodegradable petroleum-based polymers in food packaging. However, when compared to nonbiodegradable petroleum-based polymers, the use of biopolymers in food packaging has several disadvantages, such as inferior mechanical and barrier properties.[12,34]

13.7 SUSTAINABLE APPROACH TOWARD ENVIRONMENT

To improve biodegradable polymers, our surrounds must be more concerned about environmental health. Such packaging requires more research to improve its properties, storage time, dietary guidelines, and biosafety. The barrier qualities, in addition to this, needed to be improved. Degradable bioplastics generated from starch and cellulose are primarily being developed for use in food packaging.[45]

Biodegradable plastics have an advantage over conventional synthetic plastics in terms of complete and safe disposal, allowing them to be used in a wide range of applications such as agriculture, biomedical, and food packaging. Biodegradable materials used to wrap food products are largely produced renewable basic ingredients such as polylactide, starch/cellulose-based polymers, polybutylene succinate, and polybutylene succinate-co-adipate in the current scenario.[30]

Food deterioration caused by microbial contamination and lipid and protein oxidation is a severe concern for food safety and quality, and it can occur in packaged foods even when traditional food preservation methods such as freezing, drying, heating, salting, and fermentation have been utilized. All of these issues can be solved by employing bio-based active and smart packaging, which offers excellent prospects for improving quality, tenderness, and freshness while also providing antibacterial properties.[52]

The use of bio-derived polymers as food packaging is limited due to a number of factors. Besides the high cost of manufacture compared to petrochemical-derived plastics, land use and availability are also significant limiting issues for its production, preventing its industrial use. Although the features of bioplastics entice their usage, various drawbacks such as fragility, thermal volatility, low Tm, complex heat stability, and high water and oxygen accessibility prevent them from being a desirable food packaging material.[40]

13.8 APPLICATIONS OF BIO-BASED FOOD PACKAGING MATERIALS

Since the beginning of 1999, starch-based carrying bags have been available in the market. Since then, numerous distribution networks in Scandinavian and Mediterranean nations have used similar carrier bags with great success. The carrying bags have been introduced in places where separate organic waste collection is well established, and the bags are widely recognized as biodegradable composting bags by municipalities.

Bio-based food-packaging materials are predicted to start out as niche markets, where the bio-based materials' particular features provide a competitive advantage to the packaging concept. A few niche markets have been identified in the following sections, but as technology advances, there will undoubtedly be more opportunities for these materials.

13.8.1 FRUITS AND VEGETABLES

Several different types of biobased packaging materials have quite different gas characteristics. Most synthetic plastics have a 1:4 ratio between oxygen and carbon dioxide transmittance; however, certain biobased polymers have a 1:30 or higher ratio.[18,49] Carbon dioxide has a higher transmittance than oxygen, which is interesting in terms of packaging for high-respiring fruits and vegetables. However, some of these materials' high water vapor transmittance is not necessarily necessary for certain packaging applications.[49]

13.8.2 CHEESE

Cheese, like humans, breathes and releases carbon dioxide during storage, making the use of biobased packaging for cheese wrapping appealing. To avoid package inflation, packaging materials must have a reasonably high carbon dioxide transmittance rate due to the cheese's respiration. For these applications, both PHA and PLA materials will be useful.[49] Alternatively, different material combinations can be created.

13.8.3 CHILLED OR FROZEN PRODUCTS

PHA and PLA materials have low gas and water vapor barriers, making them suitable for a wide variety of food applications. Dairy products, in particular, have been mentioned as a likely PLA application.[49] Noncarbonated beverage bottles, cups, containers, cartons, and other items made of PHA and PLA will be wonderful. Furthermore, as an alternative to the synthetic coatings used now, they could be intriguing as coatings for milk, juice, and other cardboard boxes. The use of PLA and PHA containers as juice packing has yielded favorable results, with comparable vitamin C degradation and little or no sorption of fragrance components into the packaging materials (scalping).[18,49]

Thus, bio-based packagings are predicted to be highly successful for items with a short shelf-life and no strong demands on gas barrier qualities or water vapor barriers.

13.9 CURRENT TRENDS AND FUTURE SCOPE

Bio-based feedstocks are not a new notion in the chemical industry. For more than a decade, they have proven industrially practicable on a huge scale. Because the price of oil was so low and the growth of oil-based products gave so many opportunities, bio-based products were not prioritized at the time.[7] Several factors drove the development of bio-based polymers and products, including fossil fuel supply constraints and unpredictability, environmental concerns, and technological advancements.

Technology turned its focus to cellulose-based feedstocks as the argument over food vs. fuel heated up, focusing on waste from the wood and paper industries, the food sector, and even stems and leaves and solid municipal waste streams. More of these technologies are being developed to align with the above-mentioned waste streams; nevertheless, developing the entire spectrum of chemicals based on these technologies could take another 20 years.[28]

Even though novel bio-based polymers are being manufactured on a large scale, various factors still need to be determined in order for biobased polymers to have a long-term viability. The increase of first-generation bio-based fuel production will create unsustainable demands on biomass resources, endangering both biochemical and biopolymer production as well as food supply.[28] Several efforts are underway to produce useable sugars from cellulose-based feedstocks for biofuels, biochemicals, and biopolymers.[23]

13.10 SUMMARY

Because of advancements in biotechnologies and public awareness, biobased polymers are now routinely found in a wide range of applications, from commodity to high-tech. Despite the appealing applications of bio-based polymers, their use in food packaging is limited due to their high manufacturing costs and limited functional qualities when compared to competitive petro-based variants. The most important and critical activity for the preservation and marketing of food products is packaging. Petroleum-based food packaging materials have been used for decades due to their nonbiodegradability, which has serious negative consequences. Biopolymer-based biodegradable food packaging materials could be a viable replacement for nonbiodegradable plastics. The advancement of innovative biopolymer and food packaging technologies results in more consumer-friendly products.

KEYWORDS

- **biodegradability**
- **biopolymer**
- **edible films**
- **migration**
- **polysaccharides**
- **primary packaging**
- **sorption**

REFERENCES

1. Ali, R. R.; Rahman, W.; Kasmani, R. M. Tapioca Starch Biocomposite for Disposable Packaging Ware. *Chem. Eng. Trans.* **2013,** *32*, 1711–1716.
2. Almeida, C. B. D.; Corradini, E.; Forato, L. A.; Fujihara, R.; Lopes, J. F. Microstructure and Thermal and Functional Properties of Biodegradable Films Produced Using Zein. *Polímeros* **2018,** *28*, 30–37.
3. Arfin, T.; Fatima, S. Anticipating Behavior of Advanced Materials in Healthcare. In: *Advanced Biomaterials and Biodevices*; Tiwari, A., Nordin, A. N., Eds.; Wiley-Scrivener: New York, 2014; pp 243–288.

4. Arfin, T.; Tarannum, A. Polymer Materials: From the Past to the Future. In: *Green Polymeric Materials: Advances and Sustainable Development*; Ahmed, S., Annu, I. S., Eds.; Nova Science Publishers: New York, 2017; pp 73–88.
5. Arfin, T.; Sonawane, K. Bio-Based Materials: Past to Future. In: *Bio-Based Materials for Food Packaging*; Springer: Singapore, 2018; pp 1–32.
6. Auras, R.; Harte, B.; Selke, S. Sorption of Ethyl Acetate and D-Limonene in Poly (Lactide) Polymers. *J. Sci. Food Agric.* **2006,** *86*, 648–656.
7. Babu, R. P.; O'connor, K.; Seeram, R. Current Progress on Bio-Based Polymers and Their Future Trends. *Progress Biomater.* **2013,** *2* (1), 1–16.
8. Bharathi, S. V.; Moses, J. A.; Anandharamakrishnan, C. Interaction Phenomena between Packaging and Product. In: *Bio-based Materials for Food Packaging*; Springer: Singapore, 2018; pp 33–56.
9. Bourtoom, T. Edible Protein Films: Properties Enhancement. *Food Res. J.* **2009,** *16*, 1–9.
10. Campos, C. A.; Gerschenson, L. N.; Flores, S. K. Development of Edible Films and Coatings with Antimicrobial Activity. *Food Bioprocess Technol.* **2011,** *4* (6), 849–875.
11. Dyshlyuk, L.; Babich, O.; Belova, D.; Prosekov, A. Comparative Analysis of Physical and Chemical Properties of Biodegradable Edible Films of Various Compositions. *J. Food Process Eng.* **2017,** *40* (1), E12331.
12. Espitia, P. J.; Otoni, C. G. Nanotechnology and Edible Films for Food Packaging Applications. *Bio-Based Mater. Food Pack.* **2018,** 125–145.
13. Fahma, F.; Sugiarto, Sunarti, T. C. Thermoplastic Cassava Starch-PVA Composite Films with Cellulose Nanofibers from Oil Palm Empty Fruit Bunches as Reinforcement Agent. *Int. J. Polym. Sci.* **2017**.
14. Ferreira, A. R. V.; Alves, V. D.; Coelhoso, I. M. Polysaccharide-Based Membranes. *Food Pack. App. Membranes* **2016,** 2016, 20–29.
15. Fischer, S.; Vlieger, De, J.; Kock, T.; Gilberts, J. Green Composites: The Materials of the Future—A Combination of Natural Polymers and Inorganic Particles. *Unpublished Presentation at Proceedings of the Food Biopack Conference at Coperhagen – Denmark*; 27–29 August 2000; p 29.
16. Gutiérrez, T. J. Biodegradability and Compostability of Food Nanopackaging Materials. In: *Composites Materials for Food Packaging*; Cirillo, G., Kozlowski, M.A., Spizzirri, U.G., Eds.; John Wiley & Sons: New York, 2018; pp 269–296.
17. Hansen, N. M.; Plackett, D. Sustainable Films and Coatings from Hemicelluloses: A Review. *Biomacromolecules* **2008,** *9* (6), 1493–1505.
18. Haugaard, V. K.; Udsen, A. M.; Mortensen, G.; Høegh, L.; Petersen, K.; Monahan, F. Potential Food Applications of Biobased Materials. *EU Concerted Action Project—Starch Stärke* **2001,** *53* (5), 189–200.
19. Hernández-Muñoz, P.; Catala, R.; Gavara, R. Effect of Sorbed Oil on Food Aroma Loss Through Packaging Materials. *J. Agric. Food Chem.* **1999,** *47*, 4370–4374.
20. Hernández-Muñoz, P.; Catala, R.; Gavara, R. Food Aroma Partition Between Packaging Materials and Fatty Food Simulants. *Food Addit. Contam.* **2001,** *18* (7), 673–682.
21. Hu, B. Biopolymer-Based Lightweight Materials for Packaging Applications. In: *Lightweight Materials from Biopolymers and Biofibers* (Volume 1175); Yang, Y., Xu, H., Yu, X., Eds.; *J. Am. Chem. Soc.* **2014**, 239–255.
22. Jafarzadeh, S.; Jafari, S. M.; Salehabadi, A.; Nafchi, A. M.; Kumar, U. S. U.; Khalil, H. A. Biodegradable Green Packaging with Antimicrobial Functions Based on The

Bioactive Compounds from Tropical Plants and Their By-Products. *Trends Food Sci. Technol.* **2020,** *100*, 262–277.
23. Jun, X.; Bao-Hua, G. Microbial Succinic Acid, Its Polymer Poly (Butylene Succinate), and Applications. *Microbiol. Monogr.* **2010,** *14*, 347–388.
24. Malathi, A. N.; Santhosh, K. S.; Udakumar, N. Recent Trends of Biodegradable Polymer: Biodegradable Films for Food Packaging and Application of Nanotechnology in Biodegradable Food Packaging. *Curr. Trends Technol. Sci.* **2014,** *3* (2), 73–79.
25. Malmir, S.; Montero, B.; Rico, M.; Barral, L.; Bouza, R. Morphology, Thermal and Barrier Properties of Biodegradable Films of Poly (3-Hydroxybutyrate-Co-3-Hydroxyvalerate) Containing Cellulose Nanocrystals. *Comp A Appl. Sci. Manuf.* **2017,** *93*, 41–48.
26. Mangaraj, S.; Yadav, A.; Bal, L. M.; Dash, S. K.; Mahanti, N. K. Application of Biodegradable Polymers in Food Packaging Industry: A Comprehensive Review. *J. Pack. Technol. Res.* **2019,** *3* (1), 77–96.
27. Marsh, K.; Bugsu, B. Food Packaging-Roles, Materials, and Environmental Issues. *J. Food Sci.* **2007,** *72*, 39–55.
28. Michael, C.; Dirk, C.; Harald, K.; Jan, R.; Joachim, V. *Policy Paper on Bio-Based Economy: Level Playing Field for Bio-Based Chemistry and Materials*; Huerth, Germany: Nova-Institute GmbH; November 2011; p 9.
29. Mogarkar, P. R.; Arfin, T. Chemical and Structural Importance of Starch Based Derivatives and its Applications. In: *Natural Polymers: Derivatives, Blends and Composites*; Ikram, S., Ahmed, S., Eds., Vol. II; Nova Science Publishers: New York, 2017; pp 73–88.
30. Mostafa, N. A.; Farag, A. A.; Abo-Dief, H. M. Production of Biodegradable Plastic from Agricultural Wastes. *Arabian Journal of Chemistry*, **2015,** *11* (4), 546–553.
31. Muller, J.; González-Martínez, C.; Chiralt, A. Combination of Poly (Lactic) Acid and Starch for Biodegradable Food Packaging. *Materials (Basel)*, **2017,** *10* (8), 952.961.
32. Nir, M. M.; Ram, A. Sorption and Migration of Organic Liquids in Poly(Ethylene Terephthalate). *Polym. Eng. Sci.* **1996,** *36* (6), 862–868.
33. Othman, S. H. Bio-nanocomposite Materials for Food Packaging Applications: Types of Biopolymer and Nano-Sized Filler. *Agric. Agric. Sci. Procedia* **2014,** *2*, 296–303.
34. Otoni, C. G.; Avena-Bustillos, R. J.; Azeredo, H. M. C.; Lorevice, M. V.; Moura, M. R.; Mattoso, L. H. C.; Mchugh, T. H. Recent Advances on Edible Films Based on Fruits and Vegetables: A Review. *Compreh. Rev. Food Sci. Food Saf.* **2017,** *16* (5), 1151–1169.
35. Otoni, C. G.; Avena-Bustillos, R. J.; Olsen, C. W.; Bilbao-Sáinz, C.; Mchugh, T. H. Mechanical and Water Barrier Properties of Isolated Soy Protein Composite Edible Films as Affected by Carvacrol and Cinnamaldehyde Micro and Nanoemulsions. *Food Hydrocol.* **2016,** *57* (Suppl. C), 72–79.
36. Otoni, C. G.; Espitia, P. J. P.; Avena-Bustillos, R. J.; Mchugh, T. H. Trends in Antimicrobial Food Packaging Systems: Emitting Sachets and Absorbent Pads. *Food Res. Int.* **2016,** *83* (Suppl. C), 60–73.
37. Pandit, P.; Nadathur, G. T.; Maiti, S.; Regubalan, B. Functionality and Properties of Bio-Based Materials. *Bio-Based Mater. Food Pack.* **2018,** 2018, 81–103.
38. Pavlath, A. E.; Voisin, A.; Robertson, G. H. Pectin-Based Biogradable Water Insoluble Films Application of Polymers in Foods. *Die Makromolekulare Chemie (Macromolecular Chemistry), Symposium, Dallas—TX,* **1999**; Unpublished Paper, p 9.

39. Pawar, P. A.; Purwar, A. H. Bioderadable Polymers in Food Packaging. *Am. J. Eng. Res.* **2013,** *2* (5), 151–164.
40. Peelman, N.; Ragaert, P.; De Meulenaer, B. Application of Bioplastics for Food Packaging. *Trends Food Sci. Technol.* **2013,** *32* (2), 128–141
41. Razzaq, H. A. A.; Pezzuto, M.; Santagata, G. Barley B-Glucan-Protein Based Bioplastic Film with Enhanced Physicochemical Properties for Packaging. *Food Hydrocoll.* **2016,** *58*, 276–283.
42. Safa, H. L.; Bourelle, F. Sorption-Desorption of Aromas on Multi-Use PET Bottles: A Test Procedure. *Pack. Technol. Sci.* **1999,** *12*, 37–44.
43. Sintim, H. Y.; Bary, A. I.; Hayes, D. G.; English, M. E.; Schaeffer, S. M.; Miles, C. A.; Flury, M. Release of Micro-and Nanoparticles from Biodegradable Plastic During in Situ Composting. *Sci. Total Environ.* **2019,** *675*, 686–693.
44. Souza, V. G. L.; Fernando, A. L.; Pires, J. R. A.; Rodrigues, P. F.; Lopes, A. A.; Fernandes, F. M. B. Physical Properties of Chitosan Films Incorporated with Natural Antioxidants. *Ind. Crops Products* **2017,** *107*, 565–572.
45. Srivastava, P.; Bano, K.; Zaheer, M. R.; Kuddus, M. Biodegradable Smart Biopolymers for Food Packaging: Sustainable Approach toward Green Environment, In: *Bio-Based Materials for Food Packaging*; Springer: Singapore, 2018; pp 197–216.
46. Storz, H.; Vorlop, K. D. Bio-Based Plastics: Status, Challenges and Trends. *Appl. Agric. Forestry Res.* **2013,** *63*, 321–332.
47. Tawfik, M. S.; Devliegherea, F.; Huyghebaert, A. Influence of D-Limonene Absorption on the Physical Properties of Refillable PET. *Food Chem.* **1998,** *61* (1–2), 157–162.
48. Verma, R.; Vinoda, K. S.; Papireddy, M. Toxic Pollutants from Plastic Waste. *Procedia Environ. Sci.* **2016,** *35*, 701–708.
49. Weber, C. J.; Haugaard, V.; Festersen, R.; Bertelsen, G. Production and Applications of Biobased Packaging Materials for the Food Industry. *Food Addit. Contam.* **2002,** *19* (S1), 172–177.
50. Weizman, O.; Dotan, A.; Nir, Y.; Ophir, A. Modified Whey Protein Coatings for Improved Gas Barrier Properties of Biodegradable Films. *Polym. Adv. Technol.* **2017,** *28* (2), 261–270.
51. Wu, X.; Liu, Y.; Wang, W.; Han, Y.; Liu, A. Improved Mechanical and Thermal Properties of Gelatin Films Using a Nano Inorganic Filler. *J. Food Process. Eng.* **2017,** *40* (3), E12469.
52. Yu, P. H.; Chua, H.; Huang, A. L. Conversion of Food Industrial Wastes into Bioplastics. *Appl. Biochem. Biotechnol.* **1998,** *70*–72, 603–614.

INDEX